Recent Advances in Polymer Processing

Recent Advances in Polymer Processing

Editors

Ying Yan
Yiqi Wang

MDPI

Basel • Beijing • Wuhan • Barcelona • Belgrade • Novi Sad • Cluj • Manchester

Editors

Ying Yan
Mechanical Engineering
Dalian University
of Technology
Dalian
China

Yiqi Wang
Mechanical Engineering
Dalian University
of Technology
Dalian
China

Editorial Office
MDPI
St. Alban-Anlage 66
4052 Basel, Switzerland

This is a reprint of articles from the Special Issue published online in the open access journal *Polymers* (ISSN 2073-4360) (available at: www.mdpi.com/journal/polymers/special_issues/frontiers_polymer_processing).

For citation purposes, cite each article independently as indicated on the article page online and as indicated below:

Lastname, A.A.; Lastname, B.B. Article Title. *Journal Name* **Year**, *Volume Number*, Page Range.

ISBN 978-3-7258-0250-0 (Hbk)
ISBN 978-3-7258-0249-4 (PDF)
doi.org/10.3390/books978-3-7258-0249-4

Contents

Preface

Dear Esteemed Colleagues,

It is with great enthusiasm and anticipation that we introduce this Special Issue dedicated to polymer processing, with a specific focus on the machining processes of polymers. Polymers, with their widespread applications in industries ranging from aerospace to optical and biological engineering, have proven indispensable due to their remarkable properties: low density, corrosion resistance, low coefficient of friction, and scalability through mass production.

In the intricate landscape of polymer utilization, the integrity of the finished surface emerges as a pivotal factor in guaranteeing the quality of manufactured components. Recognizing this significance, our Special Issue aims to converge the latest research and review papers that delve into the frontiers of polymer processing. Specifically, we direct our attention towards the machining processes of polymers, where the intersection of processing parameters, surface quality, and material removal mechanisms holds the key to unlocking the full potential of these versatile materials.

Our motivation for assembling this collection lies in the firm belief that the true potential of polymers can be harnessed through meticulous attention to their processing capabilities. With the right methods and parameters, we can elevate the quality of polymers, paving the way for advancements in diverse applications.

This endeavor would not be possible without the collaborative spirit of the scientific community. We express our deepest gratitude to all those who have contributed to the realization of this Special Issue, and we look forward to the wealth of knowledge and insights that will be shared within its pages.

Ying Yan and Yiqi Wang
Editors

polymers

MDPI

Article

Crazing Initiation and Growth in Polymethyl Methacrylate under Effects of Alcohol and Stress

Ying Yan *, Yujia Sun, Jiyang Su, Bo Li and Ping Zhou

State Key Laboratory of High-Performance Precision Manufacturing, Department of Mechanical Engineering, Dalian University of Technology, Dalian 116024, China
* Correspondence: yanying@dlut.edu.cn

Abstract: Polymer crazing is typically a precursor to damage and considerably reduces the mechanical performance of polymer materials. The concentrated stress caused by machines and the solvent atmosphere created during machining exacerbates the formation of crazing. In this study, the tensile test method was employed to examine the initiation and progression of crazing. The research focused on polymethyl methacrylate (PMMA), both regular and oriented, and the impact of machining and alcohol solvents on the formation of crazing. The results showed that the alcohol solvent influenced PMMA through physical diffusion, whereas machining primarily affected crazing growth via residual stress. Treatment reduced the crazing stress threshold of PMMA from 20% to 35% and produced a threefold increase in its sensitivity to stress. The findings revealed that oriented PMMA exhibited 20 MPa higher resistance to crazing stress compared with regular PMMA. The results also indicated that the extension of the crazing tip and thickening were in conflict, with the crazing tip of regular PMMA severely bending under tensile stress. This study provides valuable insight into the initiation of crazing and the methods of its prevention.

Keywords: crazing; swelling; tensile test; plastic deformation

check for updates

Citation: Yan, Y.; Sun, Y.; Su, J.; Li, B.; Zhou, P. Crazing Initiation and Growth in Polymethyl Methacrylate under Effects of Alcohol and Stress. *Polymers* **2023**, *15*, 1375. https://doi.org/10.3390/polym15061375

Academic Editor: Zbigniew Bartczak

Received: 7 February 2023
Revised: 28 February 2023
Accepted: 8 March 2023
Published: 9 March 2023

1. Introduction

Polymer materials, including polymethyl methacrylate (PMMA), are widely used due to their favorable material properties. PMMA, in particular, is favored in the production of optical lenses and fighter cockpit covers as it is lighter, more resistant to corrosion, and boasts better optical and mechanical properties than quartz glass [1–6]. Crazing, however, is a ubiquitous mechanism of plastic deformation that occurs in brittle polymeric materials, including PMMA, and is caused by local plastic deformation under tensile stress. This form of damage, characterized by a tensile-stress-induced subscale crack, negatively impacts PMMA's optical properties, reducing light transmittance, affecting contrast, and causing optical distortion, which eventually leads to mechanical fracture [7]. Traditional shaping methods are not capable of producing PMMA products with the high surface accuracy and low surface roughness that is required for complex free-form surfaces, necessitating mechanical processing. However, the high level of local stress concentration during the machining and solvent cleaning processes exacerbates the formation of crazing, reducing the stress threshold for crazing initiation and triggering its development [8,9].

Exposure to a solvent atmosphere over extended periods of time leads to a reduction in the threshold for crazing and an increase in the density of crazing in PMMA. Roger P. Kambour and E.J. Kramer proposed that crazing at the crack tip in PMMA could undergo as much as 100% elastic deformation under stress before the crack propagation resumed. This deformation is estimated to account for as much as 40% of the nominal Griffith energy of crack propagation. The thickening process of methanol-induced crazing, which can result in thickness increases from 40% to 140%, absorbs a substantial amount of energy and can lead to material damage [10,11]. Numerous studies have been conducted to evaluate the impact of the solvent atmosphere on PMMA. For instance, E. H. Andrews et al. employed a creep experiment to investigate the effect of different alcohol solvents on PMMA and found that physical diffusion, rather than chemical reactions, occurred due to the large molecular weight difference between PMMA and the alcohol solvents [12]. This caused local cavities to form in the PMMA molecular chains, reducing its mechanical properties and crazing stress resistance. In another study, Kramer, Edward J. measured the kinetics of craze growth in PMMA under contact with liquid methanol and found that solvent crazing velocity was limited by the hydrodynamic transport of the solvent through a porous craze structure [13]. N. L. Thomas and A. H. Windle also studied the characteristics of methanol diffusion in PMMA and found that a concentration gradient was generated, leading to internal local stress that exceeded the stress required for crazing to occur [14]. Luo et al. investigated the diffusion between PMMA and solvent molecules and discovered that diffusion caused volume expansion and surface distortion in PMMA, leading to stress concentration [15]. The crack expansion rate of polyethylene under different environmental conditions was studied by Maximilian Thuy et al., who showed that environmental factors played a role in the crack expansion rate [16]. The effects of natural weathering on the mechanical properties of PMMA, ABS, and ASA were studied by Sánchez-Calderón, I et al., who reported that the natural environment and light considerably reduced the mechanical properties of PMMA [17]. The diffusion effect of solvents such as sedimentary oil and TPP on the crazing growth of PMMA was studied by Masoud Razavi et al., who found that the diffusion of these solvents promoted molecular chain aggregation and the nucleation of PMMA, leading to voids in the molecular network and high rates of crazing, cracking, and brittle fracture [18]. Despite these studies, research is lacking on the changes experienced by PMMA after diffusion and the growth of crazing, a suitable theoretical model describing the effect of the solvent atmosphere on the crazing process is also lacking.

The machining process induces concentrated stress, affecting the initiation and development of crazing. H. Michler studied the size, shape, and internal structure of crazing in PMMA, PS, PC, and other machined materials before and after annealing [19]. Non-annealed PMMA, when subjected to an external force quickly, nucleated, forming micro-fibrous and microcavity structures. PMMA exhibited strong strain softening under tensile stress, indicating crazing. Gao et al. monitored the progression of crazing damage using acoustic emission and finite element simulation and found that growth was uneven and nonlinear, and the constitutive models of crazing and microcracking were obtained [20]. Razavi et al. and Wang Y et al. also investigated the impact of mechanical forces on the growth of crazing and cracking in PS, PMMA, and other materials. Machining was found to improve PMMA creep deformation, resulting in irreversible deformation and the expansion of cracks with associated crazing [18,21]. The surface damage caused by machining can promote the crazing formation process by accelerating the development and coalescence of micro-voids, which can ultimately lead to material fracture [22]. This limitation substantially affects the development of polymer material machining, as processing defects and vibrations induced by machining can further contribute to the destruction of polymer materials during their use [23].

Theoretical studies on the initiation of crazing are limited. Lang RW et al. proposed a criterion for crazing initiation based on elastic strain. Kinoch AJ et al. and Jie et al. proposed criteria based on the maximum principal stress and equivalent shear stress, respectively, as per the nucleation theory [24–26]. S. Socrate et al. used finite element simulation and obtained the critical stress of polystyrene (PS) for the maximum principal stress criterion [27]. Wu et al. then used molecular dynamics simulations to evaluate the initiation of crazing in a PS material using this critical stress [28].

In this study, the dynamic evolution of crazing during stretching experiments on regular and oriented PMMA was analyzed. The regular and oriented PMMA samples used in the experiment underwent various treatments, including machining and exposure to solvent environments. The mechanisms of crazing initiation, growth, thickness increase, and crack development under tensile stress were also explored. To comprehend the microscopic process of crazing growth in PMMA, the crazing characteristics of both oriented and regular PMMA were compared, and the impact of the PMMA crazing state was investigated.

2. Materials and Experiments

2.1. Materials

In this study, regular and oriented PMMA were tested using the tensile test method [29]. The experiment used alcohol (Shunxing Chemical Company, Shenyang, China), regular PMMA, and oriented PMMA as the sample materials. Regular PMMA was created through injection processing using PMMA particles with a molecular weight of about two million (PMMAIRL409 002 IRL Series, Mitsubishi Group, Tokyo, Japan). Oriented PMMA was created by heating regular PMMA materials above the glass transition temperature (Tg), stretching a plane in a specific direction, and cooling to obtain anisotropic physical and mechanical properties. As shown in Equation (1) [30], the drawn ratio was used to describe oriented PMMA.

$$\lambda = \sqrt{\frac{h_2}{h_1}} \tag{1}$$

Here, h_1 and h_2 are the height of the vertical stretching plane before and after oriented stretching, respectively. The type of oriented PMMA used in this experiment was YB-DM-10, which has a tensile ratio of 68%. The specific acquisition process is detailed in the literature [30]. The partial mechanical properties of regular and oriented PMMA are listed in Table 1.

Table 1. Partial material properties of regular and oriented PMMA.

Property	Regular PMMA	Oriented PMMA
Tensile Strength at 23 °C (MPa)	77.8	94.6
Tensile Strength at 100 °C (MPa)	1.84	28.4
Bending strength (MPa)	130	>205
Impact strength (kJm^{-2})	22.6	57.0
Break elongation (%)	4.0	7.7

2.2. Experiments

The size of regular and oriented PMMA specimens is depicted in Figure 1. Subject to a variety of treatments, including machining, solvent environment, and loading conditions, twelve groups of experiments were conducted with five samples per group based on the Taguchi method, as detailed in Table 2 [31]. The specimens were machined using JDHGT600T 3-Axis High-Speed Machining Centers(JDHTG Series, Beijing Jing Diao, Beijing, China) with a spindle speed of 6000 r/min and feed rate of 60 mm/min, followed by polishing to ensure better surface quality. The samples were then fully immersed in an alcohol solution and sealed for 20 days to facilitate the complete diffusion between the alcohol and PMMA. Groups 3, 7, and 10 were sealed in the air for 20 days. After 20 days of exposure to either an alcohol solution or air, uniaxial tensile stress was applied to PMMA in groups 3–4 and 7–12 using an electronic universal material testing machine (Instron 5982, Instron, Boston, MA, USA). The crazing growth state was recorded under varying loads using a polarizing microscope, whereas the stretched surface state was analyzed using a confocal microscope (VK-X250, Keyence, Osaka, Japan) and a scanning electron microscope (SU5000, HITACHI, Tokyo, Japan).

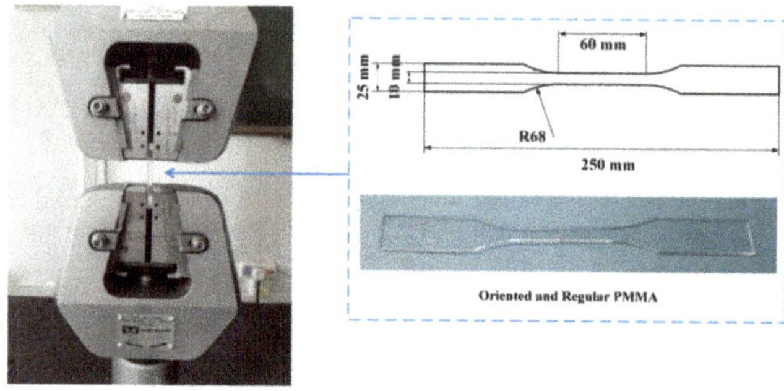

Figure 1. Experimental instruments and the size of regular and oriented PMMA specimens.

Table 2. Experimental parameters.

Group No.	Material	Machining	Solvent	Load
Group 1	regular PMMA	Non-milling	95% alcohol by volume	Standing
Group 2	oriented PMMA	Non-milling	95% alcohol by volume	Standing
Group 3	regular PMMA	Non-milling	Air	Uniaxial tension
Group 4	oriented PMMA	Non-milling	95% alcohol by volume	Uniaxial tension
Group 5	regular PMMA	Milling	95% alcohol by volume	Standing
Group 6	oriented PMMA	Milling	95% alcohol by volume	Standing
Group 7	regular PMMA	Milling	Air	Uniaxial tension
Group 8	regular PMMA	Milling	75% alcohol by volume	Uniaxial tension
Group 9	regular PMMA	Milling	95% alcohol by volume	Uniaxial tension
Group 10	oriented PMMA	Milling	Air	Uniaxial tension
Group 11	oriented PMMA	Milling	75% alcohol by volume	Uniaxial tension
Group 12	oriented PMMA	Milling	95% alcohol by volume	Uniaxial tension

3. Results and Discussion

3.1. *Effect of Solvent Atmosphere on Crazing Growth*

3.1.1. Effect of Solvent Atmosphere on Crazing Initiation without Tensile Stress

In this study, the influence of the solvent environment on crazing initiation in the absence of tensile stress was evaluated in Groups 1–2 and 5–6. We observed no crazing on the surface of PMMA without external stress; however, a complex swelling phenomenon related to localized stress was observed. The experiments showed that nonmilled regular PMMA and oriented PMMA in Groups 1–2 were immersed in alcohol and showed no crazing on the surface; instead, a swelling phenomenon appeared on both regular and oriented PMMA, which was evenly distributed in strips. Figure 2a shows that the swelling points on the surface of regular PMMA were dense and unevenly distributed, with numerous smaller-scale swelling points on the surface of larger-scale swelling points and a higher density for the derived swelling points. In contrast, Figure 2b shows that oriented PMMA primarily exhibited strip swelling points that were 60–80 μm in length with a small portion of 1–3 μm which spotted the swelling points. The arrangement of swelling points in oriented PMMA was regular and generally parallel, resulting in a more directional and consistent swelling phenomenon.

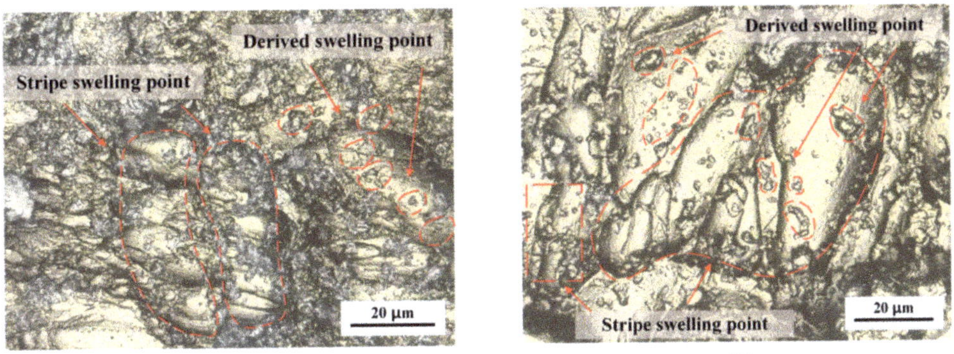

(**a**) (**b**)

Figure 2. Swelling of PMMA surface without machining for (**a**) Regular PMMA; (**b**) Oriented PMMA.

As depicted in Figure 3, the surface morphology of PMMA exhibited no crazing, but an increase in swelling points could be observed after machining. Machining caused concentrated stress points on the PMMA surface, influencing the crazing and swelling processes. The swelling points on PMMA in Groups 5–6, compared with the strip swelling points in Groups 1–2, were disordered in distribution. The height of the swelling points on the machined PMMA surface was elevated, and the swelling points were larger. The surface of the regular PMMA swelling points was rough and characterized by a complex morphology. Smaller derived swelling points, less than 1 μm in size, were observed to be intertwined with the machined surface and distributed around the central point. At the central point, crazing occurred in the same direction.

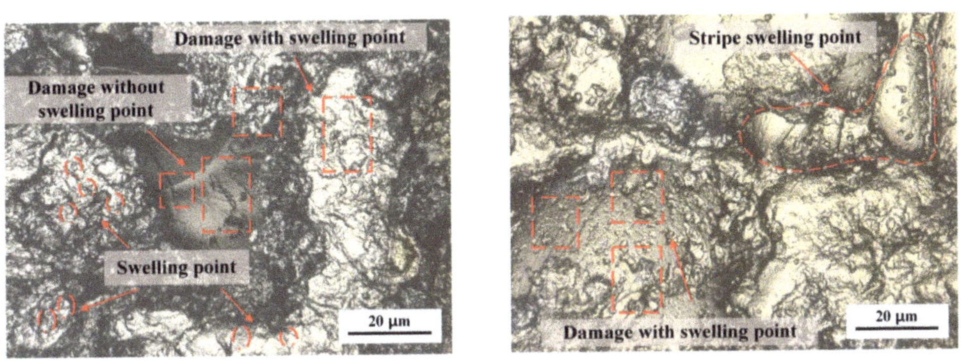

(**a**) (**b**)

Figure 3. Swelling of PMMA surface after machining with (**a**) Regular PMMA (**b**) Oriented PMMA.

In the regions of low molecular chain density in PMMA, micro-voids existed between the chains, and swelling mainly occurred in these micro voids. When PMMA came into contact with alcohol molecules, the two substances interacted through diffusion. Lisaka K proposed that the molecular weight directly affected the diffusion rate during this process [32]. Because a single PMMA molecule has a molecular weight thousands of times higher than that of an alcohol molecule, the diffusion rate of PMMA molecular chains was relatively static compared with that of alcohol molecules. Gradually, alcohol molecules were diffused into the micro-voids of PMMA, increasing the activity and kinetic energy of the PMMA molecular chains. This promoted their movement and the expansion of the region near the micro-voids, leading to the formation of swelling points. Brown H R and Kramer E J found that the internal tensile stress was the key factor leading to craze initiation by observing a mixed system of polymer and methanol [33]. The swelling points had poor surface morphology and caused concentrations and stress gradients within the matrix material surrounding them. Kausch, H.H. Dettenmaier discussed the molecular orientation at the initial stage of crazing and found that a regular molecular chain arrangement helped to establish the energy balance and reduced the possibility of the orientation mechanism transforming into crazed fiber [34]. PMMA with a more uniform molecular chain distribution, such as oriented PMMA, would be less susceptible to the formation of micro-voids during interactions with alcohol, reducing the likelihood of crazing and the associated crazing stress.

3.1.2. Effect of Solvent Atmosphere on Crazing Initiation with Tensile Stress

A comparison was made between the initiation and progression of crazing in Groups 3–4 under different solvent atmospheres, with and without 95% alcohol by volume. The term "stress at crazing" was used to refer to the stress level at which crazing began to occur, and the "sensitivity of crazing to stress" was defined as the ratio of the increase in crazing length to the increase in applied stress. The paper notes that all tensile stress was applied vertically.

Figure 4 presents the observation of crazing under tensile stress conditions. The results revealed that the crazing in oriented PMMA occurred perpendicular to the direction of tensile stress with a size ranging from 300 to 400 μm. The forward expansion of the crazing tip was primarily responsible for the growth in crazing, resulting in a slight increase in thickness. The width of the crazing was approximately 1 μm, and some derived crazing was observed. Regular PMMA also showed crazing perpendicular to the direction of tensile stress with a size ranging from 200 to 300 μm, but there was no derived crazing observed around the crazing. The formation of crazing and derived crazing on the surface of PMMA is linked to the entanglement network formed by the molecular chain. Kramer E J proposed a positive correlation between chain contour length *le* between entanglements and the extension ratio λ [35]. After directional stretching, the *le* of the oriented PMMA was much larger than that of regular PMMA, which also accounted for the longer crazing growth length in the former. Furthermore, Kramer E J and Donald A M jointly proposed that materials with a regular molecular chain arrangement were more likely to experience a large-scale entanglement motion during crazing formation [36]. As crazing was more easily generated with plane strain than plane stress, more derived crazing occurred around the oriented PMMA, which was primarily caused by the plastic deformation of the main crazing. The SEM results indicated that the crazing tip in oriented PMMA had a strong tendency to extend forward, and this extension resulted in the breaking of the supporting microfibers and the thickening of crazing along with the formation of derived crazing. Additionally, the crazing tip in regular PMMA did not show any forward extension, and the breaking of the supporting micro-fibers caused the crazing to be relatively regular.

The results of the alcohol treatment on PMMA under tensile stress are depicted in Figure 5. The crazing tips of both oriented and regular PMMA were observed to bend after alcohol exposure, causing a deviation from their perpendicular alignment with the direction of applied stress. There was a noticeable increase in the thickness of the crazing, with the end of the crazing on regular PMMA breaking, and this resulted in the formation of micro-cracks. The size of crazing in the oriented PMMA was measured between 300 and 400 μm, whereas the crazing on regular PMMA was characterized by a staggering pattern, with lengths of 400–500 μm and widths of 5–10 μm. The orientation of the crazing in the oriented PMMA changed only in size and growth direction. As the crazing tip continued to extend, thickening and fracture occurred. The internal micro-fibrils within the crazing were still visible. Furthermore, the center of the crazing in regular PMMA was found to be much larger than that of oriented PMMA, with a thicker and stronger tendency to give rise to additional crazing.

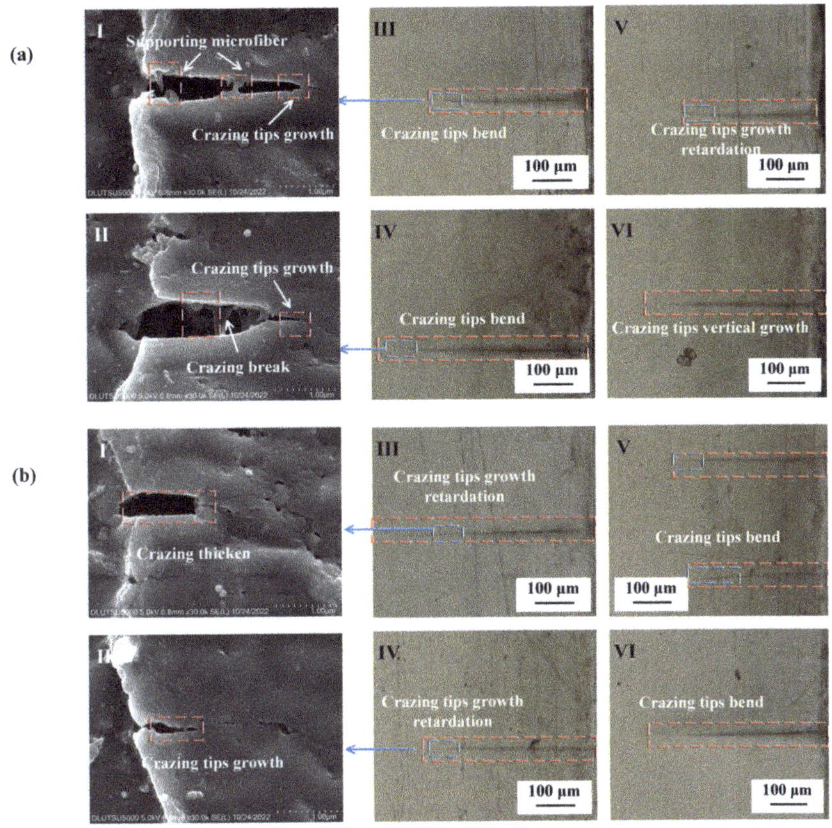

Figure 4. PMMA crazing growth state under tensile stress only. (**a**) Oriented PMMA. **I**: Crazing tip growth and supporting microfiber; **II**: crazing tip growth and break; **III–V**: crazing tip bend; **VI**: crazing tip growth retardation. (**b**) Regular PMMA. **I**: Crazing thickening and not supporting microfiber; **II**: crazing tip growth and slowly thickening; **III**,**IV**: crazing tip growth retardation; **V**,**VI**: crazing tips circuitously growth.

Figure 6 presents the relationship between the crazing length and stress, both before and after alcohol treatment, in regular and oriented PMMA obtained using a polarizing microscope. The results are summarized in Table 3, which displays the stress at which crazing occurred alongside the sensitivity of the crazing to stress. The stress value required for crazing to occur in regular PMMA decreased from 31.4 to 29.8 MPa after 95% alcohol treatment, representing a decrease of 1.6 MPa, which is an insignificant change. In contrast, the stress value in oriented PMMA fell from 51.4 to 43.4 MPa, highlighting a decrease of 8.0 MPa. This demonstrates that oriented PMMA exhibited superior resistance to crazing stress and reduced sensitivity to alcohol compared with regular PMMA. The growth of the crazing was observed to be stable.

The sensitivity of crazing to stress in regular PMMA increased from 0.049 mm/MPa after 95% alcohol treatment, and a linear relationship between the crazing length and tensile stress could still be observed. By contrast, the sensitivity of crazing to stress on the surface of oriented PMMA increased from 0.015 to 0.055 mm/MPa after alcohol treatment, and the relationship between the crazing length and tensile stress was no longer linear. A localized increase in sensitivity was observed when crazing occurred within a narrow range of stress (48.5–50.3 MPa), with a sensitivity of 0.071 mm/MPa, which was higher than that observed under other stress states (0.023 mm/MPa).

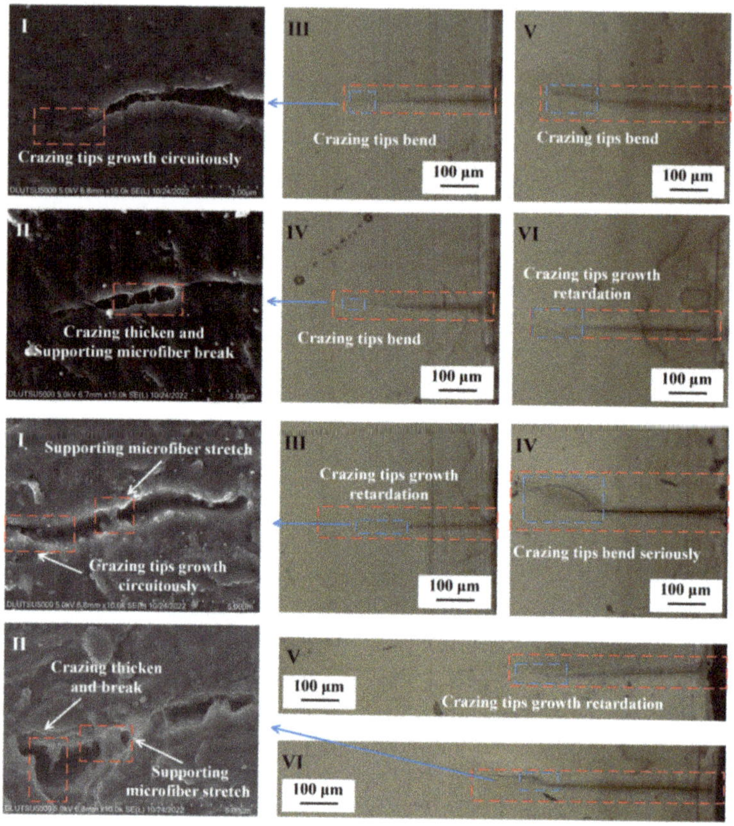

Figure 5. PMMA crazing growth state under tensile stress after alcohol treatment. (**a**) Oriented PMMA. **I**: Crazing tip circuitously grow; **II**: crazing slowly thickens and supports microfiber breaks; **III–V**: crazing tips bend; **VI**: growth retardation of crazing tips. (**b**) Regular PMMA. **I**: crazing tips circuitously grow, and supporting microfiber stretches; **II**: supporting microfiber stretches and breaks; **III**: growth retardation of crazing tips; **IV**: crazing tips serious bend. **V**: growth retardation of crazing tips; **VI**: crazing tips bend, ultimately perpendicular to the direction of tensile stress.

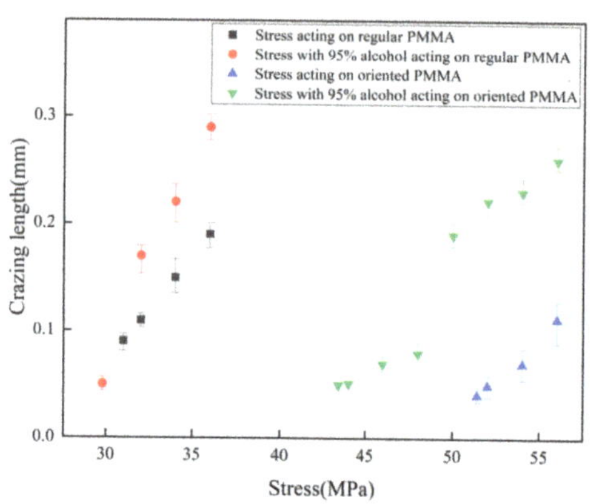

Figure 6. Effect of alcohol on crazing growth of PMMA without machining.

Table 3. Crazing growth state of PMMA with 95% alcohol.

Materials and Treatment	Stress at Crazing	Sensitivity of Crazing to Stress
Regular PMMA	31.4 MPa	0.02 mm/MPa
Regular PMMA with 95% alcohol	29.8 MPa	0.049 mm/MPa
Oriented PMMA	51.4 MPa	0.015 mm/MPa
Oriented PMMA with 95% alcohol	43.4 MPa	0.055 mm/MPa

Following alcohol treatment, the local expansion of the PMMA surface resulted in local plastic deformation and concentrated stress, lowering the material's anti-crazing stress. The effect of concentrated stress caused the crazing to rapidly grow when it passed through the swelling points during the growth process, indicating a high sensitivity to stress. In the regular PMMA, the distribution of the molecular chains was nonuniform, and there were microcavities between the molecular chains. Crazing was caused by the combination of microcavities and a break in the molecular chain. The concentrated stress caused by alcohol treatment resulted in a complex stress gradient with different stress gradient directions and sizes. During the growth process, the tip of the crazing bent was due to the combined effect of tensile and concentrated stresses. Additionally, the concentrated stress could also promote the generation of derived crazing. After directional stretching, the molecular chain distribution of the oriented PMMA became more uniform. The result of the concentrated stress was stress gradients that were close to each other in all directions. After alcohol treatment, the crazing tip of the oriented PMMA did not bend as much as that of regular PMMA. The crazing tip still had a strong tendency to extend forward. However, crazing would continue to rapidly grow rapidly when passing through the concentrated stress points, demonstrating the crazing's high sensitivity to tensile stress.

3.2. *Effect of Concentrated Stress on Crazing Initiation*

Figure 7 depicts the effect of machining on the crazing of the PMMA surface under tensile stress. The length of the crazing increased to between 400 and 700 μm, and its thickness increased to 10–100 μm after machining. A large amount of derived crazing was produced on both sides of the main crazing, and the direction of the crazing remained perpendicular to the direction of tensile stress. The tip of the crazing continued to extend forward, but at a reduced rate, as the length increased and the thickness grew. The tip was observed as bent, and the end of the crazing broke into a broad area. The interior of the crazing comprised numerous regular microcavities of varying sizes, and the tip continued to form microcavities and merge. The crazing on the surface of regular PMMA was severely bent; the total length of the main crazing was 500–900 μm, and the thickness was larger than 100 μm. The crazing growth direction was no longer perpendicular to the tensile stress direction, but the interior of the main crazing was still composed of a large amount of small crazing that was perpendicular to the tensile stress direction. The distance from the end of the crazing to the tip of the crazing was the measured crazing length. A large amount of derived crazing appeared on both sides, and the density of crazing in the regular PMMA was substantially higher than that of the oriented PMMA. The microcavities in the interior of the crazing had lost their regularity. To form the crack, the crazing on the surface of regular PMMA broke and merged with the larger-scale cavities.

The results of the study, shown in Figure 8 and Table 4, indicate that machining caused a decrease in the stress value at which crazing occurred in both regular and oriented PMMA. The decrease was 4.6 MPa for regular PMMA and 9.2 MPa for oriented PMMA, with the latter appearing more sensitive to machining. The sensitivity of crazing to stress in regular PMMA increased from 0.02 to 0.065 mm/MPa and, in oriented PMMA, from 0.015 to 0.04 mm/MPa. In the absence of machining, a linear relationship existed between the crazing and tensile stress, and the sensitivity was stable. However, after machining, crazing was more sensitive to stress at the initial stages of growth and experienced rapid growth as tensile stress increased. As the tensile stress continued to increase, the sensitivity of crazing to stress decreased, and its growth slowed, eventually breaking and transforming into a microcrack when reaching a length of 0.3–0.4 mm. The growth of the crazing was limited by the increasing thickness of the crazing and resulted in reduced sensitivity to stress.

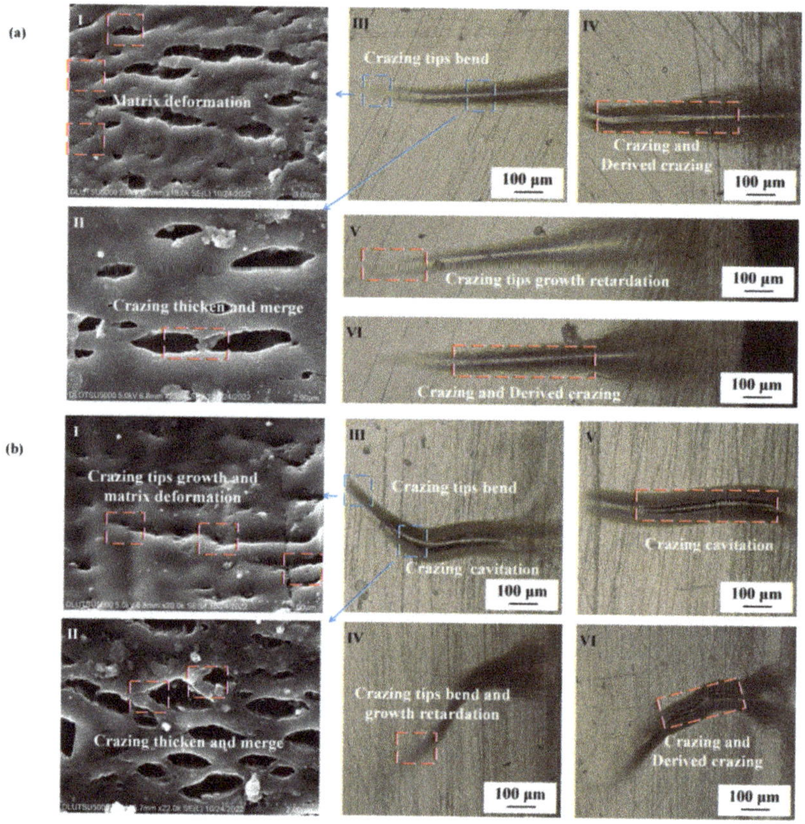

Figure 7. Crazing growth state of PMMA after milling under tensile stress. (**a**) Oriented PMMA. **I**: Crazing tip growth retardation and matrix deformation; **II**: crazing slowly thickening and supporting microfiber breaking; **III**,**IV**: crazing tip bending, causing derived crazing; **V**: crazing tip growth retardation. **VI**: large amount of derived crazing. (**b**) Regular PMMA. **I**: Crazing tip growth and matrix deformation; **II**: crazing thickening and merging; **III**: crazing tip bending and cavitation; **IV**: crazing tip bending and growth retardation; **V**: crazing cavitation; **VI**: resulting large amount of derived crazing.

Table 4. Crazing growth state of PMMA before and after machining.

Materials and Treatment	Stress at Crazing	Sensitivity of Crazing to Stress
Regular PMMA	31.4 MPa	0.02 mm/MPa
Regular PMMA after milling	26.8 MPa	0.065 mm/MPa
Oriented PMMA	51.4 MPa	0.015 mm/MPa
Oriented PMMA after milling	42.2 MPa	0.04 mm/MPa

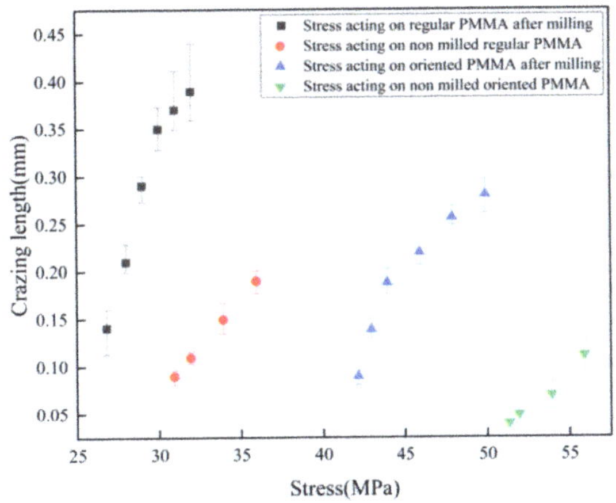

Figure 8. PMMA crazing growth state before and after machining.

For brittle PMMA materials, the coexistence and competition between shear and craze are the main damage mechanisms. On the undamaged surface, plastic deformation at the craze tip can partially inhibit the crazing process. However, surface damage induced by machining substantially reduced the stress that was required for craze initiation in materials [37]. Bucknall C. B. proposed the theoretical model of surface defects, which suggested that at the damage point, a surface defect was analogous to embedding a particle that could continuously exert stress on the surrounding material [38]. The initiation of crazing could be considered a frustrated fracture process rather than a yielding mechanism, as the presence of the surface defects created localized stress concentrations that could promote the formation and propagation of multiple crazing. This phenomenon was especially notable in brittle materials such as PMMA, where the interplay between shear plastic deformation and crazing failure was characterized by competition and coexistence, and the presence of surface damage further exacerbated this complex behavior. Residual stress was prevalent at the damage points in this experiment, and this stress state facilitated the development of frustrated fracturing under tensile loading, leading to a notable reduction in the anti-cracking stress of both oriented and regular PMMA materials. Calculations showed that a strain energy release rate (G_nasc) of less than 1 J/m was sufficient to generate a typical nascent craze with a thickness of 20 nm, which explained the low-stress requirement of only 20–30 MPa, which was able to cause crazing.

Under tensile stress, the crazing on the surface of oriented PMMA remained perpendicular to the direction of the stress, whereas regular PMMA experienced considerable bending. Saad-Gouider N proposed, through simulation, that the existence of surface defects could cause multiple crazes to form in different directions [39]. The rate and direction of multiple craze growth were affected by the size and direction of the stress gradient at the surface defect point. In this experiment, the anisotropic stress distribution at the surface defect of regular PMMA caused the bending phenomenon of crazing when the resultant force direction and the external stress direction were deflected.

3.3. Effects of Concentrated Stress and Solvent Atmosphere on PMMA Crazing Initiation

Figure 9 shows the state of PMMA surface crazing growth caused by the combined action of alcohol and machining. As shown in Figure 9a, the crazing on the surface of oriented PMMA was broken, and the cavity after the break was 500 µm long and 150 µm wide. The method for determining the crazing length was the same as that shown in Figure 7. The forward extension of the crazing tip is usually bent in two different directions. At the bending crazing tip, the crazing still grew forward, and the supporting microfiber at the crazing tip was generated and broke. As shown in Figure 9b, the scale of cavities on the surface of the regular PMMA was smaller, but the direction of the forward extension of the crazing tip was more random. The crazing in the cavity's front was also broken, but the crazing was mostly perpendicular to the tensile stress. The tip of the crazing was no longer perpendicular, and many of the supporting micro-fibers were broken.

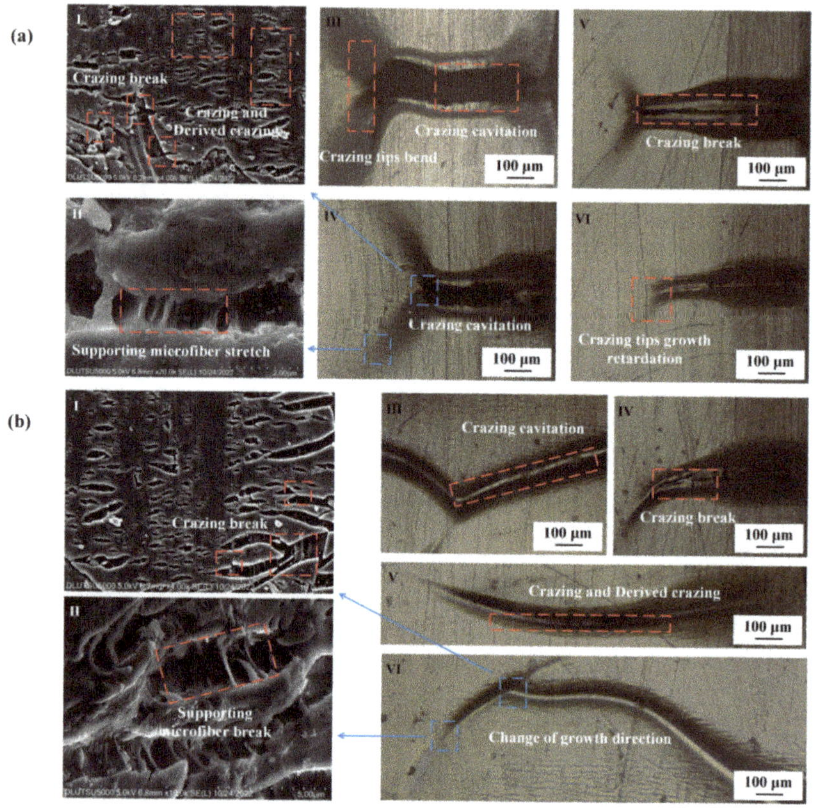

Figure 9. PMMA crazing state under the combined action of alcohol and machining. (**a**) Oriented PMMA. **I**: Crazing tip growth and matrix deformation; **II**: supporting microfiber stretching; **III**,**IV**: crazing cavitation; **V**,**VI**: crazing tip growth retardation. (**b**) Regular PMMA. **I**: Crazing breaking; **II**: supporting microfiber stretching and breaking; **III**: crazing cavitation, causing derived crazing; **IV**: crazing break; **V**: crazing thickening, causing derived crazing; **VI**: crazing tip growth retardation, and the growth direction suddenly changing.

Figure 10 illustrates the state of the crazing growth in both oriented and regular PMMA after being subjected to treatment with varying concentrations of alcohol (75% and 95% by volume) and air. The results presented in Table 5 reveal the stress levels at which crazing occurred, as well as its sensitivity to stress. Following machining, the stress required to induce crazing in oriented PMMA decreased from 42.2 to 31.8 and 31.4 MPa after being treated with 75% and 95% alcohol, respectively. A similar decrease was observed in regular PMMA, with the stress dropping from 26.8 to 23.4 and 22.8 MPa after treatment with the two alcohol concentrations, respectively. We found that treating PMMA with alcohol substantially reduced the tensile stress that was required to cause crazing, and the concentration of alcohol used had no effect on the crazing stress value.

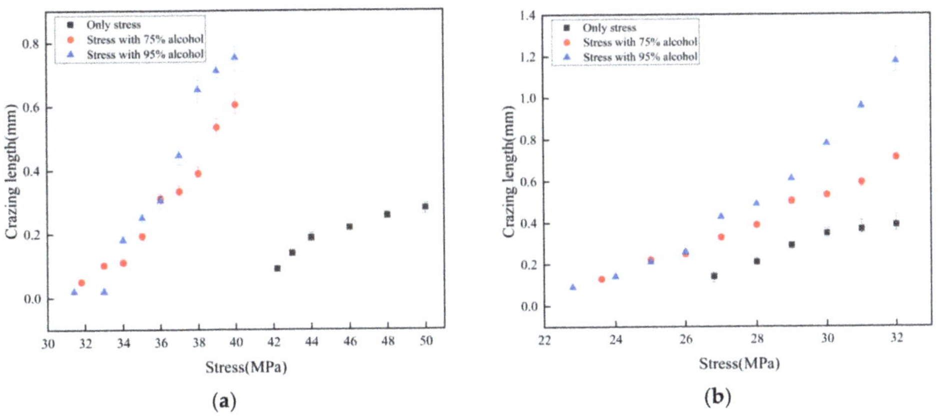

Figure 10. Crazing growth state under the combined action of alcohol and machining: (**a**) Oriented PMMA; (**b**) Regular PMMA.

Table 5. The crazing growth state of PMMA with the combined action of alcohol and machining.

Material and Treatment	Stress at Crazing	Sensitivity of Crazing to Stress
Oriented PMMA	42.2 MPa	0.02 mm/MPa
Oriented PMMA with 75% alcohol	31.8 MPa	0.035 mm/MPa
Oriented PMMA with 95% alcohol	31.4 MPa	0.055 mm/MPa
Regular PMMA	26.8 MPa	0.065 mm/MPa
Regular PMMA with 75% alcohol	23.4 MPa	0.077 mm/MPa
Regular PMMA with 95% alcohol	22.8 MPa	0.12 mm/MPa

The sensitivity of crazing to stress in PMMA after machining was influenced by the alcohol concentration. As the alcohol concentration increased from 75% to 95%, the stress sensitivity of oriented PMMA nonlinearly increased from 0.02 to 0.035 mm/MPa and then to 0.055 mm/MPa. The higher the alcohol concentration in oriented PMMA, the more sensitive crazing was to stress, resulting in the faster development of cracks. In the regular PMMA, the sensitivity of crazing to stress increased from 0.065 to 0.077 mm/MPa after treatment with 75% alcohol. The increase in sensitivity was observed as the tensile stress increased to 28.5 MPa, indicating that the presence of alcohol enhanced the development of crazing in PMMA.

Both alcohol solvents and machine damage can promote crazing initiation and progression. The physical diffusion of the alcohol solvent lowers the crazing stress threshold, whereas machining raises the stress of PMMA by creating concentrated stress points. When stress is superimposed on the surface of PMMA, the crazing of PMMA becomes extremely sensitive to tensile stress. The forward extension of the crazing tip and the thickening of the crazing simultaneously occur in the crazing growth process of oriented PMMA, and a dynamic balance can be achieved throughout the crazing growth process. When the two factors are applied to the surface of regular PMMA, the stress gradient limits the thickening of crazing. The forward extension of the crazing tip is continuously strengthened as the tensile stress increases. Additionally, the direction of the crazing tip growth is uncontrollable due to the different positions of the concentrated stress generated by the two factors. The network structure of an oriented PMMA molecular chain is more uniform after directional stretching, which effectively reduces the stress gradient that is generated by the concentrated stress and causes crazing to grow in a fixed direction.

4. Conclusions

The tensile experiments were used to analyze the crazing processes of oriented and regular PMMA after exposure to solvent atmospheres and machining. We determined that the solvent atmosphere could affect the initiation and growth of crazing in PMMA through the concentration and stress gradients generated by physical molecular diffusion. Machining also contributed to the crazing process by creating localized areas of concentrated stress. Both treatments had the potential to substantially decrease the anti-crazing stress in PMMA and enhance its sensitivity to stress during the growth of crazing. Maintaining uniform stress and strain is an effective means of reducing the crazing caused by machining and solvent environments. We also delved into the molecular chain change process and the behavior of PMMA under various influences during the crazing process and ultimately concluded the following:

- Physical diffusion causes swelling points on the surface of PMMA due to the effect of the solvent atmosphere. Regular PMMA surface swelling points are distributed in 1–3 μm spots, whereas oriented PMMA surface swelling points are distributed in 20–40 μm strips. At the swelling points, concentration and stress gradients arise; this increases the growth rate of the crazing tip and reduces the stress on PMMA against crazing.
- Machining primarily affects the growth of crazing through residual stress. The application of tensile stress leads to localized plastic deformation around areas of concentrated stress, resulting in the formation of micro-cavities. These micro-cavities eventually merge to form crazing.
- The tip of the crazing in regular PMMA severely bends due to the uneven distribution of molecular chains. Orientated PMMA, however, can effectively alleviate the residual stress that is caused by machining through directional stretching. The direction of tensile stress is always perpendicular to the direction of crazing growth.
- The critical stress threshold for PMMA crazing is diminished by both machining and exposure to solvent atmospheres. On average, the initiation stress of crazing was reduced by 20%, and the sensitivity of crazing to stress was elevated by 0.35–0.45 mm/MPa.

Author Contributions: Conceptualization, Y.Y. and P.Z.; methodology, Y.Y.; software, Y.S.; validation, Y.Y. and P.Z.; formal analysis, Y.S.; investigation, Y.S.; resources, Y.S. and B.L.; data curation, Y.S. and J.S.; writing—original draft preparation, Y.Y.; writing—review and editing, Y.Y.; visualization, P.Z.; supervision, P.Z.; project administration, P.Z.; funding acquisition, Y.Y. and P.Z. All authors have read and agreed to the published version of the manuscript.

Funding: This research was funded by the National Natural Science Foundation of China (No. 51975094, 51875078, 51991372), and the National Key Research and Development Program of China (No. 2019YFA0709102, No. 2020YFA0714502).

Institutional Review Board Statement: Not applicable.

Data Availability Statement: Data sharing is not applicable to this article.

Conflicts of Interest: The authors declare no conflict of interest.

References

1. Katti, K.S. Biomaterials in Total Joint Replacement. *Colloids Surf. B Biointerfaces* **2004**, *39*, 133–142. [CrossRef] [PubMed]
2. Wu, C.; Jin, Z.; Wang, H.; Ma, H.; Wang, Y. Design and Fabrication of a Nanofluidic Channel by Selective Thermal Oxidation and Etching Back of Silicon Dioxide Made on a Silicon Substrate. *J. Micromechanics Microeng.* **2007**, *17*, 2393–2397. [CrossRef]
3. Ramakrishna, S.; Mayer, J.; Wintermantel, E.; Leong, K.W. Biomedical applications of polymer-composite materials: A review. *Compos. Sci. Technol.* **2001**, *61*, 1189–1224. [CrossRef]
4. Zaghloul, M.Y.; Zaghloul, M.M.Y.; Zaghloul, M.M.Y. Influence of Stress Level and Fibre Volume Fraction on Fatigue Performance of Glass Fibre-Reinforced Polyester Composites. *Polymers* **2022**, *14*, 2662. [CrossRef]
5. Zaghloul, M.M.Y.; Zaghloul, M.Y.M.; Zaghloul, M.M.Y. Experimental and Modeling Analysis of Mechanical-Electrical Behaviors of Polypropylene Composites Filled with Graphite and MWCNT Fillers. *Polym. Test* **2017**, *63*, 467–474. [CrossRef]

6. Nabhan, A.; Taha, M.; Ghazaly, N.M. Filler Loading Effect of Al_2O_3/TiO_2 Nanoparticles on Physical and Mechanical Characteristics of Dental Base Composite (PMMA). *Polym. Test* **2023**, *117*, 107848. [CrossRef]
7. Berger, L.L.; Kramer, E.J. The Effect of Temperature on the Transition from Crazing to Shear Deformation in Crosslinked Polystyrene. *J. Mater. Sci.* **1988**, *23*, 3536–3543. [CrossRef]
8. Khoshaim, A.B.; Elsheikh, A.H.; Moustafa, E.B.; Basha, M.; Showaib, E.A. Experimental Investigation on Laser Cutting of PMMA Sheets: Effects of Process Factors on Kerf Characteristics. *J. Mater. Res. Technol.* **2021**, *11*, 235–246. [CrossRef]
9. Hoogenboom, R.; Becer, C.R.; Guerrero-Sanchez, C.; Hoeppener, S.; Schubert, U.S. Solubility and thermoresponsiveness of PMMA in alcohol-water solvent mixtures. *Aust. J. Chem.* **2010**, *63*, 1173–1178. [CrossRef]
10. Kambour, R.P. Mechanism of Fracture in Glassy Polymers. III. Direct Observation of the Craze Ahead of the Propagating Crack in Poly(Methyl Methacrylate) and Polystyrene. *J. Polym. Sci. Part A-2 Polym. Phys.* **1966**, *4*, 349–358. [CrossRef]
11. Kramer, E.J.; Krenz, H.G.; Ast, D.G. Mechanical Properties of Methanol Crazes in Poly(Methyl Methacrylate). *J. Polym. Sci. Polym. Phys. Ed.* **1978**, *16*, 349–366. [CrossRef]
12. Andrews, E.H.; Levy, G.M.; Willis, J. Environmental crazing in a glassy polymer: The role of solvent absorption. *J. Mater. Sci.* **1973**, *8*, 1000–1008. [CrossRef]
13. Kramer, E.J.; Bubeck, R.A. Growth Kinetics of Solvent Crazes in Glassy Polymers. *J. Polym. Sci. Polym. Phys. Ed.* **1978**, *16*, 1195–1217. [CrossRef]
14. Thomas, N.L.; Windle, A.H. Diffusion mechanics of the system PMMA-methanol. *Polymer* **1981**, *22*, 627–639. [CrossRef]
15. Luo, W.; Yang, T.; Zhang, P. Research Progress on meso damage evolution of polymers. *Prog. Mech.* **2001**, *31*, 264–275.
16. Thuy, M.; Pedragosa-Rincón, M.; Niebergall, U.; Oehler, H.; Alig, I.; Böhning, M. Environmental Stress Cracking of High-Density Polyethylene Applying Linear Elastic Fracture Mechanics. *Polymers* **2022**, *14*, 2415. [CrossRef]
17. Yildirim, F.F.; Sezer Hicyilmaz, A.; Yildirim, K. The Effects of the Weathering Methods on the Properties of the ABS, ASA and PMMA Polymers. *Polym. Test* **2022**, *107*, 107484. [CrossRef]
18. Razavi, M.; Cheng, S.; Huang, D.; Zhang, S.; Wang, S.Q. Crazing and yielding in glassy polymers of high molecular weight. *Polymer.* **2020**, *197*, 122445. [CrossRef]
19. Michler, G.H. Correlation between craze formation and mechanical behaviour of amorphous polymers. *J. Mater. Sci.* **1990**, *25*, 2321–2334. [CrossRef]
20. Gao, G.; Xing, Y. Monitoring the evolution of crazing damage in an area under stress concentration via acoustic emission. *Int. J. Precis. Eng. Manuf.* **2018**, *19*, 561–568. [CrossRef]
21. Wang, Y.; Yu, B.; Wu, H.; Lu, K.; Wang, D. Dynamic Crack Growth Characteristics of Half-Circular Disk PMMA Specimens with Prefabricated Penetration Defects under Impact Loading. *Polym. Test* **2020**, *91*, 106684. [CrossRef]
22. Krenz, H.G.; Ast, D.G.; Kramer, E.J. Micromechanics of Solvent Polystyrene Crazes in Polystyrene. *J. Mater. Sci.* **1976**, *11*, 2198–2210. [CrossRef]
23. Wang, Y.; Wang, S.; Liu, K.; Gan, Y.; Han, L.; Jiang, S. Effect of Indirect Cryogenic Cooling on the Machining Accuracy and Tool Vibration in the Turning of Polysulfone. *J. Manuf. Sci. Eng.* **2021**, *144*, 061004. [CrossRef]
24. Lang, R.W.; Manson, J.A.; Hertzberg, R.W.; Schirrer, R. Craze development in poly (methyl methacrylate) during stable fatigue crack propagation. *Polym. Eng. Sci.* **1984**, *24*, 833–842. [CrossRef]
25. Jie, M.; Li, Y.; Deng, Z. Crazing damage field at crack tip of amorphous polymer. *J. Appl. Mech.* **1998**, *15*, 7–12.
26. Argon, A.S. Physical Basis of Distortional and Dilational Plastic Flow in Glassy Polymers. *J. Macromol. Sci. Part B* **1973**, *8*, 573–596. [CrossRef]
27. Socrate, S.; Boyce, M.C.; Lazzeri, A. A micromechanical model for multiple crazing in high impact polystyrene. *Mech. Mater.* **2001**, *33*, 155–175. [CrossRef]
28. Wu, Z.; Zhang, B. Molecular dynamics simulation of polymer crazing at different strain rates. *Polym. Mater. Sci. Eng.* **2012**, *28*, 169–172.
29. Zhai, G.; Ding, Y.; Ma, Z.; Wei, Z.; Li, X.; Liu, B. Novel Triaxial Experimental Investigation on Compressive Behavior of Hollow Glass Microspheres Composites under Varied Temperature Environments. *Polym. Test* **2022**, *115*, 107745. [CrossRef]
30. Wang, X.K.; Wei, S.C.; Xu, B.S.; Chen, Y.; Yan, X.; Xia, H.H. Transparent organic materials of aircraft cockpit canopies: Research status and development trends. *Mater. Res. Innov.* **2015**, *19* (Suppl. 10), S10-199–S10-206. [CrossRef]
31. Thirumalai, R.; Senthilkumaar, J.S.; Selvarani, P.; Ramesh, S. Machining Characteristics of Inconel 718 under Several Cutting Conditions Based on Taguchi Method. *Proc. Instig. Mech. Eng. C J. Mech. Eng. Sci.* **2013**, *227*, 1889–1897. [CrossRef]
32. Iisaka, K.; Hayama, M.; Fukami, A. Dependence of Solvent Crazing in Bisphenol—A Polycarbonate on Molecular Weight of Organic Liquids. *J. Macromol. Sci. Part B* **1988**, *27*, 385–405. [CrossRef]
33. Ntal Erate, Q.A.; Brown, H.R.; Kramer, E.J. Effect of Surface Tension on the Stress in Effect of Surface Tension on the Stress in Environmental Crazes. *Polymer* **1981**, *22*, 687–690.
34. Kausch, H.H.; Dettenmaier, M. Polymer Bulletin Some Molecular Aspects of Craze Formation. *Polym. Bull.* **1980**, *3*, 565–570. [CrossRef]
35. Donald, A.M.; Kramer, E.J. Effect of Molecular Entanglements on Craze Microstructure in Glassy Polymers. *J. Polym. Sci. Part A-2 Polym. Phys.* **1982**, *20*, 899–909. [CrossRef]
36. Kramer, E.J. *Micromechanical Measurements of Polymers by Transmission Electron Microscopy*; ACS Publications: Philadelphia, PA, USA, 1983.

37. Aegon, A.S.; Hannoosh, J.G. Initiation of Crazes in Polystyrene. *Philos. Mag.* **1977**, *36*, 1195–1216. [CrossRef]
38. Bucknall, C.B. New Criterion for Craze Initiation. *Polymer* **2007**, *48*, 1030–1041. [CrossRef]
39. Saad-Gouider, N.; Estevez, R.; Olagnon, C.; Séguéla, R. Calibration of a Viscoplastic Cohesive Zone for Crazing in PMMA. *Eng. Fract. Mech.* **2006**, *73*, 2503–2522. [CrossRef]

polymers

MDPI

Article

Simulation of Cone-Jet and Micro-Drip Regimes and Printing of Micro-Scale Patterns on PET Substrate

Dazhi Wang [1,2,3,*], Zeshan Abbas [1], Liangkun Lu [1], Shiwen Liang [2], Xiangyu Zhao [1], Pengfei Xu [1], Kuipeng Zhao [1], Liujia Suo [1,2], Yan Cui [1,2], Penghe Yin [1,2], Bin Tang [4], Jin Xie [4], Yong Yang [5] and Junsheng Liang [1,2]

1 Key Laboratory for Micro/Nano Technology and System of Liaoning Province, Dalian University of Technology, Dalian 116024, China; hopenotout1214@mail.dlut.edu.cn (Z.A.); llk@mail.dlut.edu.cn (L.L.); zxy4195@dlut.edu.cn (X.Z.); xupfupc@163.com (P.X.); kuipengzhao@mail.dlut.edu.cn (K.Z.); suolj@dlut.edu.cn (L.S.); yanc@dlut.edu.cn (Y.C.); phyin@dlut.edu.cn (P.Y.); engineeringjsliang@dlut.edu.cn (J.L.)
2 Ningbo Institute, Dalian University of Technology, Ningbo 315000, China; liangsw_nbi@dlut.edu.cn
3 Key Laboratory for Precision and Non-Traditional Machining Technology of Ministry of Education, Dalian University of Technology, Dalian 116024, China
4 Laboratory of Precision Manufacturing Technology, CAEP, Mianyang 621900, China; john46311@hotmail.com (B.T.); xiejin_nwpu@163.com (J.X.)
5 Ningbo Yongxin Optics Co., Ltd., Ningbo 315000, China; yy@yxopt.com
* Correspondence: d.wang@dlut.edu.cn; Tel.: +86-411-84707170; Fax: +86-411-84707940

Abstract: The fabrication of various micro-patterns on polymer insulating substrates is a current requirement in micro-electromechanical system (MEMS) and packaging sectors. In this paper, we use electrohydrodynamic jet (E-Jet) printing to create multifaceted and stable micro-patterns on a polyethylene terephthalate (PET) substrate. Initially, simulation was performed to investigate optimized printing settings in phase field physics for the usage of two distinct functional inks. A series of simulation experiments was conducted, and it was determined that the following parameters are optimised: applied pressure of 40 kPa, high pulse voltage of 1.95 kV, low dc voltage of 1.60 kV, duty cycle of 80%, pulse frequency of 60 Hz, printing height of 0.25 mm, and printing speed of 1 mm/s. Then, experiments showed that adjusting a pressure value of 40 kPa and regulating the SEMICOSIL988/1 K ink to print micro-drops on a polymer substrate with a thickness of 1 mm prevents coffee staining. The smallest measured droplet size was 200 μm. Furthermore, underfill (UF 3808) ink was driven with applied pressure to 50 kPa while other parameters were left constant, and the minimum size of linear patterns was printed to 105 μm on 0.5-mm-thick PET substrate. During the micro-drip and cone-jet regimes, the consistency and diameter of printed micro-structures were accurately regulated at a pulse frequency of 60 Hz and a duty cycle of 80%.

Keywords: MEMS devices; PET substrate; direct writing method; drop-on-demand; micro-patterns

check for updates

Citation: Wang, D.; Abbas, Z.; Lu, L.; Liang, S.; Zhao, X.; Xu, P.; Zhao, K.; Suo, L.; Cui, Y.; Yin, P.; et al. Simulation of Cone-Jet and Micro-Drip Regimes and Printing of Micro-Scale Patterns on PET Substrate. *Polymers* **2022**, *14*, 2683. https://doi.org/10.3390/polym14132683

Academic Editor: Hai-Feng (Frank) Ji

Received: 2 June 2022
Accepted: 28 June 2022
Published: 30 June 2022

1. Introduction

The fabrication of different micro/nano-structures on polymer insulating substrates using E-Jet printing technology has generated widespread interest due to its extensive applications in flexible electronic devices [1,2], the biomedical sector [3], optoelectronic fields [4] and other different disciplines. The use of E-Jet printing technology for micro/nano-patterns have many advantages compared to traditional photolithography and ink-jet printing, i.e., a three-dimensional structuring [5], a high-resolution patterning technique [6], high efficiency of resource dissipation [7], high-speed printing with nano-sized structures [8], and a contactless method [9].

The E-Jet printing technology is usually categorized into direct writing and drop-on-demand type, depending on the mechanisms used to generate continuous-line structures and micro-droplets [10,11]. The technology has recently attracted attention for printing micro/nano-sized patterns because it can print much smaller micro-structures than the needle diameters [12]. Furthermore, many versatile materials, such as metal, organic, and even biological material, can be used without thermal damage. The needle structure used in this technology is simpler than that used in other types. In the E-Jet method, the charged fluid is extracted from the needle tip when the electrical force locally exceeds the surface tension of functional fluid [13]. There are different spraying modes with various E-Jet structures and breaking mechanisms, depending on the applied pressure, flow rate, applied voltages, liquid properties and needle configuration [7]. As a result, E-Jet printing has various limitations and drawbacks that should be addressed. For example, at a low pressure, E-Jet printing has poor print rates that can interfere with micro-structure resolution. Material is consumed or lost during cone-jet disruptions, and the air stream surrounding the interface becomes unstable. There are difficulties when printing droplets of the same size at high pressure and increasing the coffee stain effect.

To date, many studies have presented different high-resolution linear patterns and DoD micro-structures on insulating substrates. Xu et al. emphasized the meniscus dynamics and jetting properties of the E-Jet printing technique. The research was applied to overcome needle obstruction problems during the pulsing drop regime [14]. In a review employing the E-Jet 3D printing technology, Wu et al. sought to produce alternative scaffolds for several specific biological applications [15]. Zou et al. created micro/nanopatterns on polymer substrates by combining E-Jet printing with a wet metal etching method. In the realm of printed electronics fabrication technology, this process is regarded as simple and effective [16]. Liu et al. investigated the droplet trajectory caused by different factors, such as needle size and needle angle [17]. Yang et al. studied E-Jet printing under the influence of various parameters, such as voltage, flow rate and print distance, in a simulation for fabrication micro-structures [18]. Wang et al. simulated and printed micro-drops on the polymer substrate using E-Jet printing to reduce the coffee ring effect under specific conditions [19]. The micro-structures are printed using E-Jet technology by generating a stable cone-jet mode [14,15] and the micro-dripping mode [16–19]. In both cases, the charged liquid cone-jet strikes the substrate surface to form a linear line before the jet disintegrates into a multi-jet shape [5]. Similarly, the micro-structures produced using the continuous cone-jet formation only form continuous lines and micro-droplets because the ejected ink cone-jet is continuously deposited on the substrate surface [20]. The micro-dripping jet is reported to produce micro-droplets using the DoD E-Jet method. The micro-dripping mode occurs when the pulse voltage is applied [21], or a lower DC voltage is used [17]. Similarly, the modulated AC-pulse was used to print micro-scale droplets on flexible insulating substrates by deploying the E-Jet printing technique. Hence, the silver lines and droplets were printed on the insulating substrate, which has a diameter of 20 μm [22]. In our previous work, Abbas et al. performed a numerical simulation of E-Jet and printing micro-droplets on a flexible substrate. The study calculated the optimized parameters for the development of a DoD E-Jet to print stable micro-structures on a PET substrate [2]. In another study, Abbas et al. introduced a numerical simulation of stable cone-jet formation [23] and printed direct writing structures on flexible substrate. The study offered a 2-phase-field method to optimize parameters for stable cone-jet morphology on a PET substrate utilizing a set of copper control electrodes [24–26]. Subsequently, to the best of our knowledge, there are a few research studies in the literature on the variation in droplet fabrication in terms of stability and consistency printed on the polymer substrates by regulating pulse voltage. The use of the combined needle is also limited to discussing cone-jet and micro-drip regimes for printing on PET substrates. In addition, any visual results on controlling the micro-structures' size and the resolution of the printed droplets and linear patterns still face challenges when using the E-Jet printing technology due to the unfavourable properties of insulating materials [27–29].

The objective of this research is to use different functional fluids to regulate and sense the bulging and coffee stain effect in multifarious micro-structures produced on a flexible PET substrate. The goal is to introduce a new development in E-Jet printing process and to fabricate micro-patterns that can be used for the application of MEMS devices. After a series of tests, the simulation was run to obtain previously unheard-of values. Within the cone-jet and micro-drip regimes, the resolution and diameters of the printed linear and microscopic patterns were regulated. The experimental work analyses and illustrates the properties and overhead projection display of the printed micro-structures under the constant variation in DC pulse voltages during a periodic pulsed cone-jet regime by correlating these factors (e.g., projection display in microstructures) with the simulation results. As a result, the printed micro-structures are considered essential to be used in the application of different M/NEMS devices by employing E-Jet printing technology.

2. Materials and Methods

2.1. Simulation of Electrohydrodynamic Jet

In this simulation investigation, the phase field approach was used to generate stable cone-jet and micro-drip regimes. The phase field approach was exploited to determine the cone-jet's morphological properties. As a result, many types of forces, such as viscous force, gravitational force, surface tension, and electrical force, play a critical role in the stability of the cone-jet regime. The phase field simulation has already shown that E-Jet printing factors, such as flow rate, pulse frequency, and applied voltage, influence the cone-jet shape and droplet sizes [12]. The phase field approach is preferred in this study to control the influence of different functional fluids on the creation of stable cone-jet and micro-drip regimes in the quest to obtain flexible micro-fabrication on polymer substrates such as PET. During the simulation process and under the influence of electric and hydrodynamic fields, the conservation of mass and Navier–Stokes equations are solved. Then the electrical shearing force and the electrical field jointly induced high viscous shearing force and internal pressure on the inner functional ink. Figure 1 shows a schematic depiction of the distribution forces operating around the needle interface. The simulation is employed in a variety of phases, including fluid and air, which are believed to be incompressible and immiscible [19]. Equation (1) expresses the phases of functional ink.

$$\nabla . u_{ij} = 0 \tag{1}$$

where u_{ij} denotes two distinct functional ink flow rates.

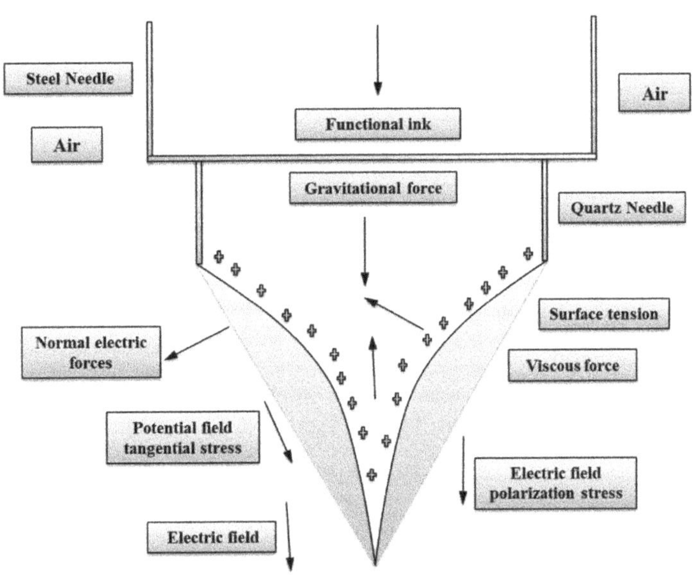

Figure 1. Forces acting around Taylor cone in the E-Jet printing process.

The Taylor–Melcher leaky dielectric model underpins the phase field approach. At the needle apex, the influence of different forces was detected. The Navier–Stokes equation, which is provided in Equation (2), was used to construct the balance of boundary conditions such as electrostatic and hydrodynamic fields [30].

$$\rho \frac{\partial \vec{u}}{\partial t}.(\vec{u}.\nabla).\vec{u} = \nabla p. + F_{st} + F_{es} + F_v + F_g + F_i \tag{2}$$

where F_{st}, F_{es}, F_v, F_g, and F_i are the surface tension, electrical force, volume force, gravity force, and pressure force created by inner functional ink, respectively. As a result, a continuity equation representing a driving force larger than zero was solved during the phase field approach to construct the interface between the liquid and air phases. Equation (1) shows the equations for the combined needle interface (3) and (4).

$$\begin{aligned} &\tfrac{\partial \phi_i}{\partial t} - \nabla.\vec{\mu}\phi_i = (\nabla.\tfrac{\gamma\lambda}{\sum_T}).\Delta\psi, \phi = phipf \\ &\lambda = \tfrac{3\varepsilon phi\sigma}{\sqrt{8}}, \gamma = X \end{aligned} \tag{3}$$

$$\begin{cases} \frac{\partial \phi_i}{\partial t} - \nabla.\vec{\mu}\phi_i = (\nabla.\frac{M_o}{\sum_T}).\Delta\phi_i \\ \phi_i = \frac{2\sum_T}{\theta}\sum_i \prod i \mp j - [\frac{1}{\sum_j}(0 \le \phi \le 1)](\frac{1}{2} - 2\varepsilon c \Delta\theta\psi) \end{cases} \\ \phi_A = phiA, \phi_B = phiB \tag{4}$$

where ϕ describes the transition state phase field variables based on Cahn–Hilliard equations [31]. The electric field is created by establishing the electrostatic charge physics surrounding the needle outlet wall, which results in body forces, as shown in Equation (5)

$$\vec{F}_{es} = q\vec{E} - \frac{1}{2}E^2\varepsilon + \nabla G\varepsilon_0 \tag{5}$$

where G is a driving force, which is also known as chemical potential and given in Equation (6)

$$G = \lambda\left[-(\nabla^2.\varphi\varphi + \frac{\varphi.(\varphi - 1)}{\theta^2})\right] = \frac{\delta\lambda}{\delta\theta^2}(\psi\varphi) \tag{6}$$

where λ is the combined energy density of the two separate functional fluids, and similarly δ is the width of a metal needle, also known as an interface thickness. The contact angle θ^2 between the polymer substrate and the functional ink is determined by computing the body forces.

2.2. Optimized Micro-Fabrication Design Considerations

COMSOL multi-physics software was used to simulate the cone-jet and micro-drip regimes. The combined needle system was created to validate the experimental design for the micro-fabrication process. The 2D axisymmetric drawing was primarily generated to build an accurate model to pursue suitable simulation results, as illustrated in Figure 2. The geometry of the combined needle created in this simulation model is shown in Figure 2a. Similarly, Figure 2b depicts the model's boundary conditions and tighter, user-controlled meshing, which can have a considerable influence on simulation fidelity. To execute the E-Jet printing model in the phase field domain, an axisymmetric geometry was created. As can be observed, the cone-jet regime in E-Jet technology involves significant electrostatic physics, which influence the charge density distribution around the combined needle. The findings were achieved at the optimal settings, where a series of simulation tests were carried out to determine the appropriate parameters for two distinct functional inks of UF 3808 and SEMICOSIL988/1K. The simulation study focuses on the regime transitions at different time intervals of a two-phase interface. Table 1 assumes and defines the boundary conditions of the electric charge field and the two fluid fields. Furthermore, the functional restrictions of liquid interfacial tension and air pressure applied to the steel needle during the cone-jet production phase had a significant influence, lowering the diameter of the jet to become even smaller than the inner diameter of a needle.

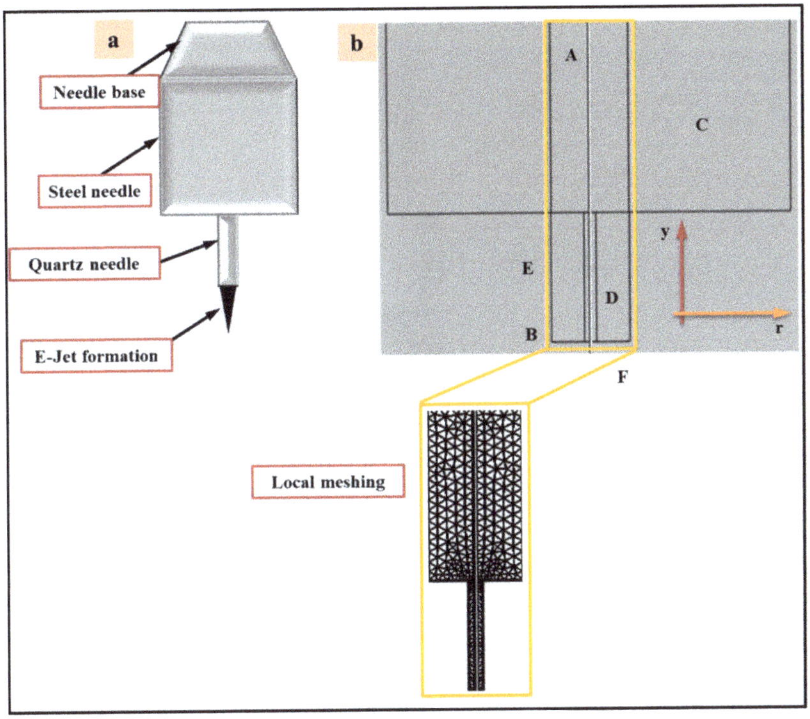

Figure 2. The numerical simulation of E-Jet printing (**a**) combined needle structure (**b**) The geometric model and basic boundary conditions and the local refined meshing scheme.

Table 1. The boundary conditions for electric charge field and various fluid fields.

Boundary Condition	Electric Charge Field	Fluid Field
A: Steel needle inlet	$\phi = V_0$	$u = Q_{metal}/A_{metal}$
B: Quartz needle inlet	$\phi = V_0$	$u = Q_{quartz}/A_{quartz}$
C: Wall of steel needle	$\phi = V_0$	$u = 0$
D: Wall of quartz needle	$\phi = V_0$	$u = 0$
E: Axisymmetric model	$\phi r = 0$	$u_r = 0$
F: Outlet of needle	$\phi = 0$	$p = 0$

A series of simulation tests were carried out in parallel using two distinct functional inks to establish the optimal printing parameters of multidimensional micro-structures. Then, during the micro-drip regime, the applied pressure of 40 kPa, DC high-pulse voltage of 1.95 kV, duty cycle of 80%, pulse frequency of 60 Hz, printing height of 0.25 mm, and printing speed of 1 mm/s were chosen as optimum parameters. During the simulation, the functional ink of SEMICOSIL988/1K was employed to establish a micro-drip regime and unique parameters were found. Following that, we ran a series of simulations with UF 3808 as the functional ink to produce the cone-jet regime. We discovered this by holding the entire parameters constant and raising the applied pressure to 50 kPa at 1.60 kV. This shows that utilising the value of the applied dc voltage may continually maintain the cone-jet regime and then generate the suitable Taylor cone form. The flexible micro-structures were then directly printed on the polymer PET substrate during the experimental investigation.

The phase field model created in this study is a helpful tool for studying the E-Jet printing process on polymer substrates. The E-Jet approach is a potential way of creating micro-scale structures for M/NEMS equipment. Figure 3 depicts the specific simulated outcomes under the improved settings for the cone-jet and micro-drip regimes.

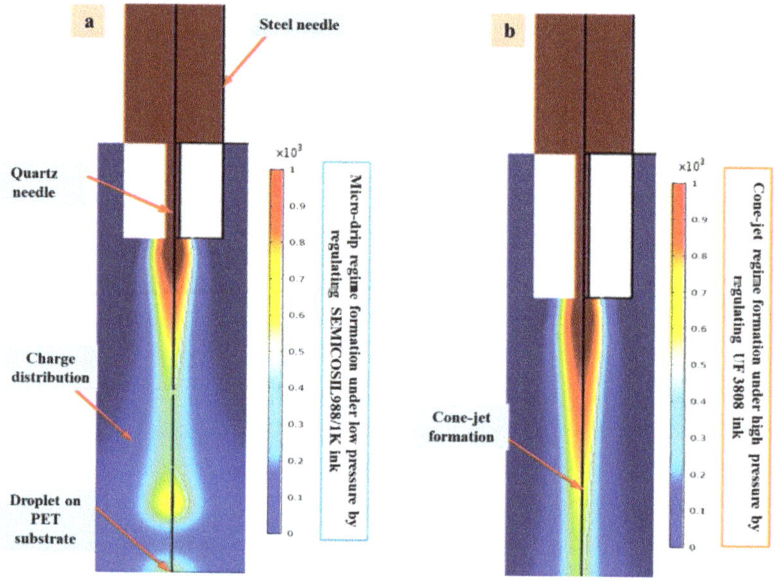

Figure 3. (**a**) Photograph of specific simulated outcomes obtained under optimum settings during the micro-drip regime (**b**) image of specific simulated results obtained with optimal settings during the cone-jet regime.

2.3. Experimental Details

The experimental setup was established in an existing study based on E-Jet printing technology, which consists of a combined needle system (steel and quartz), an air pump dispensing controller (PHD ULTRATM, Harvard Appsratus, Holliston, MA, USA), a DC pulse power supply, a computer-controlled X-Y-Z movement station and a microscopic vision system [2]. The inner and outer diameters of the steel needle were 365 μm and 700 μm, respectively. Similarly, the inner and outer diameters of the quartz needle were maintained at 50 μm and 365 μm, respectively. The distance between the combined needle and the substrate was maintained at 1 mm. Similarly, a flow diagram of the E-Jet printing process with pulse voltage implementation is shown in Figure 4. The combined needle system was connected to a DC positive pulse voltage power supply and a further inlet of needle base was connected to the air pump.

In this work, two different functional inks of UF 3808 and SEMICOSIL988/1K were used at the semiconductor packaging level to explore the influence of various process parameters and stability of the cone-jet on PET substrate through experiments. The physical properties of functional inks (Henkel Corporations Group Ltd., Stamford, CT, USA) were achieved by the suppliers. However, this material can provide excellent mechanical properties for electronic devices, and is widely used in flexible printed electronics. It is easily cured by light. The dynamic viscosity of the used inks is 450,000 cps. They have excellent bonding strength and resistance to heat and humidity cycles, and can provide a robust strength to insulating substrates. The physical performance parameters of the functional inks are given in Table 2. We aimed to print stable micro-structures on the flexible insulating substrates (i.e., PET), which were 0.5 mm and 1 mm thick. The DC positive pulse voltage power supply was used to create a considerable electric field between the steel needle and ground electrode at constant parameters. The air pump was used to provide a hydrodynamic force to push functional liquid towards the needle outlet. Voltage signals were produced using a positive-polarity, high-voltage amplifier (Smart Material GmbH, Löbtauer Str., Dresden, Germany), which was connected to a function generator (FG-7002C, EZ digital, Binh Thanh District, Ho Chi Minh City, Vietnam), and were monitored using a high-speed camera (Fastcam SA4, Photron, Westfield, MA, USA). The peak in the dc pulse voltage was 1.95 kV and pulse frequency was 1000 Hz. The duration of pulse frequency was set as a half of a period. The substrate was connected to a ground electrode power supply. The substrate was moved at a speed of 100 mm/s to a computer-controlled moving station. The schematic and experimental E-Jet printing setup used to generate a stable cone-jet is shown in Figure 5.

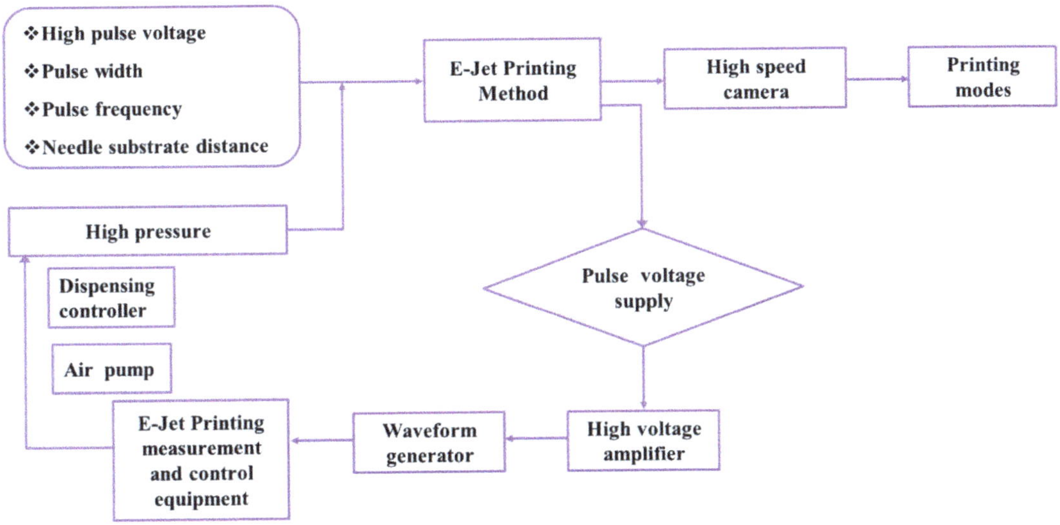

Figure 4. Flow diagram of the E-Jet printing process with pulse voltage implementation.

Table 2. Physical characteristics of the different functional inks used in existing work.

Functional Ink	Density ($g{\cdot}cm^{-3}$)	Dynamic Viscosity ($m{\cdot}Pa{\cdot}s$)	Surface Tension (N/m)	Storage Modulus (N/mm^2)	Dielectric Constant
UF 3808	1.16	450,000	0.031	260	3.24
SEMICOSIL988/1K	1.1	450,000	0.045	350	2.38

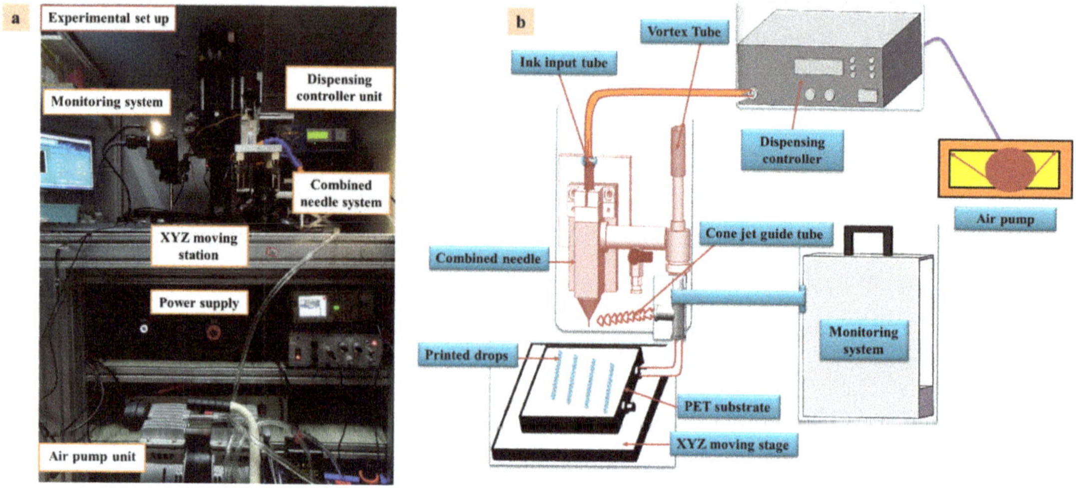

Figure 5. The (**a**) experimental setup (**b**) schematic diagram of the E-Jet printing process.

3. Results and Discussion

Stable Micro-Patterns on Flexible Substrate

The DC positive pulse voltage was selected and periodic electric field force was applied to control the formation and interruption of the E-Jet flow, to realize the DoD printing of droplet patterns on the flexible substrate. Figure 6 is a schematic diagram of the pulse waveform, where Vh represents the high-voltage pulse, Vl represents the low-voltage pulse, and the difference between Vh and Vl represents the voltage amplitude. In one cycle, the ratio of the bandwidth Tp of a high-voltage pulse to the total period Td of the pulse voltage is called a duty cycle. Further, the reciprocal of a pulse voltage period, Td, is called the pulse frequency *f*, which represents the number of times the high-voltage pulse signal is generated per unit time. Similarly, it directly disturbs the resolution and density of the micro-structures formed during the entire process. Among them, the choice of a high-voltage pulse should ensure that the end of the spray needle produces a stable micro-dripping mode or cone jet mode. The choice of a low-pressure pulse should ensure that the end of the spray needle can form a Taylor cone profile but does not produce a cone-jet. The use of pulsed electrical signals can reduce the charge of the flexible insulating substrate. Therefore, the effect of residual charges on the stability of the cone-jet and micro-droplets are considerably reduced.

A series of experiments were performed using a 0.5-mm and 1-mm-thick PET substrate and a stainless-steel quartz nozzle with an inner diameter of 50 μm. The applied pressure, high-voltage pulse, low-voltage pulse, pulse duty cycle, pulse frequency, printing height and printing speed (i.e., 40 kPa, 1.95 kV, 1.60 kV, 80%, 60 Hz, 0.25 mm and 1 mm/s) are obtained as optimized parameters for printing stable micro-structures. The formation and evolution of the electrojet on a PET substrate, achieved by increasing the different time intervals (i.e., 0 s, 0.02 s, 0.04 s, 0.08 s, 0.12 s and 0.2 s), are shown in Figure 7a. Figure 7b is the functional ink form dripping mode when no voltage is applied and Figure 7c is the cone-jet form under a high pulsed voltage. It can be seen from Figure 7a that the printing cycle of a single droplet is 0.2 s. When T = 0 s, the functional ink at the end of the quartz needle is hemispherical and begins to evolve into a conical morphology. When T = 0.12 s, the Taylor cone forms and merges, which produces cone-jetting phenomena. When T = 0.2 s, the cone-jetting is completed and the functional ink returns to a hemispherical shape.

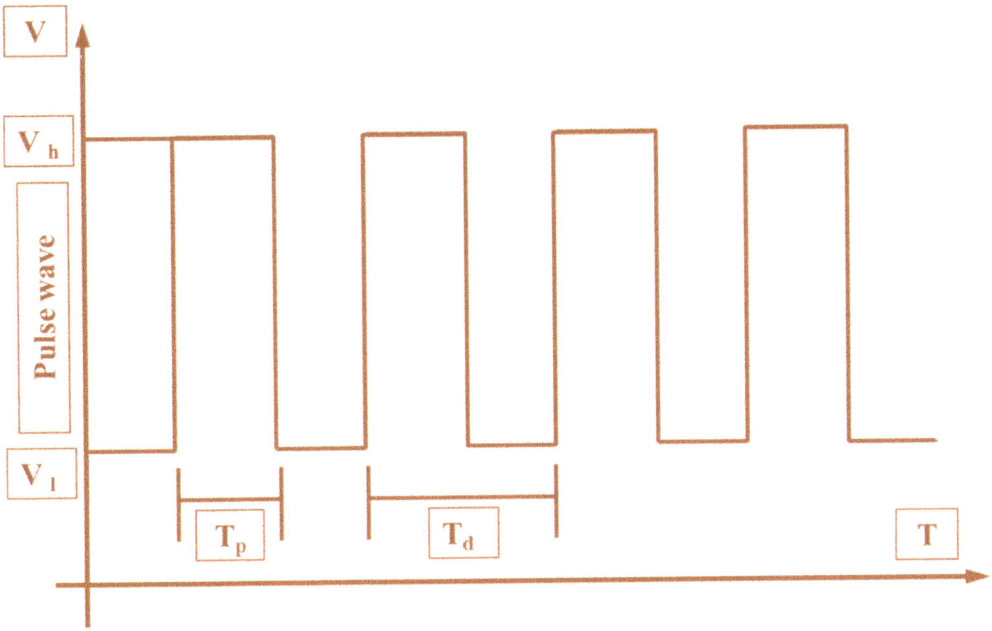

Figure 6. Schematic diagram of pulse waveform in terms of voltage across time duration.

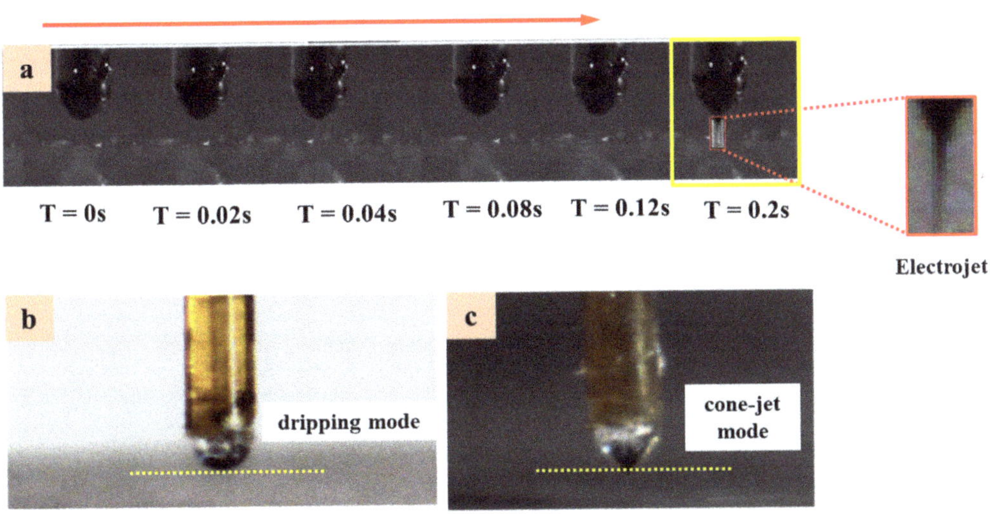

Figure 7. The (**a**) formation and evolution of electrojet in space to print on PET substrate by increasing time intervals (**b**) micro-dripping regime when no voltage at needle (**c**) cone-jet regime.

This work uses SEMICOSIL988/1K and UF3808, two different types of functional inks, to print printing micro-droplets and linear patterns on the flexible substrates. The optimized process parameters are properly adjusted to integrate the electrofluid jet for the stable droplets' morphologies. The printing control monitoring camera performs linkage control for various parameters, such as motion, pressure, voltages, and flexible insulating substrates. The micro-droplets of different sizes were printed on the flexible PET substrates with the execution of DoD E-Jet printing process as shown in Figure 8. Moreover, Figure 8a displays the different variations of several droplets in terms of their sizes that were printed in the form of array using functional ink of SEMICOSIL988/1K. The minimum droplet diameter was measured at 104 μm on the PET substrate surface. Similarly, Figure 8b shows micro-droplets being printed and the droplet diameter was measured at 85 μm. The DoD printing of droplets shows that the size of droplet is not irregular, which leads to high quality and efficient printing during the production of overhanging structures with a specific ink. The droplet morphology allows for the suppression of the adhesive flexible structure in the electronic packaging process, which avoids the coffee stain effect on droplet morphology. Due to this phenomenon, the droplet morphology does not contradict the highly insulating properties of a PET substrate and the formed structure is orderly and controllable.

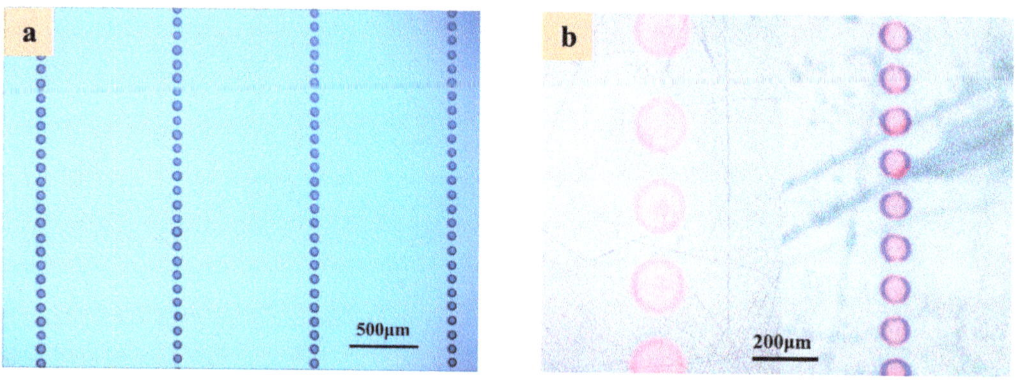

Figure 8. The DoD E-Jet printing of micro-droplets on flexible PET substrates (**a**,**b**) using functional liquid of SEMICOSIL988/1K.

When the applied pressure is increased to more than 50 kPa in the case of SEMICOSIL988/1K ink, the printed drop size rises. The research reveals that each individual droplet is not uninterrupted and implying that the interior form of the drop is not continuous due to the coffee ring effect, which is easily discernible. However, there is no visible occurrence of the coffee ring effect below 40 kPa, suggesting that the ink flow particle size is different. This generates various drop forms. In the future, we hope to increase the uniformity of the printed droplets, as well as the discontinuous behaviour in micro-structures. This will require plasma treatment procedures that are often used to increase the surface energy of polymers such as polypropylene (PP) and PET. Therefore, Figure 9 depicts many changes in the size of printed droplets due to the oscillation in their shapes. It also happens because of different printing settings for the experiments have different influence of process parameters. It also focused exclusively on the functional ink properties and then measured the minimum droplet size of 305 μm on the substrate surface.

Furthermore, Figure 10 shows the different linear patterns printed on a flexible PET substrate with a thickness of 1 mm. The applied pressure of 50 kPa was used to increase the resolution and consistency of the printed micro-structures while keeping all process parameters constant. During the printing process, parameters such as printing trajectory, printing speed, voltage and pressure were precisely linked and controlled. Figure 10a,b display the "DLUT" and "small house" patterns printed at room temperature using UF 3808 with a viscosity of 450,000 cps. The structure is uniform and continuous, and there is no glue line breakage and accumulation. The structure line space is about 200 μm between the letters, and the aspect ratio is about 0.4. The printed micro-structures indicate that the smallest feature dimension (line width) in these patterns is about 500 μm. These simple continuous linear features display that the carried charge in droplets is neutralized on these highly insulating substrates using DC pulse voltages. Otherwise, the accumulated residue charge and the resulting repulsion between droplets make continuous printing almost impossible.

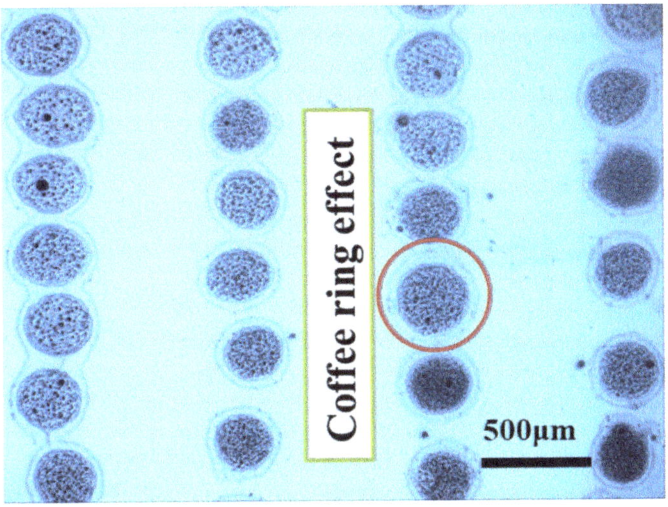

Figure 9. The unstable printed drops under 40 kPa pressure using functional liquid of SEMICOSIL988/1K.

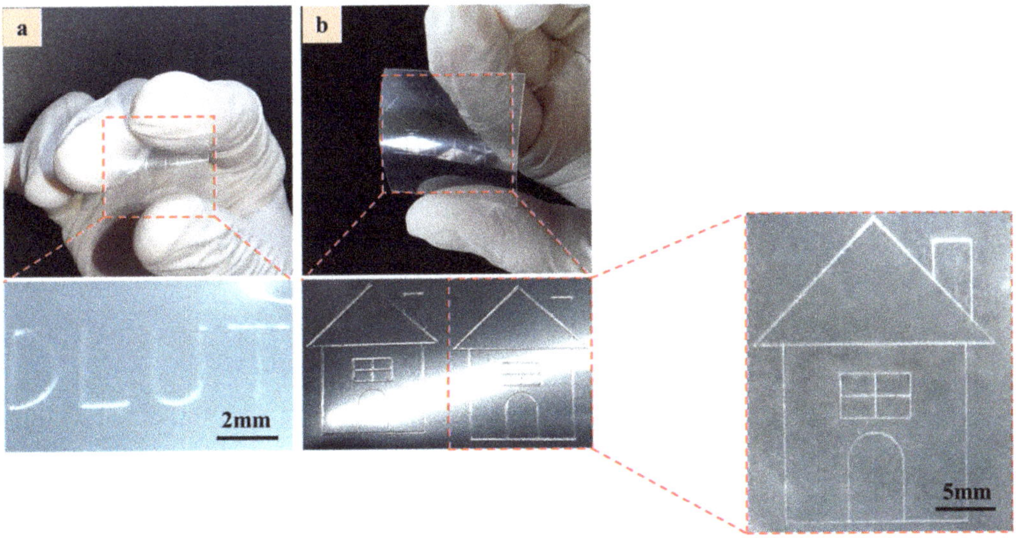

Figure 10. The printing of linear patterns on flexible PET substrates using functional liquid of UF3808 (**a**) "DLUT" pattern (**b**) "Little house" pattern.

4. Conclusions

In this paper, we used the electro-fluid jet printing platform and referred to the analysis of the simulation and experiment results. We carried out research on the influence of electrojet printing regimes, e.g., cone-jet and micro-drip, for micro-patterns on a polymer PET substrate. The influence of process parameters such as DC pulse voltage and pulse frequency on the size and morphology of the formed micro-structure indicates that the size of a micro-droplet is directly proportional to the pressure. Likewise, it is inversely proportional to the printing speed and high-pressure pulse. Primarily, the droplet size decreases and then increases with the increasing printing height. Subsequently, the influence of voltage, pressure, printing speed and printing height on the electrojet printing of flexible micro-fabrication was studied. The printing process conditions were obtained from the simulations to observe the size and shape of the micro-patterns' morphologies. By generating advantageous cone-jet and micro-drip regimes, it showed that the bulging and coffee stain effect was not predominant on a flexible PET substrate with the regulation of two different functional fluids. Finally, based on the range of influencing parameters, e.g., a metal-quartz spraying needle, functional ink viscosity and the surface energy of polymer substrate, the printed results are stable and versatile. The printing results are useful and can be applied to different polymer selections. The experimental design is deemed to be easier to implement, cost-effective, and tailored to polymer sectors. It has been concluded and demonstrated in this study that the E-Jet printing technology is a suitable method for patterning stable micro-structures on the polymer substrates. More, the polymer substrates are currently considered the most compatible material for application of biomedical devices and M/NEMS devices.

Author Contributions: Conceptualization, formal analysis, D.W.; resources, D.W.; supervision, D.W.; project administration, D.W.; funding acquisition, D.W.; Writing-original draft, Z.A.; methodology, Z.A.; software, Z.A.; data curation, Z.A.; validation, L.L.; Data curation, L.L.; Data curation, S.L.; formal analysis, X.Z.; investigation, X.Z.; data curation, P.X.; visualization, K.Z.; investigation, K.Z.; data curation, L.S.; visualization, Y.C.; investigation, Y.C.; validation, P.Y.; data curation, P.Y.; writing—review and editing, B.T.; writing—review and editing, J.X.; data curation, Y.Y.; writing-review and editing, J.L. All authors have read and agreed to the published version of the manuscript.

Funding: This research was supported by National Natural Science Foundation of China (51975104, 62074138), National Key R&D Program of China (2018YFA0703200) and Ningbo Institute of Dalian University of Technology.

Institutional Review Board Statement: Not applicable.

Informed Consent Statement: Not applicable.

Data Availability Statement: Not applicable.

Conflicts of Interest: The authors declare that they have no competing financial interest.

References

1. Sirringhaus, H.; Kawase, T.; Friend, R.H.; Shimoda, T.; Inbasekaran, M.; Wu, W.; Woo, E.P. High-resolution inkjet printing of all-polymer transistor circuits. *Science* **2000**, *290*, 2123–2126. [CrossRef] [PubMed]
2. Abbas, Z.; Wang, D.; Du, Z.; Qian, J.; Zhao, K.; Du, Z. Numerical simulation of electrohydrodynamic jet and printing micro-structures on flexible substrate. *Microsyst. Technol.* **2021**, *27*, 3125–3139. [CrossRef]
3. Hebner, T.R.; Wu, C.C.; Marcy, D.; Lu, M.H.; Sturm, J.C. Ink-jet printing of doped polymers for organic light emitting devices. *Appl. Phys. Lett.* **1998**, *72*, 519–521. [CrossRef]
4. Jayasinghe, S.N.; Townsend-Nicholson, A. Stable electric-field driven cone-jetting of concentrated biosuspensions. *Lab Chip* **2006**, *6*, 1086–1090. [CrossRef] [PubMed]
5. Huang, Y.; Duan, Y.; Pan, Y. Electrohydrodynamic direct-writing. *Nanoscale* **2013**, *5*, 12007–12017. [CrossRef] [PubMed]
6. Kim, J.; Oh, H.; Kim, S.S. Electrohydrodynamic drop-on-demand patterning in pulsed cone-jet mode at various frequencies. *J. Aerosol Sci.* **2008**, *39*, 819–825. [CrossRef]
7. Han, Y.; Dong, J. Electrohydrodynamic (EHD) printing of molten metal ink for flexible and stretchable conductor with self-healing capability. *Adv. Mater. Technol.* **2018**, *3*, 1–8. [CrossRef]
8. Wei, C.; Qin, H.; Ramírez-Iglesias, N.A.; Chiu, C.P.; Lee, Y.S.; Dong, J. High-resolution ac-pulse modulated electrohydrodynamic jet printing on highly insulating substrates. *J. Micromech. Microeng.* **2014**, *24*, 045010. [CrossRef]
9. Lee, S.; An, K.; Son, S.; Choi, J. Satellite/spray suppression in electrohydrodynamic printing with a gated head. *Appl. Phys. Lett.* **2013**, *103*, 133506. [CrossRef]
10. Abas, M.; Salman, Q.; Khan, A.M.; Rahman, K. Direct ink writing of flexible electronic circuits and their characterization. *J. Braz. Soc. Mech. Sci. Eng.* **2019**, *41*, 1–11. [CrossRef]
11. Rahman, K.; Ali, K. Fine resolution drop-on-demand electrohydrodynamic patterning of conductive silver tracks on glass substrate. *Appl. Phys. A* **2013**, *111*, 593–600. [CrossRef]

12. Choi, K.; Rahman, K.; Muhammad, N.M.; Khan, A.; Kwon, K.; Doh, Y.; Kim, H. Electrohydrodynamic Inkjet—Micro Pattern Fabrication for Printed Electronics Applications. In *Recent Advances in Nanofabrication Techniques and Applications*; InTechOpen: London, UK, 2011.
13. Han, Y.; Dong, J. Electrohydrodynamic printing for advanced micro/nanomanufacturing: Current progresses, opportunities, and challenges. *J. Micro Nano-Manuf.* **2018**, *6*, 1–20. [CrossRef]
14. Xu, W.; Zhang, S.; Xu, W. Recent progress on electrohydrodynamic nanowire printing. *Sci. China Mater.* **2019**, *62*, 1709–1726. [CrossRef]
15. Wu, Y. Electrohydrodynamic jet 3D printing in biomedical applications. *Acta Biomater.* **2021**, *128*, 21–41. [CrossRef] [PubMed]
16. Zou, W.; Yu, H.; Zhou, P.; Zhong, Y.; Wang, Y.; Liu, L. High-resolution additive direct writing of metal micro/nanostructures by electrohydrodynamic jet printing. *Appl. Surf. Sci.* **2021**, *543*, 148800. [CrossRef]
17. Liu, L.; Yang, X.; Zhang, B.; Yin, Z.; Zou, H. Trajectory analysis of the charged droplet during electrohydrodynamic jet printing. *Microsyst. Technol.* **2021**, *27*, 2935–2941. [CrossRef]
18. Yang, X.; Liu, R.; Li, L.; Yin, Z.; Chen, K.; Wang, D.F. The study of electrohydrodynamic printing by numerical simulation. *J. Electr. Eng.* **2020**, *71*, 413–418. [CrossRef]
19. Wang, D.; Abbas, Z.; Du, Z.; Du, Z.; Lu, L.; Zhao, K.; Zhao, X.; Yuan, Y.; Zong, H.; Cui, Y.; et al. Phase field simulation of electrohydrodynamic jet droplets and printing microstructures on insulating substrates. *Microelectron. Eng.* **2022**, *261*, 111817. [CrossRef]
20. Park, J.U.; Hardy, M.; Kang, S.J.; Barton, K.; Adair, K.; Mukhopadhyay, D.K.; Lee, C.Y.; Strano, M.S.; Alleyne, A.G.; Georgiadis, J.G.; et al. High-resolution electrohydrodynamic jet printing. *Nat. Mater.* **2007**, *6*, 782–789. [CrossRef]
21. Matsumoto, H.; Mizukoshi, T.; Nitta, K.; Minagawa, M.; Tanioka, A.; Yamagata, Y. Organic/inorganic hybrid nano-microstructured coatings on insulated substrates by electrospray deposition. *J. Colloid Interface Sci.* **2005**, *286*, 414–416. [CrossRef]
22. Alexander, M.S.; Paine, M.D.; Stark, J.P.W. Pulsation modes and the effect of applied voltage on current and flow rate in nanoelectrospray. *Anal. Chem.* **2006**, *78*, 2658–2664. [CrossRef] [PubMed]
23. Yogi, O.; Kawakami, T.; Mizuno, A. Properties of droplet formation made by cone jet using a novel capillary with an external electrode. *J. Electrost.* **2006**, *64*, 634–638. [CrossRef]
24. Paine, M.D.; Alexander, M.S.; Smith, K.L.; Wang, M.; Stark, J.P.W. Controlled electrospray pulsation for deposition of femtoliter fluid droplets onto surfaces. *J. Aerosol Sci.* **2007**, *38*, 315–324. [CrossRef]
25. Lee, M.W.; Kang, D.K.; Kim, N.Y.; Kim, H.Y.; James, S.C.; Yoon, S.S. A study of ejection modes for pulsed-DC electrohydrodynamic inkjet printing. *J. Aerosol Sci.* **2012**, *46*, 1–6. [CrossRef]
26. Rahman, K.; Khan, A.; Muhammad, N.M.; Jo, J.; Choi, K.H. Fine-resolution patterning of copper nanoparticles through electrohydrodynamic jet printing. *J. Micromech. Microeng.* **2012**, *22*, 1–8. [CrossRef]
27. Chen, X.; Jia, L.; Yin, X.; Cheng, J.; Lu, J. Spraying modes in coaxial jet electrospray with outer driving liquid. *Phys. Fluids* **2005**, *17*, 032101. [CrossRef]
28. Qin, H.; Wei, C.; Dong, J.; Lee, Y.S. Direct Printing and Electrical Characterization of Conductive Micro-Silver Tracks by Alternating Current-Pulse Modulated Electrohydrodynamic Jet Printing. *J. Manuf. Sci. Eng. Trans. ASME* **2017**, *139*, 1–10. [CrossRef]
29. Abbas, Z.; Wang, D.; Du, Z.; Zhao, K.; Du, Z.; Lu, L. Numerical simulation of stable electrohydrodynamic cone-jet formation and printing on flexible substrate. *Microelectron. Eng.* **2021**, *237*, 111496. [CrossRef]
30. Schnitzer, O.; Yariv, E. The Taylor-Melcher leaky dielectric model as a macroscale electrokinetic description. *J. Fluid Mech.* **2015**, *773*, 1–33. [CrossRef]
31. Jacqmin, D. Calculation of Two-Phase Navier-Stokes Flows Using Phase-Field Modeling. *J. Comput. Phys.* **1999**, *155*, 96–127. [CrossRef]

polymers

Article

Characterization of the Polyetheretherketone Weldment Fabricated via Rotary Friction Welding

Chil-Chyuan Kuo [1,2,3,4,*], Hua-Xhin Liang [1] and Song-Hua Huang [5]

1 Department of Mechanical Engineering, Ming Chi University of Technology, No. 84, Gungjuan Road, New Taipei City 24301, Taiwan
2 Research Center for Intelligent Medical Devices, Ming Chi University of Technology, No. 84, Gungjuan Road, New Taipei City 24301, Taiwan
3 Department of Mechanical Engineering, Chang Gung University, No. 259, Wenhua 1st Rd., Guishan District, Taoyuan City 33302, Taiwan
4 Center of Reliability Engineering, Ming Chi University of Technology, No. 84, Gungjuan Road, Taishan District, New Taipei City 24301, Taiwan
5 Li-Yin Technology Co., Ltd., No. 37, Lane 151, Section 1, Zhongxing Road, Wugu District, New Taipei City 24301, Taiwan
* Correspondence: jacksonk@mail.mcut.edu.tw

Abstract: Polyether ether ketone (PEEK) is frequently employed in biomedical engineering due to its biocompatibility. Traditionally, PEEK manufacturing methods involve injection molding, compression molding, additive manufacturing, or incremental sheet forming. Few studies have focused on rotational friction welding (RFW) with PEEK plastics. Based on years of RFW practical experience, the mechanical properties of the weldment are related to the burn-off length. However, few studies have focused on this issue. Therefore, the main objective of this study is to assess the effects of burn-off length on the mechanical properties of the welded parts using PEEK polymer rods. The welding pressure can be determined by the rotational speed according to the proposed prediction equation. The burn-off length of 1.6 mm seems to be an optimal burn-off length for RFW. For the rotational speed of 1000 rpm, the average bending strength of the welded parts was increased from 108 MPa to 160 Mpa, when the burn-off length was increased from 1 mm to 1.6 mm and the cycle time of RFW was reduced from 80 s to 76 s. A saving in the cycle time of RFW of about 5% can be obtained. The bending strength of the welded part using laser welding is lower than that using RFW, because only the peripheral material of the PEEK cylinder was melted by the laser.

Keywords: polyether ether ketone; rotational friction welding; burn-off length; bending strength; cycle time

check for updates

Citation: Kuo, C.-C.; Liang, H.-X.; Huang, S.-H. Characterization of the Polyetheretherketone Weldment Fabricated via Rotary Friction Welding. *Polymers* **2023**, *15*, 4552. https://doi.org/10.3390/polym15234552

Academic Editors: Ying Yan and Yiqi Wang

Received: 22 October 2023
Revised: 11 November 2023
Accepted: 24 November 2023
Published: 27 November 2023

1. Introduction

Polyether ether ketone (PEEK) has exceptional properties, such as high wear, chemical, mechanical, thermal, and radiation resistance. Thus, PEEK is widely employed in various applications, including pumps, bearings, or piston parts. PEEK is also used in medical implants. Therefore, PEEK has been frequently employed for human bone replacement due to its superior biomechanical properties [1], which have potential applications in the biomedical and aerospace industries [2,3]. PEEK exhibits outstanding thermal resilience and mechanical characteristics, showcasing its high-performance crystalline polymer nature.

Khoran et al. [4] evaluated the effects of reinforcing the fiber type on the grinding process of PEEK's composites. It was found that the tangential and normal grinding forces increased up to 67% using severe cutting conditions. Lee et al. [5] investigated the influence of jet pressure on the bond strength of PEEK. Results showed that adhesive failure is the primary mode for all groups and provides sufficient bonding durability between dental resin cement and PEEK. As a result, PEEK seems to be a promising alternative to metal in dentistry [6]. Vaddamanu et al. [7] investigated the fitting surface deformation of PEEK clasps and compared them with cobalt–chromium clasps. Results showed that the deformation of PEEK clasps' fitting surface is lower than for cobalt–chromium clasps. This finding implies that opting for PEEK as a clasp in the removable partial denture framework is a viable choice for aesthetic reasons. Bontempi et al. [8] measured the nanomechanics of medical-grade PEEK using atomic force microscopy. Results showed that two localized distributions of moduli at about 0.6 and 0.9 GPa were found. This noteworthy discovery offers insights into the ideal design of nanostructures, paving the way for advanced biomedical applications. Yang et al. [9] employed molecular dynamics to simulate local vibrational features and evaluate the local crystallinity of PEEK. The observation of the 135 cm^{-1} signal was identified, showcasing its susceptibility to intermolecular interactions. Naghavi et al. [10] proposed low-stiffness PEEK hip prostheses for minimizing the stress difference after a hip replacement. Results showed that the stiffness of the PEEK implant was about 0.276 kN/mm, showing that PEEK seems to be a promising material for implants. Zhou et al. [11] investigated the shape-memory properties and actuation performances of high-temperature PEEK. Results showed that the results obtained in this study provided a design for a PEEK actuator. Baltag et al. [12] proposed a new method for the sulfonation of PEEK with a temperature of 60 °C under ultrasonication. It was found that a maximum removal efficiency of about 99% was obtained. Saravi et al. [13] studied the fatigue behavior of PEEK implant abutments. It was found that the PEEK implant abutments seem to be feasible, compared to zirconia implant abutments. Zhang et al. [14] used the single-pulsed laser to drill PEEK. Results showed that the ejection is the primary material-removal mechanism during the slow increment stage, showing the dynamic process of keyhole evolution in the drilling of PEEK using the single-pulsed laser. Bialas et al. [15] used picosecond laser treatment to obtain the surface-infused nanogold particles with PEEK. Results showed that an increase in quantity was obtained using a laser power of 15 mW. Pedroso et al. [16] investigated the tribological performance of waterborne tribological coatings based on three binders of PEEK and polyetherketone. Results suggested that PEKK and polyetherketone seem to be attractive alternatives to PEEK.

Rotary friction welding (RFW) [17] offers various benefits, such as superior joint strength and the capability to unite dissimilar materials. Traditionally, common PEEK manufacturing methods encompass additive manufacturing [18], injection molding [19–22], milling, grinding, pulse laser drilling, or incremental sheet forming, among others [23]. It is well-known that the ultrasonic welding (UW) can be used to join PEEK plastics [24–26]. However, the distinct drawback of this approach is that it is limited to smaller and thinner PEEK components. Based on RFW practical experience, the weldment quality was affected by the burn-off length [27]. However, very few studies have focused on this issue. Based on the above two research motivations, this study aims to characterize the polyetheretherketone weldment fabricated using RFW and evaluate the impact of the burn-off length on the weld quality of PEEK polymer rods. A non-contact thermal camera was employed to investigate the temperature changes in the weld interface during RFW. Subsequent to RFW, the mechanical properties of the welded parts were examined using shore A surface hardness and three-point bending tests. Thermal properties were also investigated using differential scanning calorimetry (DSC). The culmination of this study involves proposing a technical database on RFW, explicitly focusing on the influence of the burn-off length on weld properties.

2. Materials and Methods

Figure 1 shows the research flowchart in this study. Five rotational speeds were used in this study. In this study, a PEEK polymer rod was selected as the welding specimen. The length and diameter were about 40 mm and 20 mm, respectively. It is noteworthy that the PEEK polymer rod has a melting point of around 343 °C and a decomposition temperature of approximately 575 °C [28]. The speed of the spindle was kept at constant speeds of 1000, 1350, 2000, 3000, and 4000 rpm, respectively. The cycle time of RFW under a burn-off length of 1 mm was 90 s, which involved a friction time of 30 s, a welding time of 30 s, and a cooling time under the pressure of 30 s. The axial load and feed rate are 24.1 N and 6 mm/min, respectively. The welding pressure of RFW is 0.077 MPa. To study the effects of the burn-off length on the weldment quality of PEEK polymer rods, eight different burn-off lengths of welded parts were employed in this study, i.e., 1, 1.2, 1.3, 1.4, 1.5, 1.6, 1.8, and 2 mm.

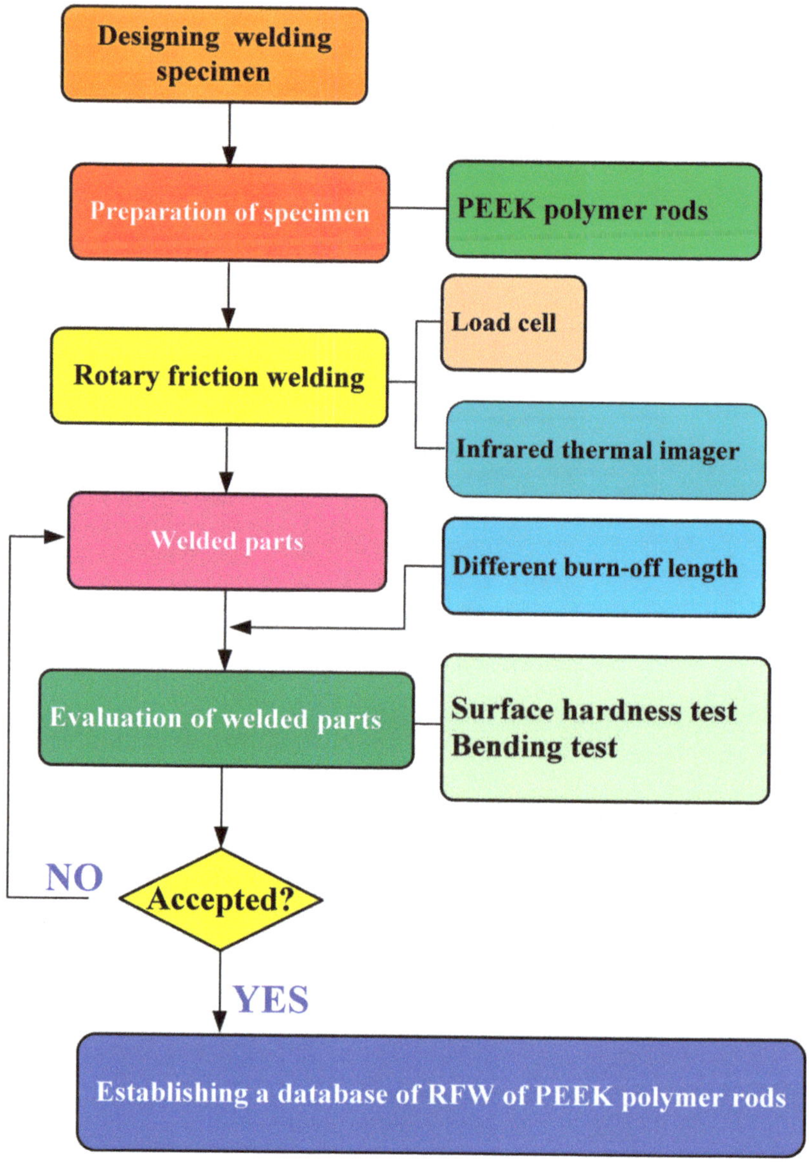

Figure 1. Research flowchart in this study.

In this study, a computer numerical control (CNC) turning machine (K-45L, Kae Jiuh, Co., Inc., New Taipei City, Taiwan) was employed to perform RFW of PEEK polymer rods, as shown in Figure 2. A non-contact thermal camera (BI-TM-F01P, Panrico trading Inc., New Taipei City, Taiwan) was used to investigate the temperature changes at the weld interface during the RFW process. A load cell (ARI742, Zhiheng Industrial Co., Inc., New Taipei City, Taiwan) was used to measure the welding force applied for RFW. Figure 3 depicts a schematic illustration of the burn-off length during RFW. The welding specimen has a round cross-section with the diameter and length of 20 mm and 40 mm. One specimen is firmly held stationary. The other specimen is fixed with a chuck and is rotated at a constant rotational speed. The frictional heat is generated at the interface of two workpieces during RFW. Figure 4 shows the experimental setup for bending, surface hardness, and DSC tests of welded parts. A 1500 W pulsed fiber laser (COMET Inc., New Taipei City, Taiwan) with a wavelength of 1062 nm was also used to join PEEK polymer rods. The output power is about 40 W, and the repetition frequency of the laser is about 80 kHz. The laser welding speed and spot size are 2 mm/s and 3 mm, respectively. The three-point bending machine (RH-30, Shimadzu Inc., Kyoto, Japan) and The Shore A durometer (MET-HG-A, SEAT Inc. New Taipei City, Taiwan) were used to evaluate the weldment quality after RFW. The movement speed of the punch is about 1 mm/s. The bending strength (σ) can be calculated using the following Equation (1).

$$\sigma = \frac{8PL}{\pi d^3} \tag{1}$$

where P is the applied load, d is the diameter of the PEEK polymer rod, and L is the length between two supports.

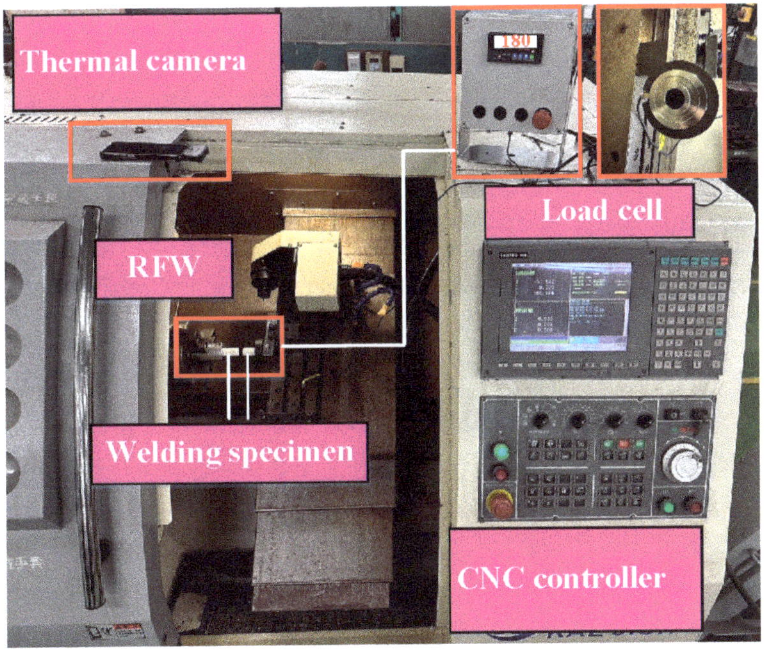

Figure 2. Computer numerical control lathe used for RFW.

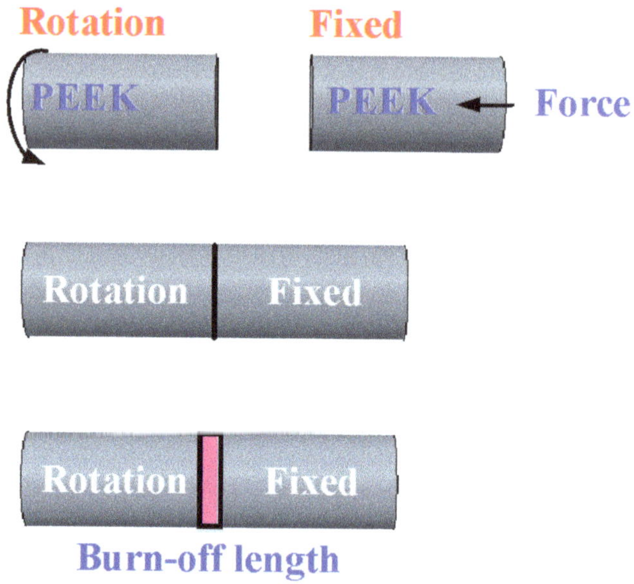

Figure 3. Schematic illustration of burn-off length during RFW.

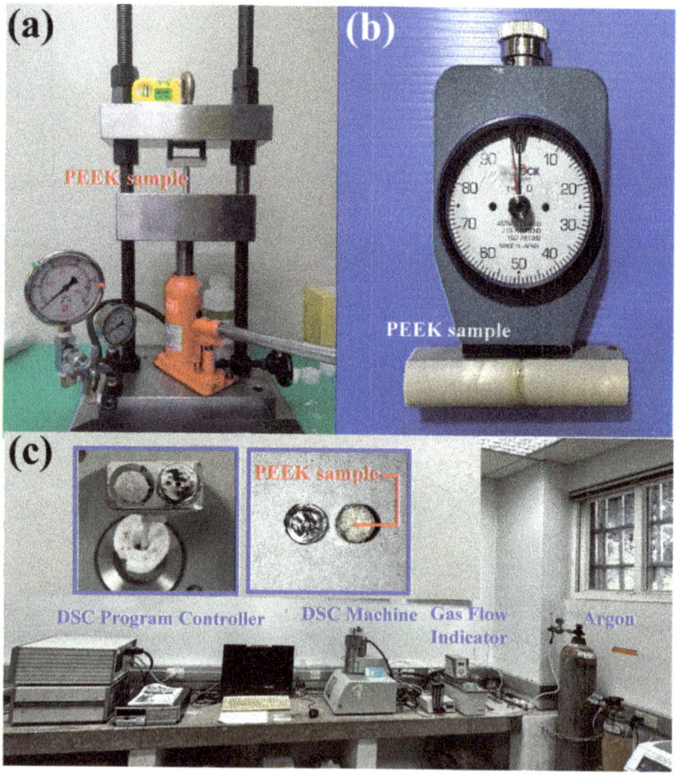

Figure 4. Experimental setup for (**a**) bending test, (**b**)surface hardness, and (**c**) DSC test of welded parts.

The scanning electron microscopy (FE-SEM) (JEC3000-FC, JEOL Inc. Tokyo, Japan) and optical microscope (OM) (Quick Vision 404, Mitutoyo Inc., Tokyo, Japan) were used to investigate the fracture surfaces after a series of bending tests. The thermal transitions of the weldments were examined using the differential scanning calorimetry (DSC) (STA 409 PC Luxx Simultaneous thermal analyzer, Netzsch-Gerätebau GmbH Inc., Selb, Germany). A mass of between 10 and 15 mg of the welded joint samples was placed in platinum crucibles for the DSC. The specimens were heated at a temperature ranging from 30 °C to 400 °C under a nitrogen gas flow rate of about 25 cc/min. Both the heating rate and cooling rate were 10 °C/min. A technical database was finally established about the characterization of PEEK weldment fabricated by RFW.

3. Results and Discussion

In this study, five specimens were produced. Based on many years of RFW practical experience, the weldment quality was significantly affected by the maximum temperature in the weld interface during RFW of the PEEK polymer rods. This study used a thermal camera to investigate the temperature in the weld joint during RFW of PEEK polymer rods. Figure 5 shows the relationship between time and the weld joint temperature under the rotational speed of 1000 rpm. The maximum temperature in the weld interface is about 377 °C. Figure 6 shows the relationship between time and the weld joint temperature under the rotational speed of 1350 rpm. The maximum temperature in the weld interface is about 380 °C. Figure 7 shows the relationship between time and the weld joint temperature under the rotational speed of 2000 rpm. The maximum temperature in the weld interface is about 382 °C. Figure 8 shows the relationship between time and the weld joint temperature under the rotational speed of 3000 rpm. The maximum temperature in the weld interface is about 383 °C. Figure 9 shows the relationship between time and the weld joint temperature under the rotational speed of 4000 rpm. However, the peak temperature in the weld interface is only about 363 °C. The possible reason is that the friction coefficient of the material is different due to the high rotation speed. It is interesting to note that these temperatures at the weld joint were higher than the melting point of PEEK polymer materials [28].

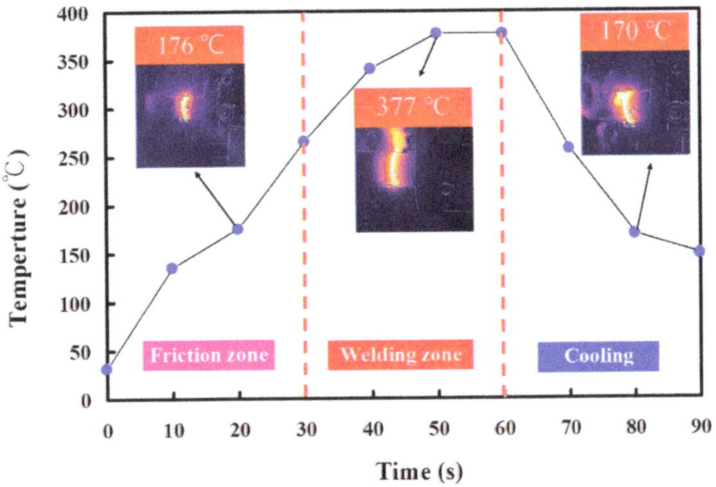

Figure 5. The relationship between time and weld joint temperature under the rotational speed of 1000 rpm.

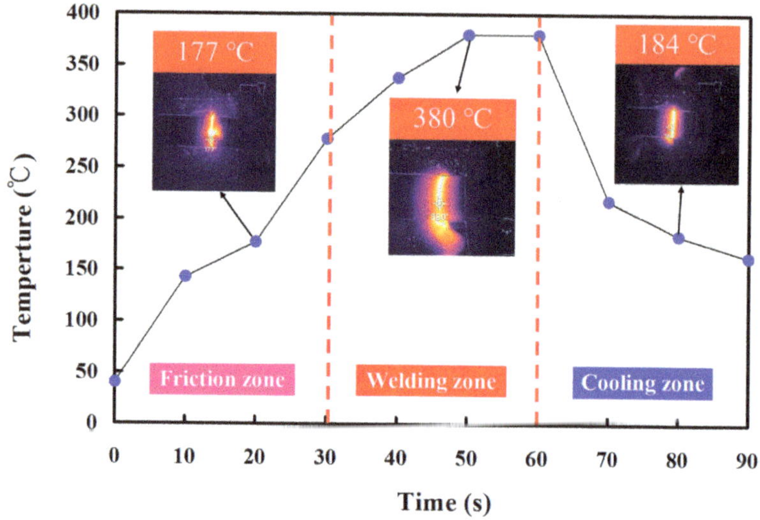

Figure 6. The relationship between time and weld joint temperature under the rotational speed of 1350 rpm.

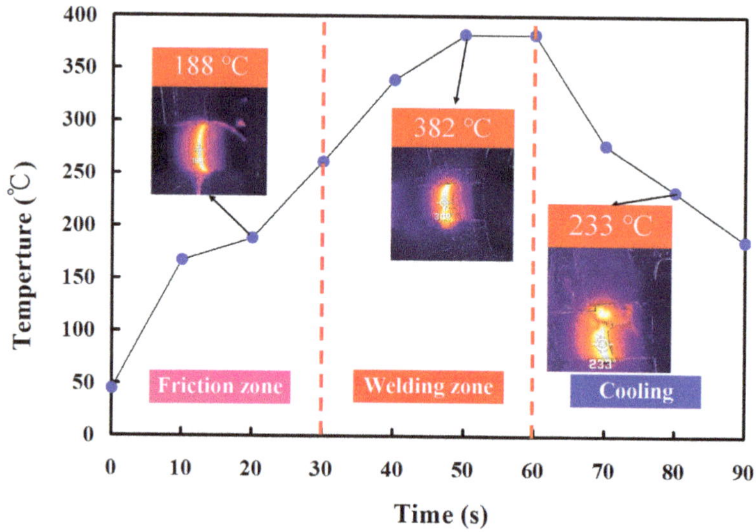

Figure 7. The relationship between time and weld joint temperature under the rotational speed of 2000 rpm.

Five test specimens and five rotational speeds were used to perform RFW in this study. To investigate the relationship between the rotational speed of RFW and welding force, the feed rate is fixed to perform the experiment. In this study, the feed rate is fixed at 6 mm/min. Figure 10 shows the relationship between the rotational speed and welding pressure during RFW of PEEK polymer rods. The welding force for rotational speeds of 1000, 1350, 2000, 3000, and 4000 rpm is 24.1, 26.1, 28.1, 31.4, 37.4 N, respectively. The welding pressure for rotational speeds of 1000, 1350, 2000, 3000, and 4000 rpm is 76, 83, 89, 99, and 119 kPa, respectively. For the feed rate of 6 mm/min, the welding pressure (y) can be determined by the rotational speed (x) according to the prediction equation of y = 0.0134x + 62.762 with a correlation coefficient of 0.9753. This study found that the weld interface after the RFW is a curved surface. Figure 11 shows the evolution mechanism of the weld interface surface after RFW of PEEK polymer rods, showing the change in the weld interface profile after RFW of the PEEK polymer rods. Compared to the center position B of the welding specimen, the edge position A of the welding specimen has a higher linear velocity under the same number of revolutions, which results in a higher temperature at edge position A and more material melting.

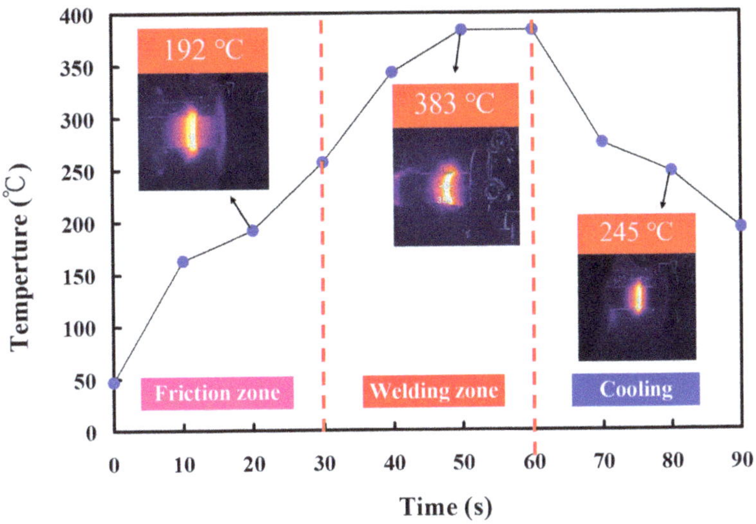

Figure 8. The relationship between time and weld joint temperature under the rotational speed of 3000 rpm.

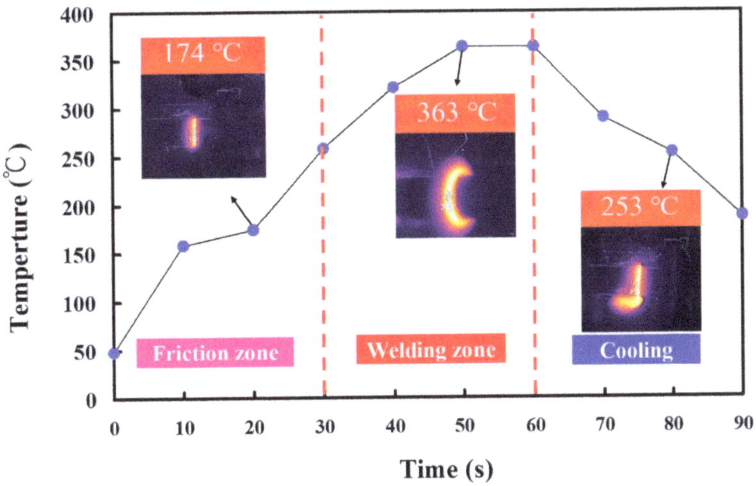

Figure 9. The relationship between time and weld joint temperature under the rotational speed of 4000 rpm.

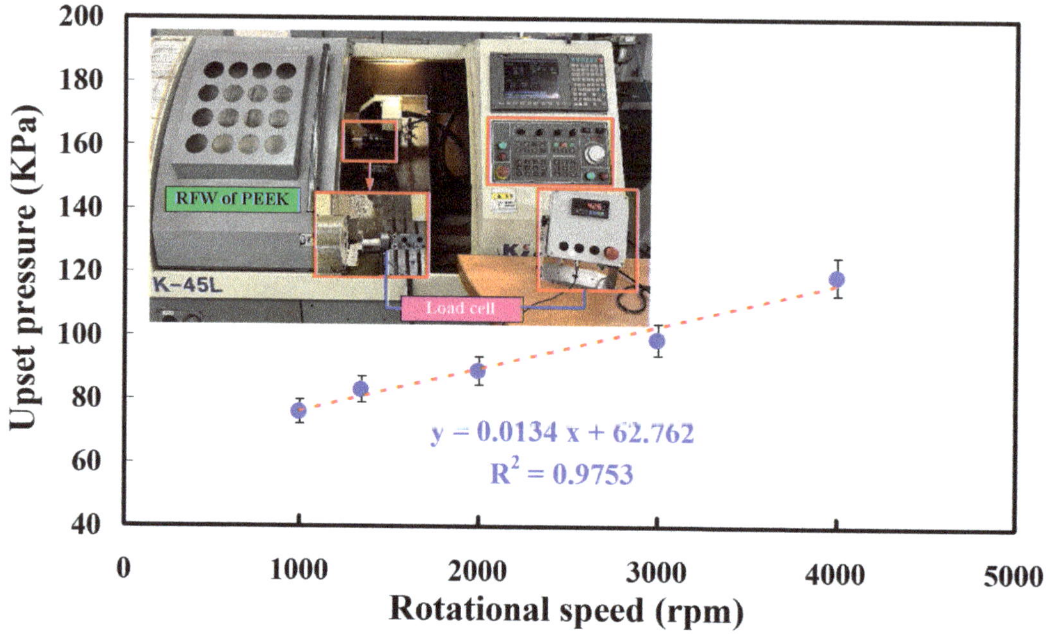

Figure 10. The relationship between rotational speed and welding pressure during RFW of PEEK polymer rods.

In this study, five specimens were performed. Figure 12 shows the sets of welded joints under different burn-off lengths. Figure 13 shows the bending strength of the welded parts after bending tests. The mean bending strength of the PEEK polymer rods is about 183 MPa. For the rotational speed of 1000 rpm, the average bending strength of the welded parts using eight different burn-off lengths for RFW is about 108 MPa, 130 MPa, 134 MPa, 148 MPa, 152 MPa, 156 MPa, 158 MPa, and 158 MPa, respectively. The cycle time of RFW is about 70 s, 72 s, 73 s, 74 s, 75 s, 76 s, 78 s, and 80 s, respectively. For the rotational speed of 1350 rpm, the average bending strength of the welded parts using eight different burn-off lengths for RFW is about 116 MPa, 132 MPa, 134 MPa, 150 MPa, 154 MPa, 156 MPa, 160 MPa, and 162 MPa, respectively. For the rotational speed of 2000 rpm, the average bending strength of the welded parts using eight different burn-off lengths for RFW is about 124 MPa, 132 MPa, 136 MPa, 152 MPa, 156 MPa, 166 MPa, 166 MPa, and 168 MPa, respectively. For the rotational speed of 3000 rpm, the average bending strength of the welded parts using eight different burn-off lengths for RFW is about 128 MPa, 138 MPa, 144 MPa, 156 MPa, 162 MPa, 170 MPa, 172 MPa, and 178 MPa, respectively. For the rotational speed of 4000 rpm, the average bending strength of the welded parts using eight different burn-off lengths for RFW is about 134 MPa, 142 MPa, 150 MPa, 162 MPa, 168 MPa, 178 MPa, 180 MPa, and 184 MPa, respectively. Based on the above results, this study found three phenomena: (a) The burn-off length of RFW has an impact on the average bending strength of the welded parts. As can be seen, the mean bending strength of the PEEK polymer rods stabilizes when the burn-off length surpasses 1.6 mm. Thus, the optimal burn-off length of 1.6 mm is suitable for cylindrical rods with a length of 40mm and a diameter of 20 mm for five rotational speeds. (b) The average bending strength of the welded parts increased from 108 MPa to 160 MPa when the burn-off length increased from 1 mm to 1.6 mm for the rotational speed of 1000 rpm. The increase in the average bending strength of the welded parts is about 48%. This result is confirmed by the thermal analyses from DSC traces.

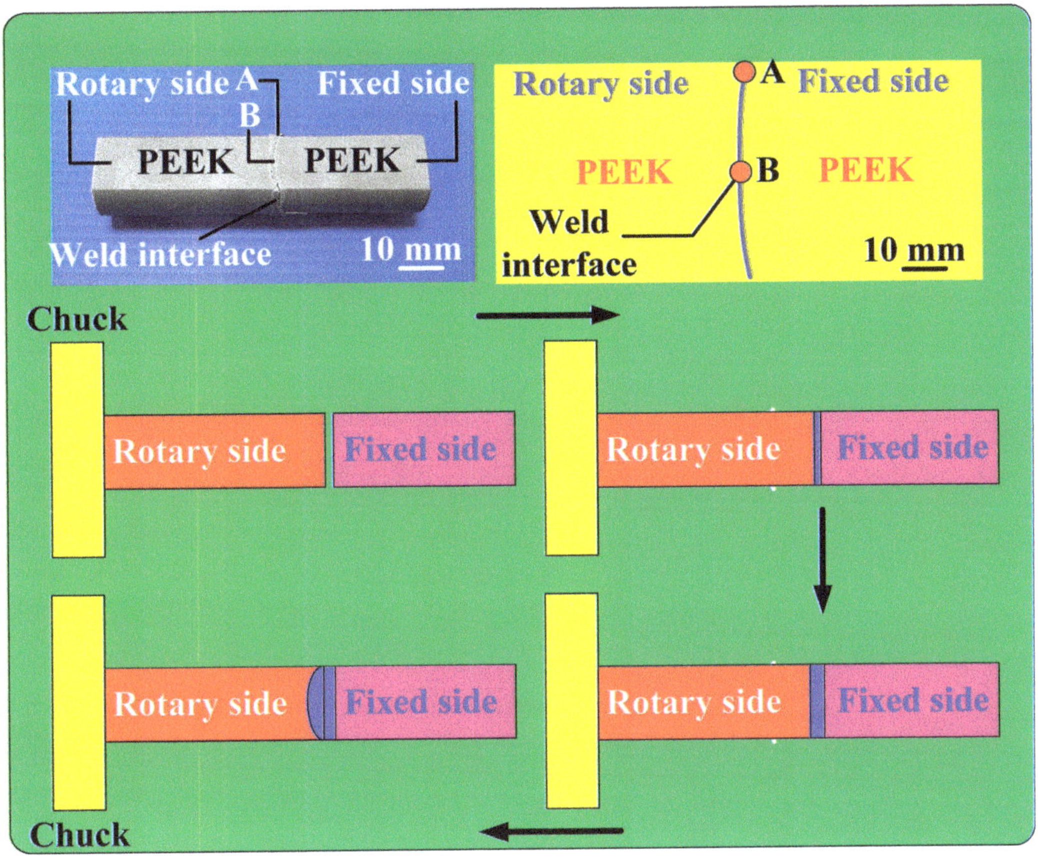

Figure 11. The evolution mechanism of weld interface surface after RFW of PEEK polymer rods.

The basis for choosing the polynomial orders is based on the correlation coefficient. A high correlation coefficient determines the polynomial orders. For the rotational speed of 1000 rpm, the bending strength of the welded parts (y) can be determined by the burn-off length (x) according to the prediction equation of $y = 0.1111x^3 - 2.8095x^2 + 24.175x + 87.857$ with a correlation coefficient of 0.9828. For the rotational speed of 1350 rpm, the bending strength of the welded parts (y) can be determined by the burn-off length (x) according to the prediction equation of $y = 0.0404x^3 - 1.4502x^2 + 16.557x + 101.43$ with a correlation coefficient of 0.975. For the rotational speed of 2000 rpm, the bending strength of the welded parts (y) can be determined by the burn-off length (x) according to the prediction equation of $y = 0.0404x^3 - 1.4502x^2 + 16.557x + 101.43$ with a correlation coefficient of 0.975. For the rotational speed of 2000 rpm, the bending strength of the welded parts (y) can be determined by the burn-off length (x) according to the prediction equation of $y = -0.2121\,x^3 + 2.2446x^2 + 1.4567x + 120.57$ with a correlation coefficient of 0.9839. For the rotational speed of 3000 rpm, the bending strength of the welded parts (y) can be determined by the burn-off length (x) according to the prediction equation of $y = -0.4762x^2 + 11.476x + 116.5$ with a correlation coefficient of 0.9935. For the rotational speed of 4000 rpm, the bending strength of the welded parts (y) can be determined by the burn-off length (x) according to the prediction equation of $y = -0.5119x^2 + 12.107x + 120.82$ with a correlation coefficient of 0.9914.

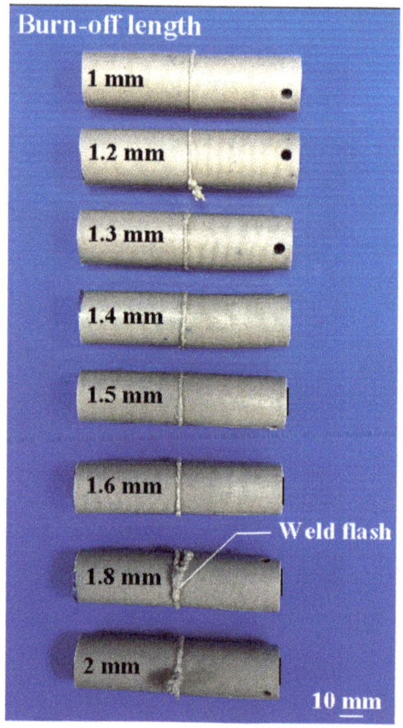

Figure 12. Sets of welded joints under different burn-off lengths.

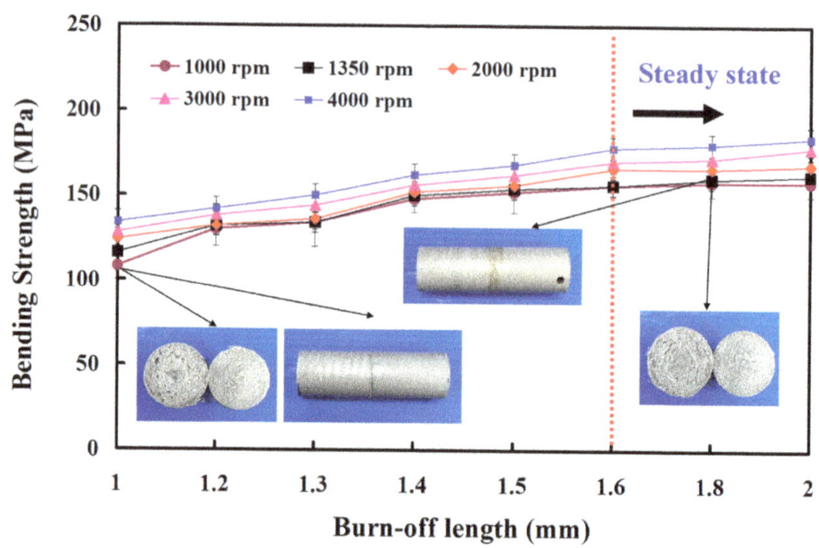

Figure 13. Bending strength of the welded parts after bending tests.

Figure 14 shows the dependence of the cycle time of RFW and burn-off length. It should be noted that the cycle time of RFW (y) can be determined by the burn-off length (x) according to the prediction equation of $y = 10x + 60$. Conventionally, the cycle time of RFW is 80 s for the burn-off length of 2 mm. According to the results obtained in this study, the burn-off length of 0.4 mm can be reduced. Thus, the cycle time of RFW can be reduced from 80 s to 76 s for five rotational speeds. The saving in the cycle time of RFW is about 5%.

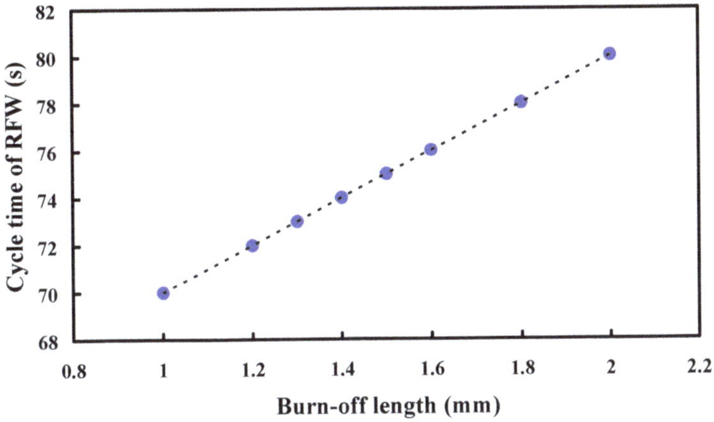

Figure 14. Dependence of the cycle time of RFW and burn-off length.

Figure 15 shows the DSC traces of welded parts. The heat capacities of welded parts fabricated using RFW with the burn-off length of 1.8 mm and 1 mm are about 0.94 mW/mg and 0.50 mW/mg, respectively [29]. This result shows the molecular orientation [30,31] in the weld interface of the welded parts fabricated using RFW with a burn-off length of 1.8 mm is higher than that of the welded parts fabricated using RFW with a burn-off length of 1 mm.

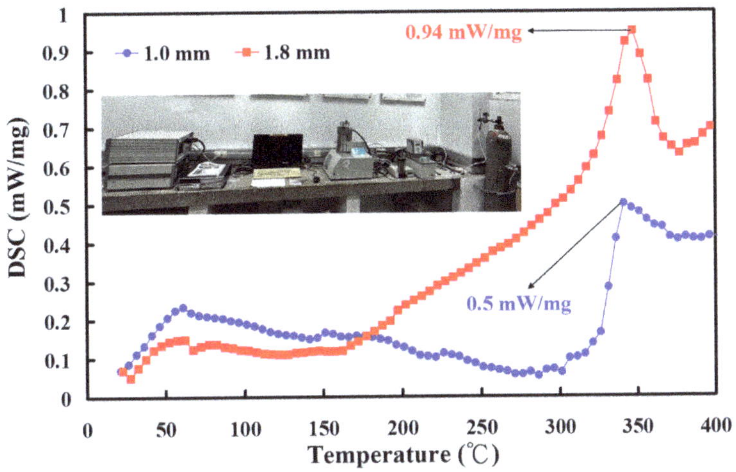

Figure 15. DSC traces of welded parts.

Figure 16 shows the fracture surfaces of the welded parts after the bending tests. Based on the analysis of the cross-section, this study found that the number of pores in the fracture section when RFW was performed with a burn-off length of 1.6 mm was significantly less than the number of pores in the fracture section when RFW was performed using a burn-off length of 1mm. The proportion of pores in the welding area was reduced from approximately 13% to 6%. Figure 17 shows the SEM micrographs of fracture surfaces of welded parts for RFW at 1000, 2000, 3000, and 4000 rpm. It should be noted that the pores were observed in the fracture surfaces of welded parts [32]. The emergence of these pores can be attributed to the probable decomposition of materials [33]. As the rotational friction welding speed rises, the maximum temperature of the weld bead also increases. Consequently, both the fluidity and centrifugal force of the molten material experience a boost. Consequently, a noticeable reduction in pores occurs, thereby enhancing the mechanical properties of the welded components.

Figure 16. Fracture surfaces of the welded parts after bending tests with (**a**) burn-off length of 1 mm and (**b**) burn-off length of 1.6 mm.

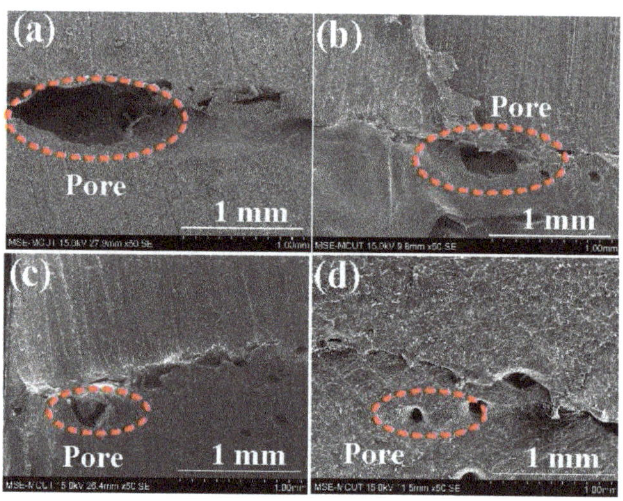

Figure 17. SEM micrographs of fracture surfaces of welded parts for RFW at (**a**) 1000, (**b**) 2000, (**c**) 3000, and (**d**) 4000 rpm.

The distinct advantage of laser welding is its precision. To evaluate whether a laser can be used to weld PEEK polymer rods, a pulsed fiber laser with an output power of about 40 W was used to weld PEEK polymer rods. Figure 18 shows a typical welded part under laser welding of PEEK polymer rods. The results showed that only the peripheral material of the PEEK cylinder was melted by the laser. The bending strength of the welded part using laser welding is lower than that using RFW. The average bending strength is about 46 MPa. Therefore, the laser is not suitable for PEEK cylindrical welding. Figure 19 shows the surface hardness of the weld joint. There are 10 measurement points at the weld interface. As can be seen, the average Shore hardness (SH) A in the weld interface is about SH 83 for five rotational speeds. The average Shore hardness A of five welded parts is about SH 82, SH 83, SH 81, SH 83, and SH 82, respectively.

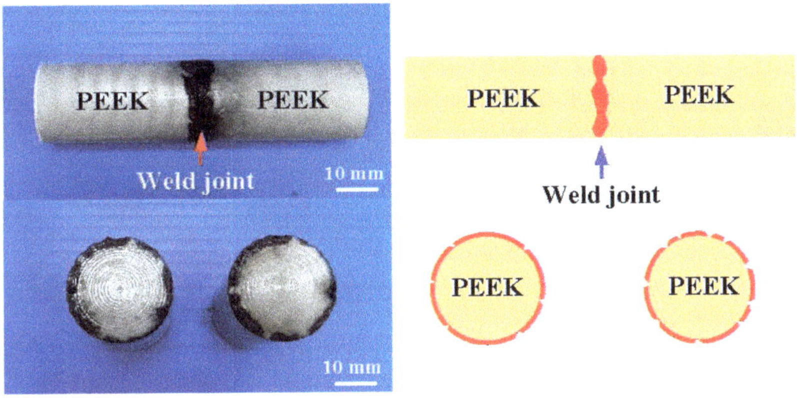

Figure 18. A typical welded part for laser welding of PEEK polymer rods.

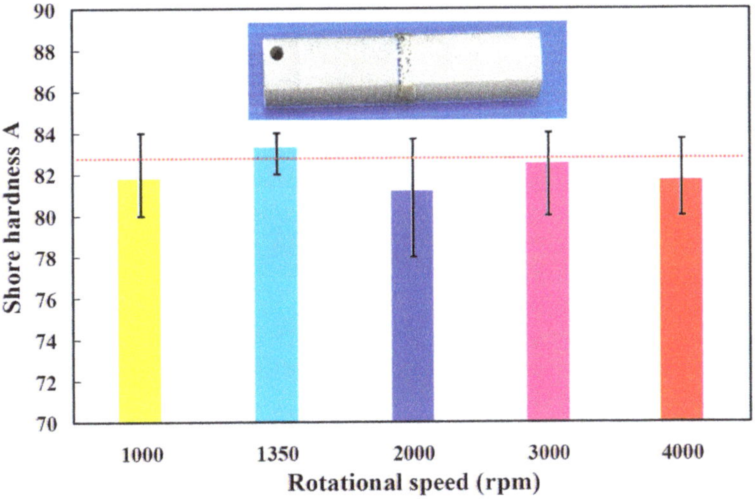

Figure 19. Surface hardness of the weld joint.

This study utilizes a CNC turning machine [34–36] to weld PEEK polymer rods. The advantage of this approach is that it is suitable to join large or thicker parts compared with UW [24–26]. In addition, the proposed method provides low environmental pollution and energy consumption compared with conventional arc welding [37]. Therefore, the proposed method has industrial applicability and complies with the sustainable development goals

7 and 12 [38–40]. The simulation software [41,42] is also recommended to predict the optimum process parameters of RFW, such as the rotational speed, feed rate, axial pressure, welding time, or friction time. In addition, integrating RFW with automation and robotics systems [43] is also a promising research topic because it could enhance the production efficiency and consistency [44]. These are interesting research topics and are currently being investigated.

4. Conclusions

Traditionally, PEEK components were predominantly produced through injection molding or compression molding. Nevertheless, there has been limited attention given to the process of RFW involving PEEK polymer rods. Based on practical RFW experience, it has been observed that the burn-off length plays a crucial role in influencing the mechanical properties of the welded components. Consequently, the main objective of this study was to examine how the burn-off length affects the mechanical properties of welded parts made from PEEK polymer rods. The primary findings derived from the experimental investigation in this study can be summarized as follows:

1. The welding pressure (y) can be determined by the rotational speed (x) according to the prediction equation of $y = 0.0134x + 62.762$ with a correlation coefficient of 0.9753.
2. The optimal burn-off length of 1.6 mm seems suitable for cylindrical rods with a length of 40 mm and a diameter of 20 mm for five rotational speeds.
3. For the rotational speed of 1000 rpm, the average bending strength of the welded parts increased from 108 MPa to 160 MPa when the burn-off length increased from 1 mm to 1.6 mm, and a saving in the cycle time of RFW about 5% can be obtained since the cycle time of RFW was reduced from 80 s to 76 s.
4. The peak temperature in the weld interface for a rotational speed of 4000 rpm is only about 363 °C, which was attributed to the friction coefficient of the material being different due to the high rotation speed.
5. The bending strength of the welded component through laser welding is inferior compared to that achieved using RFW. This discrepancy arises from the fact that the fiber laser only melts the peripheral material of the PEEK cylinder during the laser welding process.

Author Contributions: C.-C.K.: Wrote the paper, conceived and designed the analysis, and performed the analysis; H.-X.L. and S.-H.H.: Collected the data and contributed data or analysis tools. All authors have read and agreed to the published version of the manuscript.

Funding: This study received financial support from the Ministry of Science and Technology of Taiwan under contract nos. NSTC 111-2221-E-131-015-MY2, MOST 110-2221-E-131-023, and MOST 109-2637-E-131-004.

Data Availability Statement: Data are contained within the article.

Conflicts of Interest: Author Song-Hua Hunag was employed by the company Li-Yin Technology. The remaining authors declare that the research was conducted in the absence of any commercial or financial relationships that could be construed as a potential conflict of interest.

References

1. Yang, Z.; Chen, F.; Gatea, S.; Ou, H. Design of the novel hot incremental sheet forming experimental setup, characterization of formability behavior of polyether-ether-ketone (PEEK). *Int. J. Adv. Manuf. Technol.* **2020**, *106*, 5365–5381. [CrossRef]
2. Rashed, K.; Kafi, A.; Simons, R.; Bateman, S. Effects of fused filament fabrication process parameters on tensile properties of polyether ketone ketone (PEKK). *Int. J. Adv. Manuf. Technol.* **2022**, *122*, 3607–3621. [CrossRef]
3. Rashed, K.; Kafi, A.; Simons, R.; Bateman, S. Optimization of material extrusion additive manufacturing process parameters for polyether ketone ketone (PEKK). *Int. J. Adv. Manuf. Technol.* **2023**, *126*, 1067–1091. [CrossRef]
4. Khoran, M.; Azarhoushang, B.; Amirabadi, H. Evaluating the influence of reinforcing fiber type on the grinding process of PEEK's composites. *Int. J. Adv. Manuf. Technol.* **2021**, *119*, 2187–2200. [CrossRef]

5. Lee, P.-C.; Peng, T.-Y.; Ma, T.-L.; Chiang, K.-Y.; Mine, Y.; Lee, I.-T.; Yu, C.-C.; Chen, S.-F.; Yu, J.-H. Effect of Various Airborne Particle Abrasion Conditions on Bonding between Polyether-Ether-Ketone (PEEK) and Dental Resin Cement. *Polymers* **2023**, *15*, 2114. [CrossRef]
6. Miura, D.; Ishida, Y.; Shinya, A. The Effects of Different Molding Orientations, Highly Accelerated Aging, and Water Absorption on the Flexural Strength of Polyether Ether Ketone (PEEK) Fabricated by Fused Deposition Modeling. *Polymers* **2023**, *15*, 1602. [CrossRef]
7. Vaddamanu, S.K.; Alhamoudi, F.H.; Chaturvedi, S.; Alqahtani, N.M.; Addas, M.K.; Alfarsi, M.A.; Vyas, R.; Kanji, M.A.; Zarbah, M.A.; Alqahtani, W.M.S.; et al. Retentive Forces and Deformation of Fitting Surface in RPD Clasp Made of Polyether-Ether-Ketone (PEEK). *Polymers* **2023**, *15*, 956. [CrossRef]
8. Bontempi, M.; Capozza, R.; Visani, A.; Fini, M.; Giavaresi, G.; Gambardella, A. Near-Surface Nanomechanics of Medical-Grade PEEK Measured by Atomic Force Microscopy. *Polymers* **2023**, *15*, 718. [CrossRef]
9. Yang, X.; Yokokura, S.; Nagahama, T.; Yamaguchi, M.; Shimada, T. Molecular Dynamics Simulation of Poly(Ether Ether Ketone) (PEEK) Polymer to Analyze Intermolecular Ordering by Low Wavenumber Raman Spectroscopy and X-ray Diffraction. *Polymers* **2022**, *14*, 5406. [CrossRef]
10. Naghavi, S.A.; Lin, C.; Sun, C.; Tamaddon, M.; Basiouny, M.; Garcia-Souto, P.; Taylor, S.; Hua, J.; Li, D.; Wang, L.; et al. Stress Shielding and Bone Resorption of Press-Fit Polyether–Ether–Ketone (PEEK) Hip Prosthesis: A Sawbone Model Study. *Polymers* **2022**, *14*, 4600. [CrossRef]
11. Zhou, Y.; Ren, L.; Zang, J.; Zhang, Z. The Shape Memory Properties and Actuation Performances of 4D Printing Poly (Ether-Ether-Ketone). *Polymers* **2022**, *14*, 3800. [CrossRef] [PubMed]
12. Baltag, L.; Cojocaru, C.; Enache, A.-C.; Samoila, P.; Harabagiu, V. Ultrasonic-Assisted Rapid Preparation of Sulfonated Polyether Ether Ketone (PEEK) and Its Testing in Adsorption of Cationic Species from Aqueous Solutions. *Materials* **2022**, *15*, 7558. [CrossRef] [PubMed]
13. Saravi, B.; Flohr, A.; Patzelt, S.B.; Spies, B.C.; Hazard, D.; Kohal, R.J. Fatigue and Fracture Resistance Testing of Polyether Ether Ketone (PEEK) Implant Abutments in an Ex Vivo Chewing Simulator Model. *Materials* **2022**, *15*, 6927. [CrossRef] [PubMed]
14. Zhang, Y.; Yu, G.; Tian, C.; Li, Z.; Shao, J.; Li, S.; He, X. Hole Morphology and Keyhole Evolution during Single Pulse Laser Drilling on Polyether-Ether-Ketone (PEEK). *Materials* **2022**, *15*, 2457. [CrossRef] [PubMed]
15. Bialas, O.; Lis, M.; Woźniak, A.; Adamiak, M. Laser Superficial Fusion of Gold Nanoparticles with PEEK Polymer for Cardiovascular Application. *Materials* **2021**, *14*, 971. [CrossRef] [PubMed]
16. Pedroso, J.M.; Enger, M.; Bandeira, P.; Magalhães, F.D. Comparative Study of Friction and Wear Performance of PEK, PEEK and PEKK Binders in Tribological Coatings. *Polymers* **2022**, *14*, 4008. [CrossRef] [PubMed]
17. Belkahla, Y.; Mazouzi, A.; Lebouachera, S.E.I.; Hassan, A.J.; Fides, M.; Hvizdoš, P.; Cheniti, B.; Miroud, D. Rotary friction welded C45 to 16NiCr6 steel rods: Statistical optimization coupled to mechanical and microstructure approaches. *Int. J. Adv. Manuf. Technol.* **2021**, *116*, 2285–2298. [CrossRef]
18. Rinaldi, M.; Cecchini, F.; Pigliaru, L.; Ghidini, T.; Lumaca, F.; Nanni, F. Additive Manufacturing of Polyether Ether Ketone (PEEK) for Space Applications: A Nanosat Polymeric Structure. *Polymers* **2021**, *13*, 11. [CrossRef]
19. Liu, C.; Chan, K.W.; Shen, J.; Liao, C.Z.; Yeung, K.W.K.; Tjong, S.C. Polyetheretherketone Hybrid Composites with Bioactive Nanohydroxyapatite and Multiwalled Carbon Nanotube Fillers. *Polymers* **2016**, *8*, 425. [CrossRef]
20. Zhang, Y.; Hansen, H.N.; Sørensen, S. Replication of micro-pillars by PEEK injection moulding with CrN-coated Ni tool. *Int. J. Adv. Manuf. Technol.* **2015**, *80*, 383–388. [CrossRef]
21. Griffiths, C.A.; Bigot, S.; Brousseau, E.; Worgull, M.; Heckele, M.; Nestler, J.; Auerswald, J. Investigation of polymer inserts as prototyping tooling for micro injection moulding. *Int. J. Adv. Manuf. Technol.* **2009**, *47*, 111–123. [CrossRef]
22. Deringer, T.; Drummer, D. In situ curing and bonding of epoxy prepregs in epoxy thermoset injection molding. *Int. J. Adv. Manuf. Technol.* **2021**, *117*, 2667–2677. [CrossRef]
23. Su, C.; Lv, S.; Wang, R.; Lv, Y.; Lou, S.; Wang, Q.; Guo, S. Effects of forming parameters on the forming limit of single-point incremental forming of sheet metal. *Int. J. Adv. Manuf. Technol.* **2021**, *113*, 483–501. [CrossRef]
24. Khatri, B.; Roth, M.F.; Balle, F. Ultrasonic Welding of Additively Manufactured PEEK and Carbon-Fiber-Reinforced PEEK with Integrated Energy Directors. *J. Manuf. Mater. Process.* **2023**, *7*, 2. [CrossRef]
25. Alexenko, V.O.; Panin, S.V.; Stepanov, D.Y.; Byakov, A.V.; Bogdanov, A.A.; Buslovich, D.G.; Panin, K.S.; Tian, D. Ultrasonic Welding of PEEK Plates with CF Fabric Reinforcement—The Optimization of the Process by Neural Network Simulation. *Materials* **2023**, *16*, 2115. [CrossRef] [PubMed]
26. Bonmatin, M.; Chabert, F.; Bernhart, G.; Cutard, T.; Djilali, T. Ultrasonic welding of CF/PEEK composites: Influence of welding parameters on interfacial temperature profiles and mechanical properties. *Compos. Part A Appl. Sci. Manuf.* **2022**, *162*, 107074. [CrossRef]
27. Kunhirunbawon, S.; Suwichien, N.; Jantarasricha, T. Friction welding parameter for AA6063 using ANFIS prediction. *Int. J. Adv. Manuf. Technol.* **2023**, *128*, 2589–2597. [CrossRef]
28. Naganaboyina, H.P.S.; Nagaraju, P.; Sonaye, S.Y.; Bokam, V.K.; Sikder, P. In-house processing of carbon fiber-reinforced polyetheretherketone (CFR-PEEK) 3D printable filaments and fused filament fabrication-3D printing of CFR-PEEK parts. *Int. J. Adv. Manuf. Technol.* **2023**, *128*, 5011–5024. [CrossRef]

29. Assawakawintip, T.; Santiwong, P.; Khantachawana, A.; Sipiyaruk, K.; Chintavalakorn, R. The Effects of Temperature and Time of Heat Treatment on Thermo-Mechanical Properties of Custom-Made NiTi Orthodontic Closed Coil Springs. *Materials* **2022**, *15*, 3121. [CrossRef]
30. El-Geassy, A.A.; Halim, K.S.A.; Alghamdi, A.S. A Novel Hydro-Thermal Synthesis of Nano-Structured Molybdenum-Iron Intermetallic Alloys at Relatively Low Temperatures. *Materials* **2023**, *16*, 2736. [CrossRef]
31. Koniorczyk, P.; Zieliński, M.; Sienkiewicz, J.; Zmywaczyk, J.; Dębski, A. Experimental Studies of Thermophysical Properties and Microstructure of X37CrMoV5-1 Hot-Work Tool Steel and Maraging 350 Steel. *Materials* **2023**, *16*, 1206. [CrossRef] [PubMed]
32. Elkolali, M.; Nogueira, L.P.; Rønning, P.O.; Alcocer, A. Void Content Determination of Carbon Fiber Reinforced Polymers: A Comparison between Destructive and Non-Destructive Methods. *Polymers* **2022**, *14*, 1212. [CrossRef] [PubMed]
33. Misra, D.; Mitra, S.; Meena, H.L.; Datta, S.N.; Nayak, J.; Kalvettukaran, P.; Paul, S. Study of experimental and numerical simulation of high-energy laser processing on carbon fiber reinforced polymer. *Int. J. Adv. Manuf. Technol.* **2023**, *129*, 429–444. [CrossRef]
34. Liu, H.; Wang, C.; Liu, Z.; Liu, K.; Jiang, S.; Wang, Y. Numerical prediction of machining-induced surface residual stress for TC4 cryogenic turning. *Int. J. Adv. Manuf. Technol.* **2021**, *114*, 131–144. [CrossRef]
35. Zhang, Y.; Xu, X. Machine learning surface roughnesses in turning processes of brass metals. *Int. J. Adv. Manuf. Technol.* **2022**, *121*, 2437–2444. [CrossRef]
36. Bagga, P.J.; Makhesana, M.A.; Darji, P.P.; Patel, K.M.; Pimenov, D.Y.; Giasin, K.; Khanna, N. Tool life prognostics in CNC turning of AISI 4140 steel using neural network based on computer vision. *Int. J. Adv. Manuf. Technol.* **2022**, *123*, 3553–3570. [CrossRef]
37. Miao, J.; Li, Y.; Ren, B.; Dong, Z.; Zou, W.; Chang, C.; Chang, Y. Current status of research on numerical simulation of droplet transfer in CO2 gas–shielded welding. *Int. J. Adv. Manuf. Technol.* **2023**, *128*, 1–15. [CrossRef]
38. Abdul-Rashid, S.H.; Mohamad, M.N.; Sakundarini, N.; Ghazilla, R.A.R.; Thurasamy, R. Modelling sustainable manufacturing practices effects on sustainable performance: The contingent role of ownership. *Int. J. Adv. Manuf. Technol.* **2022**, *122*, 3997–4012. [CrossRef]
39. Tian, S.; Xie, X.; Xu, W.; Liu, J.; Zhang, X. Dynamic assessment of sustainable manufacturing capability based on correlation relationship for industrial cloud robotics. *Int. J. Adv. Manuf. Technol.* **2021**, *124*, 3113–3135. [CrossRef]
40. Klimant, P.; Koriath, H.J.; Schumann, M.; Winkler, S. Investigations on digitalization for sustainable machine tools and forming technologies. *Int. J. Adv. Manuf. Technol.* **2021**, *117*, 2269–2277. [CrossRef]
41. Ribeiro, C.A.C.; Ferreira, J.R.; Lima e Silva, S.M. Thermal influence analysis of coatings and contact resistance in turning cutting tool using COMSOL. *Int. J. Adv. Manuf. Technol.* **2022**, *118*, 275–289. [CrossRef]
42. Schorr, L.; Johnson, B.; McFall, J.; Shepherd, D.; Hadimani, R.L. High-temperature gripper for collaborative robots in additive manufacturing. *Int. J. Adv. Manuf. Technol.* **2023**, *128*, 1291–1303. [CrossRef]
43. Erkol, H.O.; Bailey, M.; Palardy, G.; Barbalata, C. Predicting composite laminates roughness: Data-driven modeling approaches using force sensor data from robotic manipulators. *Int. J. Adv. Manuf. Technol.* **2023**, *128*, 1801–1813. [CrossRef]
44. Feddoul, Y.; Ragot, N.; Duval, F.; Havard, V.; Baudry, D.; Assila, A. Exploring human-machine collaboration in industry: A systematic literature review of digital twin and robotics interfaced with extended reality technologies. *Int. J. Adv. Manuf. Technol.* **2023**, *129*, 1917–1932. [CrossRef]

polymers

Article

Optimization of Dimensional Accuracy and Surface Roughness of SLA Patterns and SLA-Based IC Components

Aishabibi Mukhangaliyeva, Damira Dairabayeva, Asma Perveen and Didier Talamona *

Department of Mechanical and Aerospace Engineering, School of Engineering and Digital Sciences, Nazarbayev University, Astana 010000, Kazakhstan; aishabibi.mukhangaliyeva@alumni.nu.edu.kz (A.M.); damira.dairabayeva@alumni.nu.edu.kz (D.D.); asma.perveen@nu.edu.kz (A.P.)

* Correspondence: didier.talamona@nu.edu.kz

Abstract: Rapid investment casting is a casting process in which the sacrificial patterns are fabricated using additive manufacturing techniques, making the creation of advanced designs possible. One of the popular 3D printing methods applied in rapid investment casting is stereolithography because of its high dimensional precision and surface quality. Printing parameters of the used additive manufacturing method can influence the surface quality and accuracy of the rapid investment cast geometries. Hence, this study aims to investigate the effect of stereolithography printing parameters on the dimensional accuracy and surface roughness of printed patterns and investment cast parts. Castable wax material was used to print the sacrificial patterns for casting. A small-scale prosthetic biomedical implant for total hip replacement was selected to be the benchmark model due to its practical significance. The main results indicate that the most significant stereolithography printing parameter affecting surface roughness is build angle, followed by layer thickness. The optimum parameters that minimize the surface roughness are 0.025 mm layer thickness, 0° build angle, 1.0 support density index, and across the front base orientation. As for the dimensional accuracy, the optimum stereolithography parameters are 0.025 mm layer thickness, 30° build angle, 0.6 support density index, and diagonal to the front base orientation. The optimal printing parameters to obtain superior dimensional accuracy of the cast parts are 0.05 mm layer thickness, 45° build angle, 0.8 support density index, and diagonal to the front model base orientation. With respect to the surface roughness, lower values were obtained at 0.025 mm layer thickness, 0° build angle, 1.0 support density index, and parallel to the front base orientation.

Keywords: rapid investment casting; stereolithography; dimensional accuracy; surface roughness; Taguchi; optimization; printing parameters

Citation: Mukhangaliyeva, A.; Dairabayeva, D.; Perveen, A.; Talamona, D. Optimization of Dimensional Accuracy and Surface Roughness of SLA Patterns and SLA-Based IC Components. *Polymers* **2023**, *15*, 4038. https://doi.org/10.3390/polym15204038

Academic Editors: Ying Yan and Yiqi Wang

Received: 16 August 2023
Revised: 3 October 2023
Accepted: 3 October 2023
Published: 10 October 2023

1. Introduction

Investment casting (IC), also referred to as lost wax casting or precision casting, is a near-net shape process of manufacturing complex metal geometries. IC can be used to cast non-ferrous and ferrous metals and produces parts with high dimensional accuracy and excellent surface finish. IC is widely used in numerous industries including automotive, aerospace, medicine, military, jewelry, and energy sectors [1].

The conventional IC process is shown in Figure 1. First, a metal mold for a wax pattern is manufactured through the fabrication of a metal die. The molten wax is then injected into the metal mold to create a wax pattern. A casting tree is assembled with multiple wax patterns. Afterward, the casting tree is dipped into a ceramic slurry containing alumina and colloidal silica to create the interior walls of a ceramic mold. The obtained mold is heated in an oven to melt all wax content. The molten metal is poured into the ceramic mold and allowed to solidify and cool before the destruction of the ceramic shell [2].

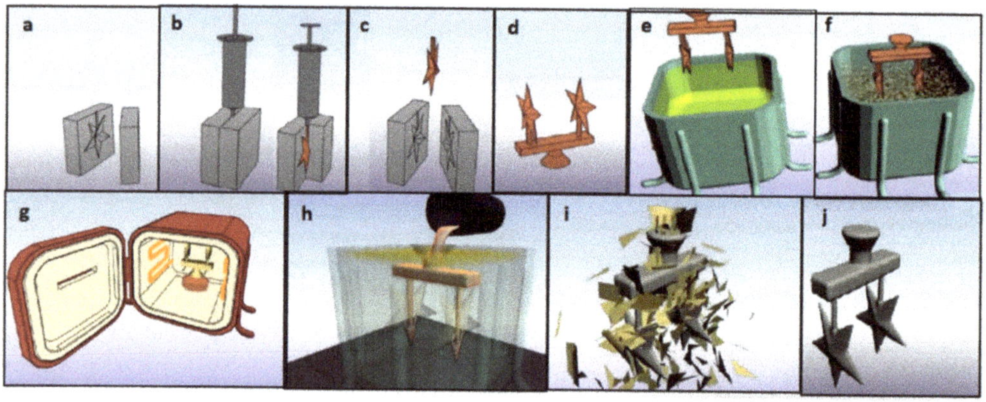

Figure 1. Conventional investment casting: (**a**) Preparing a metal mold for wax injection; (**b**) injecting wax to create patterns; (**c**) ejecting wax from the mold; (**d**) assembling wax patterns into a tree; (**e**) ceramic slurry coating; (**f**) coating with stucco; (**g**) dewaxing by heating; (**h**) pouring metal; (**i**) removing a ceramic shell; (**j**) a cast part that is prepared to be severed from the assembly [1].

One of the issues faced with conventional IC is the manufacturing of wax patterns since it requires a difficult tooling process. The pattern-making stage can be simplified by using additive manufacturing (AM) to replace the IC steps such as the hard tooling of metal molds and wax injection. AM, also known as rapid prototyping, is a layer-by-layer fabrication process that emerged in the 1980s. In recent decades, AM technologies evolved significantly to be used in IC. The AM-assisted IC is called rapid investment casting (RIC) due to the reduced time needed to prepare the patterns for casting [3].

The AM technologies in RIC are used either through an indirect or direct approach. A 3D-printed mold is used to prepare the wax pattern in the indirect RIC, while the pattern is fabricated using a printer in the direct RIC [4]. The application of AM-based sacrificial patterns in IC reduces the lead time and cost significantly for individual and small-size productions when compared to the conventional process. The AM processes used for RIC include stereolithography (SLA), fused filament fabrication (FFF), selective laser sintering (SLS), and laminated object manufacturing (LOM). In order to create castings of superior quality, the AM-based patterns are required to have a low melting point, sufficient mechanical strength and toughness, low surface roughness, high dimensional accuracy, and no residual ash following burnout [5,6]. SLA is regarded as a very precise method among 3D printing techniques [7].

SLA is a vat photopolymerization technology that uses ultraviolet (UV) light to cure or solidify liquid resin, a photosensitive thermoset material. The UV radiation initiates the curing reaction of a resin which results in the cross-linking of molecular chains and the polymerization of oligomers and monomers. Depending on the printing device, there might be a bottom-up or top-down method. Indeed, desktop 3D printers, such as Formlabs 3D printers, employ the bottom-up approach, whereas industrial 3D printers use the top-down technique [8]. As shown in Figure 2, the main components of the SLA machines are the laser, mirrors, galvanometers, resin tank, and build platform. A layer of liquid photopolymer is scanned by a laser to harden the resin in the areas specified by the CAD model. After the curing of one layer, the surface is covered with a thin film of resin for further exposure to the laser beam. The platform is raised/moved up along the *z*-axis for the curing of each consecutive layer [1].

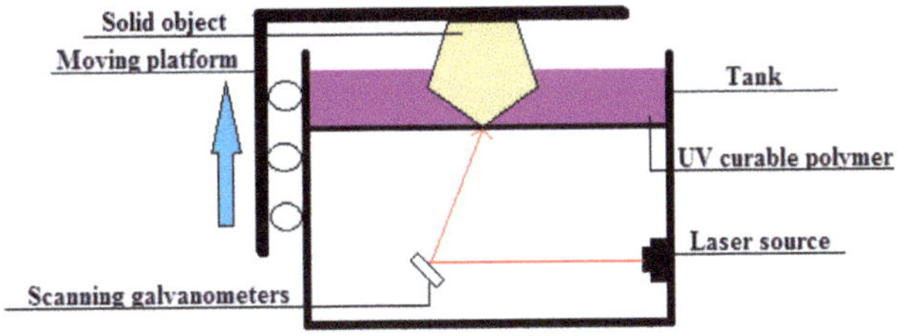

Figure 2. A general schematic of the SLA process (bottom-up).

There are several benefits of casting with SLA. Firstly, compared to traditional IC, its manufacturing cost and lead time are decreased. Secondly, waste is reduced as a result of the faster machining [9,10]. Also, it is possible to implement modifications to the model when designing without affecting the production time [1].

Most of the SLA studies considered the effect of the layer thickness, position on the build platform, and build angle on surface roughness or dimensional accuracy of SLA parts. A summary of the literature review is given in Table 1.

Table 1. Summary of SLA studies.

Source	Printing Process Parameters	Properties	Materials	Findings about Parameters Optimization
[11]	Layer thickness	Surface roughness, dimensional stability	V3 (FLGPR03) grey resin	- No effect of layer thickness on surface roughness - A 25 µm layer thickness for dimensional accuracy
[12]	Orientation in relation to the printer's build platform	Dimensional accuracy, surface roughness	Denture base polymethyl methacrylate (PMMA)	- A 90 degree angle resulted in the lowest dimensional errors, and surface roughness
[13]	Layer thickness, model structure, model base orientation, position build platform	Surface roughness	V3 (FLGPR03) grey resin	- No effect of model structure and build platform position was found
[14]	Build orientation (front, side, top), angular orientation around all the three-principal axis	Surface roughness	Visijet Sl Clear	- Surface roughness is not statistically significantly affected by altering the build orientation, However, it is influenced by changing the axis or angle - A 90° orientation along the *x* and *z*-axis resulted in lowest roughness
[15]	Printing orientation Position on build platform Layer thickness	Dimensional accuracy	Formlabs Clear Resin	Dimensional accuracy was improved when printing with - lower layer thickness (0.025 mm) - in center of the platform
[16]	Build orientation, positioning	Dimensional accuracy	Dental SG resin	- Used 45° orientation - Positioning in the center of the platform were recommended to be used

Table 1. *Cont.*

Source	Printing Process Parameters	Properties	Materials	Findings about Parameters Optimization
[17,18]	Build angle in relation to the printer's build platform	Dimensional accuracy	Dental SG Resin	- The 0 and 45-degree build angles resulted in surgical templates with the highest level of precision
[17]	Build angle in relation to the z-axis	Accuracy (trueness and precision)	Formlabs Clear Resin	- Printing at a 45-degree angle yielded superior accuracy when compared to printing at 0 or 90 degrees
[19]	Layer thickness	Dimensional accuracy	Grey and Cast resin	- No statistically significant variations in accuracy between the three different layer thicknesses
[20]	Build angle with respect to the build platform	Dimensional accuracy	Dental LT Clear Resin	- The general accuracy is not greatly affected by the print orientation
[21]	Layer thickness (0.025 and 0.050 mm) Part orientation (horizontal, slanted, vertical in relation to the build platform)	Dimensional accuracy, surface roughness	Formlabs Rigid 10K V0 and High Temperature V02	- The 0.050 mm layer thickness led to low shape quality - Surface roughness was minimized due to the horizontal orientation

Khaledi et al. and Piangsuk et al. have studied the dimensional accuracy of SLA-based RIC and milled copings used in prosthetic dentistry [22,23]. According to the outcomes, both SLA-printed and SLA-based RIC components demonstrated higher dimensional accuracy than milled components.

A prosthetic biomedical implant for total hip replacement was selected to be the benchmark model due to its practical significance based on the available literature. A hip implant is used to replace a hip joint, one of the largest ball-and-socket joints in the human body, which allows for the free movement of a leg, supports the upper part of the body, and absorbs the movement impact. As shown in Figure 3, the hip implant consists of three main parts, such as the cup, ball, and stem. The cup is a replacement for the hip socket in the pelvis bone. The ball placed in the cup is inserted into the acetabular shell, while the stem is put into the femoral canal for stability [24].

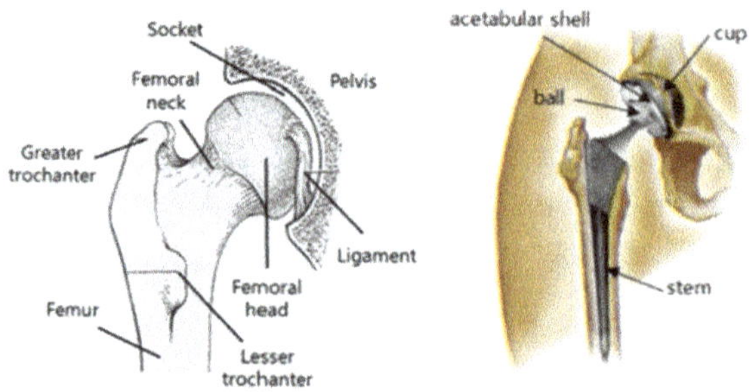

Figure 3. Parts of a hip implant [24].

Several researchers have studied the dimensional accuracy and surface roughness of hip implants produced by FFF and FFF-based IC. A study by Singh et al. investigated the effect of IC process parameters such as slurry viscosity, number of slurry layers, and dry time of the primary coating on the surface roughness of hip joints developed by a combination of IC and FFF [25]. The design was used the femoral ball and shortened stem. It was found that the most influential factor on the quality of a cast hip implant surface is the number of layers. Kumar et al. examined three hip implants with different volume-to-surface (V/A) ratios [26]. The first model consisted of only a stem, while the other two implants were designed with a femoral ball. Researchers focused on both the printing and casting parameters such as the orientation, density of the pattern, mold thickness, grade of material, type of pattern, and V/A ratio. According to the results, an orientation of 0° angle is recommended for Acrylonitrile Butadiene Styrene (ABS) patterns while a wax layer notably increased the surface quality of cast parts. A simplified model of the hip implant was designed by Tiwary et al., and the printing parameters examined in this article were extrusion speed, infill, number of top and bottom solid layers, and layer thickness [27]. It was found that the surface roughness of the FFF parts can be decreased by making the top and bottom layers solid and it was suggested that the RIC technique be applied for a low volume of production.

In the RIC, the process parameters used to print the sacrificial patterns can affect the surface roughness and dimensional accuracy of the castings. Hence, the surface finish and accuracy of the printed patterns must be improved by optimizing the printing parameters. According to the literature review, numerous studies focused on the dimensional accuracy and surface roughness of the SLA-printed parts separately. However, there is lack of studies on the effects of the SLA process parameters on the surface roughness and dimensional accuracy of both printed and cast parts and on castable wax resin material. Consequently, this study intends to examine and fill these research gaps. The aim of this work is to investigate the effect of the SLA printing parameters on the dimensional accuracy and surface roughness of the printed patterns and the cast components. The following objectives were defined to attain the aim of this study:

- Design a benchmark model of the small-scale prosthetic biomedical hip implant.
- Fabricate benchmark patterns using the SLA printing technology and assess their dimensional accuracy and surface roughness.
- Cast and measure the dimensional accuracy and surface roughness of the AM-based IC components.
- Examine the experimental results and obtain the optimized printing parameters for dimensional accuracy and surface roughness.
- Investigate the hardness, microstructure, and composition of the IC parts.

2. Materials and Methods

This section describes the experimental procedure, model design, printing materials, design of experiments, and equipment used in the study.

2.1. Experimental Procedure

This study consists of several stages as shown in Figure 4. Firstly, the virtual model of the hip implant was created using a computer-aided design software, SolidWorks 2021. The material and printing parameters were selected. Taguchi's L18 orthogonal array was chosen for the design of experiments (DOE). The SLA printer, FormLabs Form 3 (Formlabs Inc., Somerville, MA, USA), was used to print the specimens. The surface roughness and dimensional accuracy of the printed parts were measured and processed to examine the effect of parameters on these properties. Afterward, the printed samples were used to fabricate the mold for the casting. The IC parts were produced using aluminum. The surface finish and dimensional deviation of these parts were measured. The microhardness and microstructure of the IC parts were characterized.

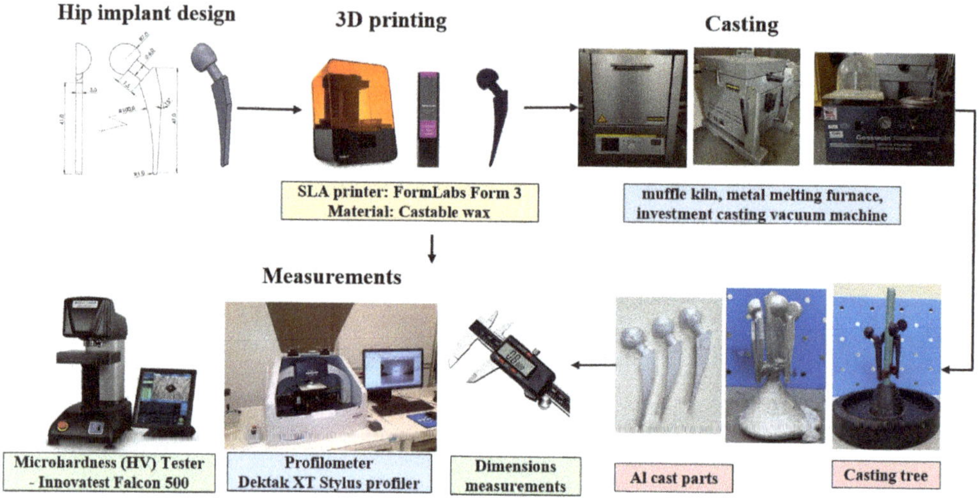

Figure 4. Experimental procedure.

2.2. Benchmark model design

The small-scale prosthetic biomedical hip implant model designed for the experiment is shown in Figure 5. The model was created using the SolidWorks 2021 software based on the literature review and resized down to an appropriate scale for 3D printing [22–24]. The main geometric features of the hip implant are the femoral head, the neck, and the stem. The model was designed to be the half of the implant since the part was required to lie flat on the surface for measurement purposes.

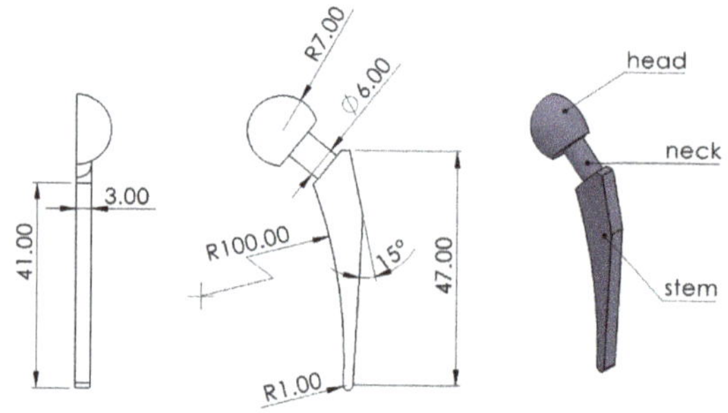

Figure 5. Geometry and feature labeling of the hip implant model.

2.3. SLA Printing of Patterns

The SLA patterns were printed using FormLabs Form 3. The printer was equipped with 250 mW laser power and low-force stereolithography (LFS) technology. The size of the laser spot was 85 μm [28].

The material used for the SLA printing of patterns was FormLabs Castable Wax V1 (FLCWPU01) resin, an acrylate photopolymer with a wax content of 20%. This resin produced the pattern for casting with a high level of detail, with print resolutions of 25 and 50 μm and a superior surface finish. The castable wax resulted in a clean burnout and almost zero ash content. The layer thicknesses that can be printed with this material are 0.025 and 0.05 mm. According to the material specifications, no-post curing is needed for the castable wax [29]. The characteristics of the resin are given in Table 2. The Preform 3.27.1 software was used to set up the parts for SLA printing.

Table 2. Resin characteristics [28].

Casting Properties		Mechanical Properties	
Wax content	20%	Young's Modulus	220 MPa
Ash content	0.0–0.1%	Tensile Strength	12 MPa
		Elongation at break	13%

2.4. Design of Experiment

The Taguchi method of parametric design of experiment (DOE) was employed to examine the impact of the selected printing parameters (see Table 3) on the dimensional accuracy and surface roughness of the hip implant model. The Taguchi method allows for a reduction in the number of experiments while still determining the sensitivity and importance of each parameter. In this study, the process parameters of SLA printing were chosen to be layer thickness, build angle (see Figure 6), support density, and model base orientation (see Figure 7). In Figure 6, model base orientations are across the front; build angles are 0° and 30° in part A and B, respectively. To obtain B, the A was rotated along the x-axis by 30°. Then, the two other cases where the build angle was 30° and model base orientation was diagonal to the front and parallel to the front were obtained by rotating B along the z-axis by 45° and 90°, respectively. The support density was chosen in the range [0.6, 1.0] because a support density of less than 0.6 was not recommended and showed an error by the Preform software when the checking model for printability while using a support density of more than 1.0; surface finish was considerably reduced due to the footprints of the excess support structures. According to the chosen DOE, there should be 3 levels of support density, therefore, 0.6, 0.8, and 1.0 were chosen.

Table 3. SLA printing parameters and control levels.

Parameters	Level 1	Level 2	Level 3
A: Layer thickness (mm)	0.025	0.050	-
B: Build angle (°)	0	30	45
C: Support density	0.6	0.8	1
D: Model base orientation	Parallel to the front	Across the front	Diagonal to the front

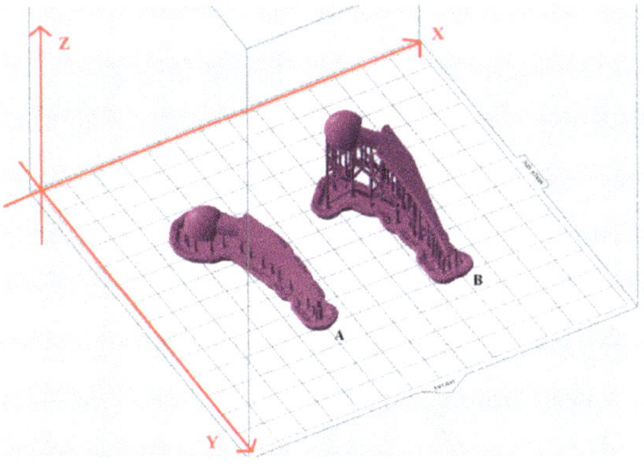

Figure 6. Build angle of 0° (**A**) and 30° (**B**), across the front model base orientation.

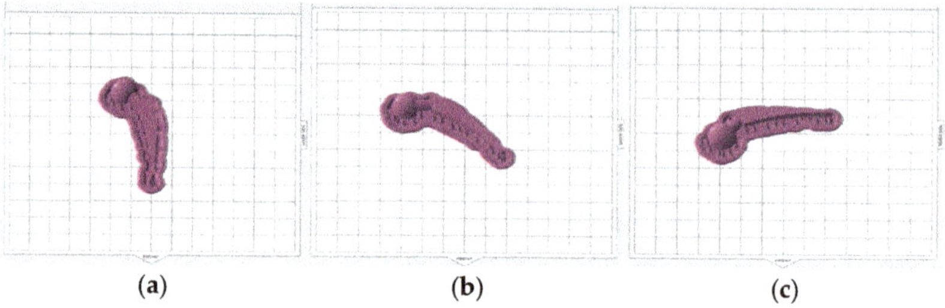

Figure 7. Model base orientations (build angle of 0°); (**a**) across the front; (**b**) diagonal to the front; (**c**) parallel to the front.

The Minitab Statistical Software Version 21.1.0 was used to create Taguchi's experimental design. L18 Orthogonal Array with 18 rows as shown in Table 4 was created for the chosen factors and levels. There is a setting for the different experimental SLA printing runs in each row. Each experimental run was repeated three times as shown in Figure 8.

Table 4. Design of experiments (DOE) for SLA prints.

No.	A	B	C	D
1	1	1	1	1
2	1	1	2	2
3	1	1	3	3
4	1	2	1	1
5	1	2	2	2
6	1	2	3	3
7	1	3	1	2
8	1	3	2	3
9	1	3	3	1
10	2	1	1	3
11	2	1	2	1
12	2	1	3	2
13	2	2	1	2
14	2	2	2	3
15	2	2	3	1
16	2	3	1	3
17	2	3	2	1
18	2	3	3	2

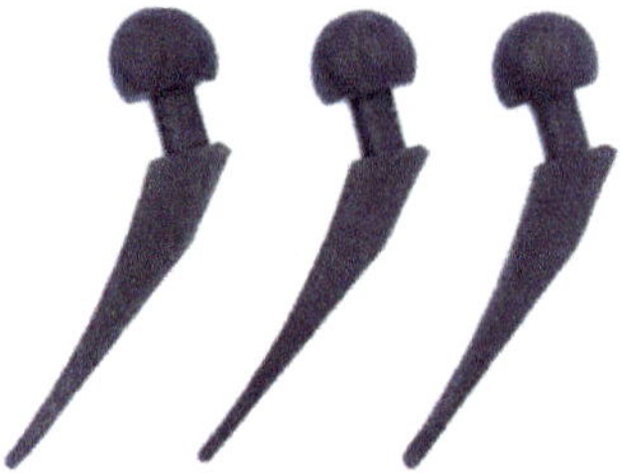

Figure 8. Hip implant printed using SLA.

2.5. Investment Casting

Aluminum wrought alloy (AMg3 (1530), Kazakhstan Metal Industrial Company, Astana, Kazakhstan) was chosen as the casting metal for this experiment since it is a lightweight, highly recyclable, inexpensive, and available metal with good strength, stiffness, and high thermal conductivity. The equipment used for casting procedures is shown in Figure 9.

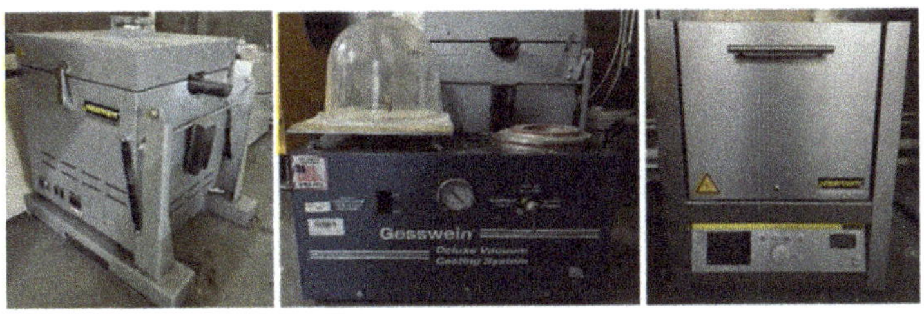

(a) (b) (c)

Figure 9. Equipment used for casting (**a**) metal melting furnace; (**b**) IC vacuum machine; (**c**) muffle kiln.

Firstly, wax rods were used to build the patterns in a casting tree, as can be seen in Figure 10a. Then, the mold creation was carried out using Ultravest investment powder from Ranson and Randolf and distilled water in a 39/100 water to powder ratio. According to the instructions for the investment powder, 690 g of powder and 269.4 g of water were mixed for an investment in a flask with a 10.2 cm height and 8.3 cm diameter. After mixing the powder with water for 2–3 min, the investment table side of the Gesswein Deluxe Vacuum Casting System machine was utilized to remove trapped air from the mixture. The mixture was then added to the flask containing the casting tree, and the vacuum was applied once again. Mold reaches its peak green intensity after 2 h, thus it was left to sit undisturbed for that time period.

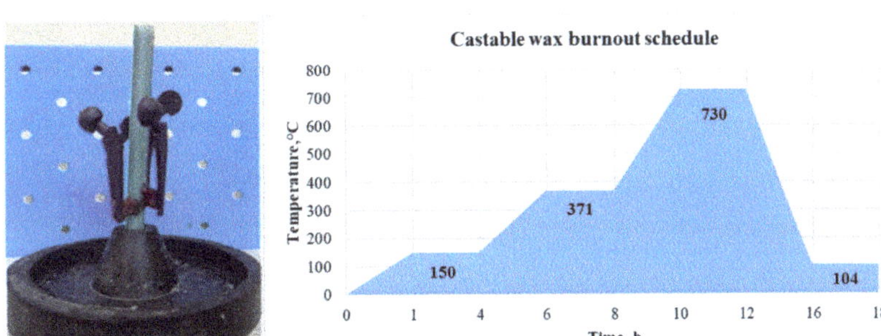

(**a**) (**b**)

Figure 10. (**a**) Casting tree; (**b**) burnout cycle of the castable wax.

For further burnout of the flask with investment, Nabertherm Muffle Furnace L15/11/B410 (Nabertherm GmbH, Lilienthal, Germany) was used. The flask was put into a kiln which was preheated to 150 °C. The burnout cycle for the selected castable wax material was approximately 15 h as illustrated in Figure 10b. Aluminum bars were placed into a Nabertherm K1/13 (Nabertherm GmbH, Lilienthal, Germany) metal melting furnace and melted at 750 °C during the final two hours of the burnout cycle. The casting parameters used were as follows: casting temperature of 750 °C, mold temperature before casting of 104 °C, pouring speed of 2.5 cm/s, and time before cooling 20 min.

The flask was taken out of the kiln and put into the casting chamber of the Gesswein Deluxe Vacuum Casting System machine (Gesswein, Bridgeport, CT, USA) after 1.5 h of gradually melting the metal. The vacuum was switched on and the molten aluminum metal was poured into the flask. After 15–20 min of metal solidification, the hot flask was cooled with water and the investment casting shells were removed from th e cast parts. Next, the cast hip implant parts were sawed off from the casting tree and no post-processing was carried out. The hip implants produced using IC are shown in Figure 11.

Figure 11. Hip implant produced using IC.

2.6. Measurements

A Vernier digital caliper was used to measure the dimensional accuracy of the printed components. The features that were measured and checked for accuracy are shown in Table 5 and Figure 5.

Table 5. SLA printing parameters and control levels.

Dimensions	Radius	Width	Length	Thickness
Features	Head	Neck	Stem	Stem
			Neck	Neck

The surface roughness was measured using a profilometer—Dektak XT Stylus (Bruker, Billerica, MA, USA) profiler. The profilometer was in contact with the sample using a stylus with a single-crystal diamond tip with a radius of 12.5 μm. The sample stage slid the sample underneath the stylus tip to draw a profile while the equipment maintained a steady stylus force. This device assessed the surface roughness by profiling the surface topography. The surface roughness was measured in the stem of the implant. The arithmetical average roughness (*Ra*) refers to the arithmetic mean of all the absolute variances from the center line over the total length and is defined by Equation (1).

$$R_a = \frac{1}{L} \int_0^L |z(x)| dx \tag{1}$$

where L is a sampling length and z is a coordinate of the curve profile [30].

The measurement setup parameters for the profilometer were the following: duration of 20 sec, length of 5000 μm, range of 524 μm, and 7 mg of stylus force. An average of three measurements was taken for each specimen.

The Vickers microhardness of the SLA-based IC was measured using the Micro-hardness (HV) Tester—Innovatest Falcon 500 (INNOVATEST Europe BV, Maastricht, The Netherlands). The hardness was measured three times for each specimen. A load of 0.5 kg was applied to the specimen with a dwelling time of 10 s. The surface of the parts was polished using the polishing machine before the test.

The microstructure was examined using the Scanning Electron Microscope JEOL JSM-IT200 (JEOL Ltd., Tokyo, Japan). The Energy Dispersive Spectroscopy (EDS) analysis was conducted to investigate the chemical composition of the specimen while imaging. The hip implants were cut using a micro wire electrical discharge machine (μWEDM) to prepare the samples for the SEM. The surface of the samples was polished using a polishing machine. The samples also underwent wet etching to remove the material from the surface.

3. Results and Discussion

This section represents the results of the experiments and analysis.

3.1. Analysis of Measurements

The S/N ratio for the smaller-is-better category was selected to minimize the surface roughness and dimensional deviation of the printed patterns. The S/N ratio is calculated using Equation (2).

$$\eta = -10log(\frac{1}{n}\sum_{i=1}^{n}\gamma_i^2) \quad (2)$$

where η is an S/N ratio, γ_i is a response for a given factor level combination, and n is a number of responses [30].

3.2. SLA Printed Patterns: Dimensional Accuracy

The dimensions of all the printed components were recorded and compared with the nominal values. The average deviations of all features are demonstrated in Table 6. Both expansion and shrinkage occurred since the average deviation from the nominal values varied between −0.21 mm and 0.12 mm. The lowest deviation corresponded to the neck length, while the stem length had the highest average deviation. Both the stem and neck thicknesses of the printed parts were larger than the nominal dimensions.

Table 6. Dimensional deviation of the SLA printed parts.

Dimension	Features	Nominal Value (mm)	Average Measured Value (mm)	Range		Average Deviation (mm)
				Max (mm)	Min (mm)	
Radius	Sphere	7.00	7.12	7.50	6.56	0.12
Width	Neck	6.00	5.81	5.92	5.64	−0.19
Thickness	Stem	3.00	3.10	3.54	2.82	0.10
	Neck	3.00	3.09	3.58	2.78	0.09
Length	Stem	48.00	47.79	48.92	47.50	−0.21
	Neck	7.50	7.57	7.89	7.38	0.07

The mean and S/N rankings of the parameters for the dimensional accuracy of each feature are given in Table A1 in Appendix A, where 1 and 4 correspond to the most and least significant printing parameter, respectively. The sum of the rankings of each parameter was calculated and, based on this, the significance of factors was revealed. Both the mean and S/N rankings gave the same outcomes. Based on their influence on the dimensional accuracy of the SLA printed parts, the most significant parameter was the build angle, followed by the support density, model base orientation, and layer thickness.

The optimal level for each parameter was determined for each feature measurement using the mean analysis. The results are demonstrated in Table A2 in Appendix A. To obtain the superior dimensional accuracy of the printed components, the number of occurring frequencies of each level for each factor was added, and the level with the highest frequency was selected as the optimal level.

3.3. SLA Printed Patterns: Surface Roughness

The roughness of all SLA samples was measured and the average surface roughness, standard deviation, and S/N ratio value for each experimental run are given in Table 7. The highest surface roughness (2.927 μm) was measured for Run 14, while the lowest value (0.882 μm) was recorded for Run 2.

Table 7. Results of surface roughness (SLA printed patterns).

Experiment Run No.	Surface Roughness Ra (μm)	Standard Deviation (μm)	S/N Ratio (dB)
1	0.922	0.060	0.702
2	0.882	0.052	1.091
3	1.023	0.118	−0.199
4	1.210	0.068	−1.655
5	1.398	0.140	−2.908
6	1.367	0.111	−2.717
7	1.466	0.235	−3.325
8	1.619	0.179	−4.183
9	2.355	0.799	−7.439
10	1.205	0.226	−1.621
11	0.987	0.095	0.115
12	1.111	0.087	−0.917
13	2.860	0.102	−9.127
14	2.927	0.306	−9.328
15	2.629	0.099	−8.395
16	2.875	0.624	−9.172
17	2.861	0.536	−9.131
18	2.028	0.662	−6.141

The main effects plot and response table for the means are shown in Figure 12 and Table 8, respectively. The most influential process parameter on the surface roughness was the build angle with a contribution of 59.11%. The dependence of the surface roughness on the build angle is monotonic since surface roughness is minimized with the decreasing build angle. The lowest surface roughness was recorded for the 0° angle, which is in line with the available findings. This can be explained by the staircase phenomenon, which is intrinsic to the layer-by-layer printing processes and more vivid on sloped surfaces [31,32].

The second most significant printing parameter was the layer thickness. The contribution of this parameter was 24.66%. The correlation between the surface roughness and layer thickness was also monotonic as the roughness decreased with a lower layer thickness. This finding agrees with the current literature [33,34]. Hence, the staircase effect can be minimized by selecting the appropriate values of build angle and layer thickness.

The model base orientation and support density showed the lowest effect on the surface roughness of the printed parts, with a contribution of 1.28% and 0.10%, respectively. The support density of 1.0 and across the front base orientation resulted in a slightly lower surface roughness. The p-values also indicate whether a particular process parameter significantly affects surface roughness. As given in Table 9, the p-values of the build angle and layer thickness were 0.0003 and 0.0022, respectively. Since both p-values are less than the significance level of 0.05, the association between these process parameters and surface roughness is statistically significant. On the contrary, the effect of support density and model base orientation was statistically insignificant. The *p*-values of support density (0.9645) and model base orientation (0.6613) were greater than the significance level.

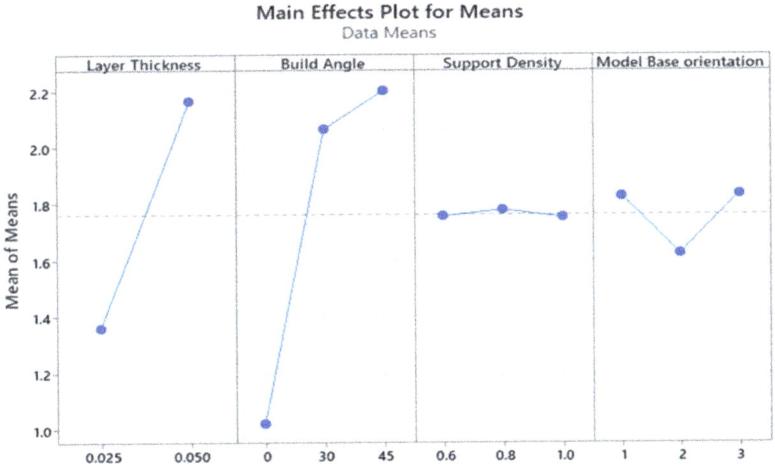

Figure 12. Main effects plot for the means for surface roughness of SLA printed parts.

Table 8. Response table for the means for surface roughness of SLA printed parts.

Level	Layer Thickness (mm)	Build Angle (°)	Support Density	Model Base Orientation
1	1.360	1.022	1.756	1.827
2	2.165	2.065	1.779	1.624
3	-	2.201	1.752	1.836
Delta	0.804	1.179	0.027	0.212
Rank	2	1	4	3

Table 9. ANOVA table for surface roughness of SLA parts.

Source	DOF	Seq SS	Adj SS	Adj MS	F	P	Contribution %
Layer thickness (mm)	1	60.803	60.803	60.803	16.6179	0.00223	24.66
Build angle (°)	2	145.755	145.755	72.877	19.9181	0.00033	59.11
Support density	2	264	264	0.132	0.0361	0.96468	0.10
Model base Orientation	2	3.155	3.155	1.578	0.4312	0.66128	1.28
Residual error	10	36.589	36.589	3.659			14.84
Total	17	246.565					100

The optimized printing parameters for the dimensional accuracy and the surface roughness are given in Table A3 in Appendix A. For the dimensional accuracy, the optimum printing parameters are a layer thickness of 0.025 mm, build angle of 30°, 0.6 support density, and diagonal to the front base orientation. To minimize the surface roughness, a 0.025 mm layer thickness, 0° build angle, 1.0 support density, and across the front base orientation are recommended to be used.

For validation, the SLA parts were printed with the optimized parameters, and their dimensional accuracy and surface roughness were measured. They gave better results than the SLA parts printed with the non-optimized parameters. The average surface roughness was equal to 0.906 μm. From the outcomes of dimensional accuracy measurements shown in Table A4 in Appendix A, it can be noted that average dimensional deviations of the SLA parts printed with the optimized parameters ranges from −0.06 to 0.15 mm.

3.4. SLA-Based IC Parts: Dimensional Accuracy

The dimensions of all SLA-based IC parts were recorded and compared with the nominal values, with the average deviations of all features given in Table 10. The average deviation from the nominal values ranged between −0.40 mm and 0.15 mm. Similar to the SLA printed patterns, the lowest deviation of the cast parts from the nominal value corresponded to the neck length, whereas the stem length had the highest average deviation. The sphere radius and neck and stem thicknesses were larger than the nominal dimensions.

Table 10. Dimensional deviation of the SLA-based IC parts.

Dimension	Features	Nominal Value (mm)	Average Measured Value (mm)	Range		Average Deviation (mm)
				Max (mm)	Min (mm)	
Radius	Sphere	7.00	7.15	7.51	6.89	0.15
Width	Neck	6.00	5.92	6.08	5.74	−0.08
Thickness	Stem	3.00	3.12	3.75	2.68	0.12
	Neck	3.00	3.10	3.58	2.51	0.10
Length	Stem	48.00	47.60	47.93	47.16	−0.40
	Neck	7.50	7.48	7.56	7.37	−0.02

The comparison between the SLA patterns and the cast parts in terms of the dimensional deviation from the nominal value is shown in Figure 13. Moreover, it is important to note that, when calculating dimensional deviation, the average measured values of both the SLA printed and the SLA-based IC parts were compared with the nominal value from the CAD file. The features such as neck thickness, stem thickness, and sphere radius had higher values of positive deviation (oversized) in the cast parts compared to the printed patterns. In the case of the neck width, the printed patterns had a larger negative deviation (undersized), which means that the expansion of the cast parts was present. This trend was present for the majority of the features. In contrast, the shrinkage of the cast parts was observed in the two length features. The stem length saw further shrinkage and had a negative deviation in both the printed and the cast parts. As for the neck length, the printed patterns were oversized, while the cast parts were undersized. This is possibly due to the shrinkage of the wax patterns during the burnout and the shrinkage of the aluminum during the metal solidification in the Z-direction as the patterns were attached to the feed sprue vertically [35].

The common assumption in IC (vacuum-assisted utilizing plaster molds) is that the casting samples will be slightly smaller than the original designs due to the phenomena of solidification shrinkage. However, there can be cases in which the casting samples are bigger than the patterns, and such occurrences can be related to several factors. When castable wax patterns are melted to form the investment mold, they may initially expand significantly and therefore the IC parts can have dimensions that are bigger than those of the original patterns [35]. Furthermore, larger castings might arise from non-uniform shrinkage caused by variations in the cooling rates of mold and metal, and particularly from the thicker to thinner areas of the part [36]. Piwonka [37] stated that the mold's deformation needs to be taken into consideration to estimate the final dimensions of the IC parts. The heating and cooling processes, including the preheating, burnout, and dewaxing, and pouring can alter the casting dimensions of the mold [38].

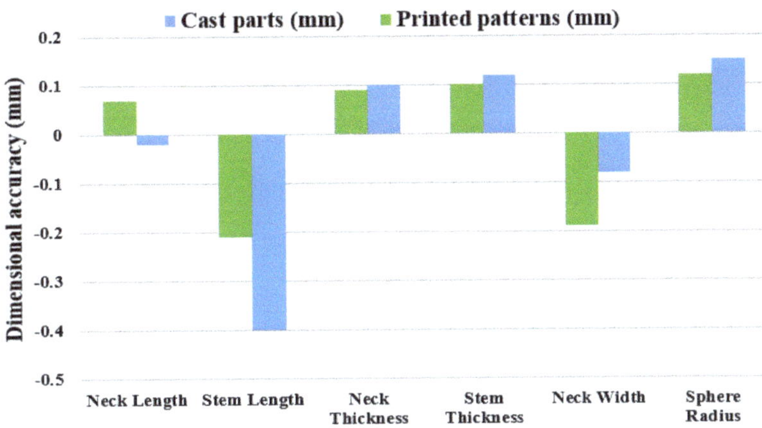

Figure 13. Comparison between the SLA patterns and the cast parts.

The mean ranking of the parameters for the dimensional accuracy of each feature is given in Table A5 in Appendix A. Based on their influence on the dimensional accuracy of the cast parts, the most significant parameter was the build angle, followed by the layer thickness. The support density and the model base orientation were the least important factors.

The optimal level for each parameter was determined for each feature measurement using the mean analysis. As given in Table A6 in Appendix A, the optimal printing parameters to obtain superior dimensional accuracy in the cast parts are a 0.050 mm layer thickness, 45° build angle, 0.8 support density, and diagonal to the front model base orientation.

3.5. SLA-Based IC Parts: Surface Roughness

The average surface roughness results of the SLA-based IC parts, standard deviation, and S/N ratio value for each experimental run are given in Table 11. The highest surface roughness (3.139 µm) was measured for Run 7, which was printed with a 0.025 mm layer thickness, 45° build angle, 0.6 support density, and across the front model base orientation. In contrast, the lowest value (1.406 µm) was recorded for Run 1 as shown in Tables 3 and 4, with printing parameters such as a 0.025 mm layer thickness, 0° build angle, 0.6 support density, and parallel the front model base orientation.

The main effects plot and the response table for means are illustrated in Figure 14 and Table 12, respectively. The build angle and the layer thickness were revealed to be the parameters with the greatest impact. The 0° angle had the lowest surface roughness, which is consistent with the data provided by Federov et al. [39]. The staircase effect, which is inherent to layer-by-layer printing methods and is more pronounced on sloping surfaces [1], can be utilized for clarifying phenomena. Layer thickness was the second most important printing parameter. Surface roughness and layer thickness have a monotonic relationship, with roughness decreasing as layer thickness decreases. The result is consistent with the findings of Jacobs [40]. Therefore, by choosing proper values for the build angle and the layer thickness, the staircase effect may be reduced.

The model base orientation and the support density showed statistically negligible effects on surface roughness according to the ranking of process parameters. The surface roughness was lowered with increasing the support density and with changing the model base orientation from diagonal to the front to parallel to the front. Lower roughness values were obtained with 1.0 support density.

Table 11. Results of surface roughness (SLA-based IC patterns).

Experiment Run No.	Surface Roughness Ra (μm)	Standard Deviation (μm)	S/N Ratio (dB)
1	1.406	0.072	−2.961
2	1.650	0.047	−4.349
3	1.504	0.522	−3.542
4	1.766	0.047	−4.939
5	2.879	0.047	−9.185
6	2.465	0.264	−7.836
7	3.139	0.737	−9.935
8	3.038	0.998	−9.653
9	2.805	0.696	−8.959
10	3.003	0.047	−9.552
11	2.330	0.516	−7.349
12	1.576	0.118	−3.951
13	2.667	0.687	−8.522
14	2.527	0.610	−8.053
15	2.658	0.674	−8.491
16	2.798	0.047	−8.936
17	2.440	0.562	−7.759
18	2.983	0.861	−9.492

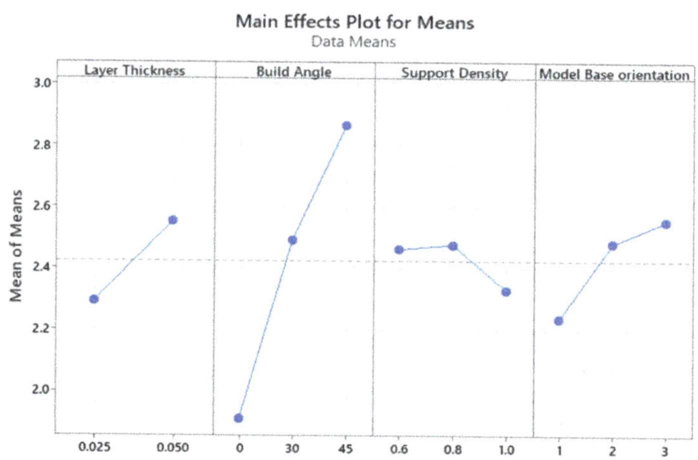

Figure 14. Main effects plot for means for surface roughness of SLA-based IC parts.

The build angle was the process parameter that had the greatest impact on the surface roughness, as shown in Table 13, with a contribution of 48.98%. The layer thickness, which contributed 6.86% of the total variance, was the second most important component. The model base orientation and support density each made 4.75% and 1.51% of the total contribution.

Table 12. Response table for means for surface roughness of SLA-based IC components.

Level	Layer Thickness (mm)	Build Angle (°)	Support Density	Model Base Orientation
1	2.305	1.912	2.463	2.265
2	2.564	2.494	2.478	2.482
3	-	2.868	2.332	2.547
Delta	0.259	0.956	0.146	0.282
Rank	2	1	4	3

Table 13. ANOVA table for surface roughness of the SLA-based IC parts.

Source	DOF	Seq SS	Adj SS	Adj MS	F	P	Contribution %
Layer thickness	1	6.414	6.414	6.414	1.81	0.208	6.86
Build angle	2	45.804	45.804	22.902	6.46	0.016	48.98
Support density	2	1.416	1.416	0.708	0.20	0.822	1.51
Model base orientation	2	4.438	4.438	2.219	0.63	0.554	4.75
Residual error	10	35.445	35.445	3.545			37.90
Total	17	93.517					100

Additionally, the p-values show whether a certain process variable substantially influences the surface roughness. The build angle has a p-value of 0.016. The correlation between this process parameter and surface roughness is statistically significant since the p-value is under the threshold of 0.05 for significance. On the other hand, due to the exceeded significance threshold, the effects of the layer thickness (0.208), support density (0.822), and model base orientation (0.554) were insignificant according to the statistical analysis.

Furthermore, it can be noted that the percentage contribution of residual error for the surface roughness of the SLA-based IC components is higher (37.90%) compared to the percentage contribution of the other sources such as layer thickness, build angle, support density, and model base orientation (see Table 13). The residual error is known as the amount of data variance left unaccounted for by the model in an ANOVA table. A large percentage contribution of residual error suggests that there could be other factors or variables impacting the response variable but not being taken into account by the model. Consequently, more research may be required to find and resolve any potential causes of bias or mistakes in the model.

The optimized printing parameters for dimensional accuracy and surface roughness of the SLA-based IC components are presented in Table A7 in Appendix A. To achieve better surface roughness results, a layer thickness of 0.025 mm, build angle of 0°, support density of 1, and parallel to the front model base orientation should be applied.

For the validation of the results, the SLA parts were printed with optimized parameters for the SLA-based IC components, and using those patterns the casting was performed. The average surface roughness of the SLA-based IC parts with the optimized printing parameters was equal to 1.462 μm. The results of dimensional accuracy measurements are demonstrated in Table A8 in Appendix A with maximum average deviation of 0.31 mm and minimum of −0.04 mm.

3.6. Microstructure and Hardness of SLA-Based IC parts

The SLA-based IC parts were characterized using the SEM, Vickers hardness test, and microstructure analysis. They had a microhardness of 101.6 HV with a standard deviation of 0.7. According to the energy dispersive spectroscopy (EDS) results, Aluminum, Copper, and Magnesium are present in the SLA-based IC samples with percentages by weight of

94.3 %, 4.25%, and 1.45%, respectively. The interface of Microhardness (HV) tester and microstructure of the SLA-based IC parts are shown in Figure 15.

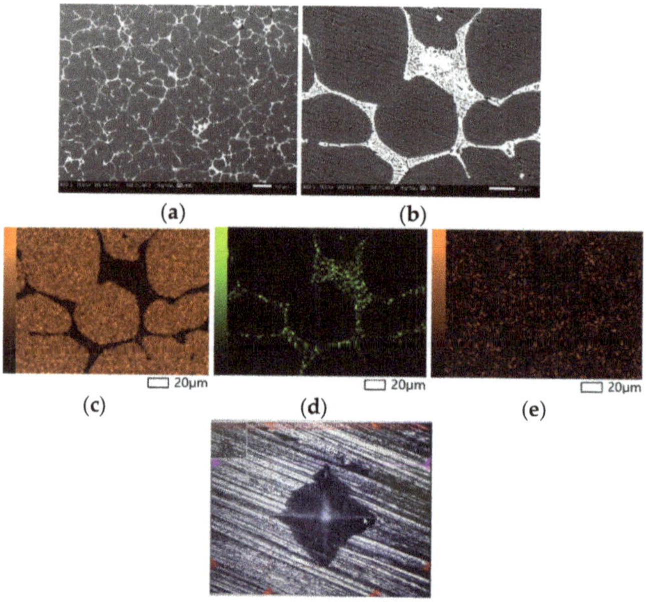

Figure 15. SEM of the SLA-based IC parts (**a**) Microstructure (×100); (**b**) Microstructure (×700); (**c**) Al; (**d**) Cu; (**e**) Mg; (**f**) Microhardness (HV) tester interface.

4. Conclusions

This study examined the effect of process parameters on the surface roughness and dimensional accuracy of the SLA patterns by using the small-scale prosthetic hip implant as a benchmark model. The Taguchi L18 array was used for DOE and fifty-four SLA patterns were printed for the measurements. The optimum parameters which reduce the surface roughness and dimensional deviation of the SLA printed parts and the SLA-based IC parts were established. The main findings of the study are as follows:

The most influential SLA printing parameter affecting the surface roughness was the build angle, followed by the layer thickness. The optimum values that minimize the surface roughness are 0.025 mm layer thickness, 0° build angle, 1.0 support density, and across the front base orientation. As for the dimensional accuracy, the optimum printing parameters are a 0.025 mm layer thickness, 30° build angle, 0.6 support density, and diagonal to the front base orientation.

The SLA-based IC components had the optimum values for the surface roughness such as 0.025 mm layer thickness, 0° build angle, 1.0 support density, and parallel to the front base orientation. As for the dimensional accuracy, the optimum printing parameters are 0.050 mm layer thickness, 45° build angle, 0.8 support density, and diagonal to the front base orientation.

Author Contributions: Investigation, Writing (original draft, review, and editing): A.M. and D.D.; Methodology, D.T. and A.P.; Review and Editing: A.P. and D.T.; Conceptualization and Supervision: D.T. and A.P.; Funding acquisition: D.T.; All authors are involved in discussions of the results. All authors have read and agreed to the published version of the manuscript.

Funding: This research was funded under the target program No. OR07665556 for the project entitled "Additive Manufacturing Systems and Metal Powders for the Kazakhstani industry", by the Ministry of Industry and Infrastructure Development of the Republic of Kazakhstan.

Institutional Review Board Statement: Not applicable.

Data Availability Statement: Data are contained within the article.

Conflicts of Interest: The authors declare no conflict of interest.

Appendix A

Table A1. Mean ranking of parameters for dimensional accuracy of SLA parts.

Feature	Layer Thickness (mm)	Build Angle (°)	Support Density	Model Base Orientation
Neck width	3	2	4	1
Neck length	3	4	1	2
Neck thickness	4	1	3	2
Stem length	2	1	3	4
Stem thickness	4	1	2	3
Sphere radius	4	1	2	3
Sum	20	10	15	15
Ranking of sums	3	1	2	2

Table A2. Optimum printing parameters for dimensional accuracy of SLA parts.

Feature	Layer Thickness (mm)	Build Angle (°)	Support Density	Model Base Orientation
Neck width	0.025	30	1	Parallel to the front
Neck length	0.050	0	0.6	Across the front
Neck thickness	0.025	30	0.6	Diagonal to the front
Stem length	0.025	30	0.6	Across the front
Stem thickness	0.025	30	0.6	Diagonal to the front
Sphere radius	0.050	30	0.6	Diagonal to the front
Optimal level	0.025	30	0.6	Diagonal to the front

Table A3. Optimized printing parameters for dimensional accuracy and surface roughness of SLA parts.

Parameters	Dimensional Accuracy (mm)		Surface Roughness (μm)	
	Level	Value	Level	Value
A: Layer thickness (mm)	1	0.025	1	0.025
B: Build angle (°)	2	30	2	30
C: Support density	1	0.6	1	0.6
D: Model base orientation	3	Diagonal to the front	3	Diagonal to the front

Table A4. Dimensional deviation of the SLA parts printed with optimized parameters.

Dimension	Features	Nominal Value (mm)	Average Measured Value (mm)	Range		Average Deviation (mm)
				Max (mm)	Min (mm)	
Radius	Sphere	7.00	7.02	7.04	7.00	0.02
Width	Neck	6.00	5.85	5.89	5.83	−0.15
Thickness	Stem	3.00	2.99	3.00	2.98	−0.01
	Neck	3.00	2.99	3.05	2.94	−0.01
Length	Stem	48.00	48.03	48.06	48.01	0.03
	Neck	7.50	7.56	7.57	7.52	0.06

Table A5. Mean ranking of parameters for dimensional accuracy of SLA-based IC parts.

Feature	Layer Thickness (mm)	Build Angle (°)	Support Density	Model Base Orientation
Neck width	4	1	2	3
Neck length	1	2	4	3
Neck thickness	2	1	3	4
Stem length	1	2	3	4
Stem thickness	4	1	3	2
Sphere radius	4	1	3	2
Sum	16	8	18	18
Ranking of sums	2	1	3	3

Table A6. Optimum printing parameters for dimensional accuracy of SLA-based IC parts.

Feature	Layer Thickness (mm)	Build Angle (°)	Support Density	Model Base Orientation
Neck width	0.025	30	0.8	Across the front
Neck length	0.025	30	0.8	Diagonal to the front
Neck thickness	0.050	45	1.0	Diagonal to the front
Stem length	0.050	0	0.8	Across the front
Stem thickness	0.050	45	1.0	Diagonal to the front
Sphere radius	0.050	45	0.6	Diagonal to the front
Optimal level	0.050	45	0.8	Diagonal to the front

Table A7. Optimized printing parameters for dimensional accuracy and surface roughness of SLA-based IC components.

Parameters	Dimensional Accuracy (mm)		Surface Roughness (μm)	
	Level	Value	Level	Value
A: Layer thickness (mm)	2	0.050	1	0.025
B: Build angle (°)	3	45	1	0
C: Support density	2	0.8	3	1
D: Model base orientation	3	Diagonal to the front	1	Parallel to the front

Table A8. Dimensional deviation of the optimized SLA-based IC parts.

Dimension	Features	Nominal Value (mm)	Average Measured Value (mm)	Range		Average Deviation (mm)
				Max (mm)	Min (mm)	
Radius	Sphere	7.00	7.05	7.13	6.94	0.05
Width	Neck	6.00	5.88	6.09	5.73	−0.12
Thickness	Stem	3.00	3.06	3.17	2.84	0.06
	Neck	3.00	3.04	3.20	2.91	0.04
Length	Stem	48.00	47.69	47.93	47.16	−0.31
	Neck	7.50	7.35	7.46	7.24	−0.15

References

1. Mukhtarkhanov, M.; Perveen, A.; Talamona, D. Application of Stereolithography Based 3D Printing Technology in Investment Casting. *Micromachines* **2020**, *11*, 946. [CrossRef]
2. Badanova, N.; Perveen, A.; Talamona, D. Concise review on Pattern making process in Rapid Investment Casting: Technology, Materials & Numerical modelling aspect. *Adv. Mater. Process. Technol.* **2022**, *8*, 966–978.
3. Sahebrao Ingole, D.; Madhusudan Kuthe, A.; Thakare, S.B.; Talankar, A.S. Rapid prototyping—A technology transfer approach for development of rapid tooling. *Rapid Prototyp. J.* **2009**, *15*, 280–290. [CrossRef]
4. Chua, C.; Feng, C.; Lee, C.; Ang, G. Rapid investment casting: Direct and indirect approaches via model maker II. *Int. J. Adv. Manuf. Technol.* **2005**, *25*, 26–32. [CrossRef]
5. Chhabra, M.; Singh, R. Rapid casting solutions: A review. *Rapid Prototyp. J.* **2011**, *17*, 328–350.
6. Lee, S.; Park, W.; Cho, H.; Zhang, W.; Leu, M.-C. A neural network approach to the modelling and analysis of stereolithography processes. *Proc. Inst. Mech. Eng. Part B J. Eng. Manuf.* **2001**, *215*, 1719–1733. [CrossRef]
7. Yarlagadda, P.K.; Hock, T.S. Statistical analysis on accuracy of wax patterns used in investment casting process. *J. Mater. Process. Technol.* **2003**, *138*, 75–81.
8. Lakkala, P.; Munnangi, S.R.; Bandari, S.; Repka, M. Additive manufacturing technologies with emphasis on stereolithography 3D printing in pharmaceutical and medical applications: A review. *Int. J. Pharm. X* **2023**, *5*, 100159.
9. Mueller, T. Stereolithography-based prototyping: Case histories of applications in product development. In Proceedings of the IEEE Technical Applications Conference and Workshops, Northcon/95, Conference Record, Portland, OR, USA, 10–12 October 1995; pp. 305–310.
10. Raja, V.; Zhang, S.; Garside, J.; Ryall, C.; Wimpenny, D. Rapid and cost-effective manufacturing of high-integrity aerospace components. *Int. J. Adv. Manuf. Technol.* **2006**, *27*, 759–773.
11. Sabbah, A.; Romanos, G.; Delgado-Ruiz, R. Impact of layer thickness and storage time on the properties of 3D-printed dental dies. *Materials* **2021**, *14*, 509. [CrossRef]
12. Shim, J.S.; Kim, J.-E.; Jeong, S.H.; Choi, Y.J.; Ryu, J.J. Printing accuracy, mechanical properties, surface characteristics, and microbial adhesion of 3D-printed resins with various printing orientations. *J. Prosthet. Dent.* **2020**, *124*, 468–475. [CrossRef]
13. Arnold, C.; Monsees, D.; Hey, J.; Schweyen, R. Surface quality of 3D-printed models as a function of various printing parameters. *Materials* **2019**, *12*, 1970. [CrossRef]
14. Aravind Shanmugasundaram, S.; Razmi, J.; Mian, M.J.; Ladani, L. Mechanical anisotropy and surface roughness in additively manufactured parts fabricated by stereolithography (SLA) using statistical analysis. *Materials* **2020**, *13*, 2496. [CrossRef]
15. Cotabarren, I.; Palla, C.A.; McCue, C.T.; Hart, A.J. An assessment of the dimensional accuracy and geometry-resolution limit of desktop stereolithography using response surface methodology. *Rapid Prototyp. J.* **2019**, *25*, 1169–1186. [CrossRef]
16. Unkovskiy, A.; Bui, P.H.-B.; Schille, C.; Geis-Gerstorfer, J.; Huettig, F.; Spintzyk, S. Objects build orientation, positioning, and curing influence dimensional accuracy and flexural properties of stereolithographically printed resin. *Dent. Mater.* **2018**, *34*, e324–e333.
17. Rubayo, D.D.; Phasuk, K.; Vickery, J.M.; Morton, D.; Lin, W.-S. Influences of build angle on the accuracy, printing time, and material consumption of additively manufactured surgical templates. *J. Prosthet. Dent.* **2021**, *126*, 658–663. [CrossRef]
18. Hada, T.; Kanazawa, M.; Iwaki, M.; Arakida, T.; Soeda, Y.; Katheng, A.; Otake, R.; Minakuchi, S. Effect of printing direction on the accuracy of 3D-printed dentures using stereolithography technology. *Materials* **2020**, *13*, 3405.
19. Resende, C.C.D.; Barbosa, T.A.Q.; Moura, G.F.; Rizzante, F.A.P.; Mendonça, G.; Zancopé, K.; das Neves, F.D. Cost and effectiveness of 3-dimensionally printed model using three different printing layer parameters and two resins. *J. Prosthet. Dent.* **2021**, *129*, 350–353.
20. McCarty, M.C.; Chen, S.J.; English, J.D.; Kasper, F. Effect of print orientation and duration of ultraviolet curing on the dimensional accuracy of a 3-dimensionally printed orthodontic clear aligner design. *Am. J. Orthod. Dentofac. Orthop.* **2020**, *158*, 889–897. [CrossRef]

21. Basile, V.; Modica, F.; Surace, R.; Fassi, I. Micro-texturing of molds via Stereolithography for the fabrication of medical components. *Procedia CIRP* **2022**, *110*, 93–98. [CrossRef]
22. Khaledi, A.-A.; Farzin, M.; Akhlaghian, M.; Pardis, S.; Mir, N. Evaluation of the marginal fit of metal copings fabricated by using 3 different CAD-CAM techniques: Milling, stereolithography, and 3D wax printer. *J. Prosthet. Dent.* **2020**, *124*, 81–86. [CrossRef]
23. Piangsuk, T.; Dawson, D.V.; El-Kerdani, T.; Lindquist, T.J. The accuracy of post and core fabricated with digital technology. *J. Prosthodont.* **2023**, *32*, 221–226.
24. Jun, Y.; Choi, K. Design of patient-specific hip implants based on the 3D geometry of the human femur. *Adv. Eng. Softw.* **2010**, *41*, 537–547.
25. Singh, R.; Singh, S.; Kapoor, P. Investigating the surface roughness of implant prepared by combining fused deposition modeling and investment casting. *Proc. Inst. Mech. Eng. Part E J. Process Mech. Eng.* **2016**, *230*, 403–410. [CrossRef]
26. Kumar, P.; Ahuja, I.S.; Singh, R. Effect of process parameters on surface roughness of hybrid investment casting. *Prog. Addit. Manuf.* **2016**, *1*, 45–53. [CrossRef]
27. Tiwary, V.K.; Padmakumar, A.; Deshpande, A.S.; Rangaswamy, N. Surface enhancement of FDM patterns to be used in rapid investment casting for making medical implants. *Rapid Prototyp. J.* **2019**, *25*, 904–914. [CrossRef]
28. Jiao, Y.; Li, X.; Zhang, X.; Li, G.; Fang, J.; Xuan, S.; Liu, L.; Wang, S.; Xie, H. Silver antibacterial surface adjusted by hierarchical structure on 3D printed porous titanium alloy. *Appl. Surf. Sci.* **2023**, *610*, 155519.
29. Castable Wax. Available online: https://formlabs-media.formlabs.com/datasheets/castable_wax_technical_data_sheet.pdf (accessed on 1 September 2022).
30. Sivaiah, P.; Chakradhar, D. Modeling and optimization of sustainable manufacturing process in machining of 17-4 PH stainless steel. *Measurement* **2019**, *134*, 142–152. [CrossRef]
31. Khodaii, J.; Rahimi, A. Improving the surface roughness in stereolithography by controlling surface angle, hatch spaces, and postcuring time. *Eng. Rep.* **2020**, *2*, e12193. [CrossRef]
32. Oropallo, W.; Piegl, L.A. Ten challenges in 3D printing. *Eng. Comput.* **2016**, *32*, 135–148.
33. Mat, M.; Ramli, F.R.; Alkahari, M.R.; Sudin, M.N.; Abdollah, M.; Mat, S. Influence of layer thickness and infill design on the surface roughness of PLA, PETG and metal copper materials. *Proc. Mech. Eng. Res. Day* **2020**, *7*, 64–66.
34. Kunjan, C. Development of process model for optimal selection of process parameters for geometric tolerances and surface roughness in stereolithography. *ADMT J.* **2017**, *9*, 103–113.
35. Badanova, N.; Perveen, A.; Talamona, D. Study of SLA Printing Parameters Affecting the Dimensional Accuracy of the Pattern and Casting in Rapid Investment Casting. *J. Manuf. Mater. Process.* **2022**, *6*, 109. [CrossRef]
36. Sabau, A.S. Alloy shrinkage factors for the investment casting process. *Metall. Mater. Trans. B* **2006**, *37*, 131–140. [CrossRef]
37. Piwonka, T.; Wiest, J. Factors affecting investment casting pattern die dimensions. *Incast* **1998**, *11*, 8–13.
38. Bansode, S.N.; Phalle, V.M.; Mantha, S.S. Optimization of process parameters to improve dimensional accuracy of investment casting using Taguchi approach. *Adv. Mech. Eng.* **2019**, *11*, 1687814019841460. [CrossRef]
39. Fedorov, K.; Fayazbakhsh, K.; Ravindran, C. Surface roughness and dimensional tolerances in A319 alloy samples produced by rapid investment casting process based on fused filament fabrication. *Int. J. Adv. Manuf. Technol.* **2022**, *119*, 4423–4437.
40. Jacobs, P.F. *Stereolithography and Other RP&M Technologies: From Rapid Prototyping to Rapid Tooling*; Society of Manufacturing Engineers: New York, NY, USA, 1995.

polymers MDPI

Article

Optimization of Printing Parameters to Enhance Tensile Properties of ABS and Nylon Produced by Fused Filament Fabrication

Andrei Yankin, Yerassyl Alipov, Ali Temirgali, Gaini Serik, Saniya Danenova, Didier Talamona and Asma Perveen *

Department of Mechanical and Aerospace Engineering, School of Engineering and Digital Sciences, Nazarbayev University, Astana 010000, Kazakhstan; andrei.iankin@nu.edu.kz (A.Y.); yerassyl.alipov@alumni.nu.edu.kz (Y.A.); ali.temirgali@alumni.nu.edu.kz (A.T.); gaini.serik@alumni.nu.edu.kz (G.S.); saniya.danenova@alumni.nu.edu.kz (S.D.); didier.talamona@nu.edu.kz (D.T.)

* Correspondence: asma.perveen@nu.edu.kz; Tel.: +7-(7172)-70-91-32

Abstract: This study aimed to identify the optimum printing parameters for the fused filament fabrication (FFF) of acrylonitrile butadiene styrene (ABS) and polyamide (nylon), to improve strength properties. For this purpose, the methodology of the paper involves an experimental study that used Taguchi's method to identify the effects of the infill pattern, infill density, and printing speed on the mechanical properties of the materials. ABS and nylon plastic parts were tested in tension to failure. Based on the results of the tensile tests, it was found that ABS material produced the highest ultimate tensile strength when printed using a tri-hexagonal infill pattern, 100% infill density, and a printing speed of 65 mm/s. On the other hand, nylon material exhibited a better performance when printed using an octet geometric structure, with identical other parameters.

Keywords: FFF; FDM; ABS; nylon; tensile test; parametric study

check for updates

Citation: Yankin, A.; Alipov, Y.; Temirgali, A.; Serik, G.; Danenova, S.; Talamona, D.; Perveen, A. Optimization of Printing Parameters to Enhance Tensile Properties of ABS and Nylon Produced by Fused Filament Fabrication. *Polymers* **2023**, *15*, 3043. https://doi.org/10.3390/polym15143043

Academic Editors: Ying Yan and Yiqi Wang

Received: 6 June 2023
Revised: 6 July 2023
Accepted: 7 July 2023
Published: 14 July 2023

1. Introduction

Additive manufacturing (AM) refers to a relatively novel approach to a rapid prototyping technique that enables the creation of an object, layer by layer, with the help of a plastic material or metal powder [1]. The capabilities of 3D-printing technologies are wide enough to produce an object of almost any shape, including shapes unlikely to be achieved via conventional methods, and avoiding the use of massive machines and high levels of manpower.

Fused filament fabrication (FFF) is the most widely used and recognized AM technique (Figure 1), primarily due to its affordability and accessibility as a desktop printer, compared to other methods. FFF printing is particularly valuable in producing cost-effective parts quickly, creating rigid models, and constructing prototypes for validation purposes [2]. As a result, the range of applications of 3D-printed parts using FFF technology and thermoplastic materials is rapidly expanding. The automotive, aerospace, medical, industrial, manufacturing, and architecture industries, among others, are increasingly incorporating FFF technology into their processes [2,3].

In FFF, a thermoplastic filament is melted and used to construct the cross-sectional geometry of an object on a build platform. Various materials can be utilized in FFF, including polycarbonate (PC), polylactic acid (PLA), acrylonitrile butadiene styrene (ABS), polyamide (nylon), and more [4–7]. The printing process involves heating the thermoplastic polymer filament to its viscous point, and then extruding it through a nozzle, in a layer-by-layer fashion, onto a glass plate. The layers adhere to one another through the molecular interaction between the extruded molten filament and the solidified layer. This thermoplastic behavior allows for the fusion of layers, and facilitates the transition from a molten state to a solid state, as the temperature decreases.

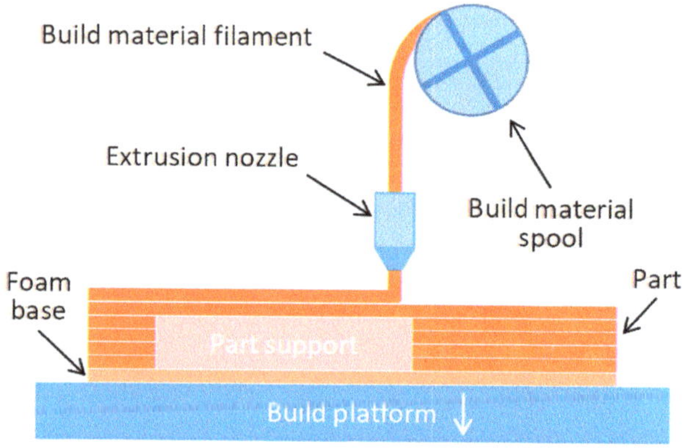

Figure 1. Fused filament fabrication process schematic.

Despite the numerous advantages of FFF printing compared to conventional manufacturing methods, certain limitations need to be addressed. The wide range of limitations includes the lower mechanical properties, and inferior surface quality, of the printed parts. One of the key challenges is to address the formation of voids between the layers in printed parts [8]. This issue arises from the relatively weak interaction between the extruded layer and the solidified part [9]. Consequently, further research is necessary to investigate and enhance the mechanical properties of the final printed object.

The process parameters impacting FFF printing include the infill density, infill patterns (internal geometric structure), extrusion temperature, nozzle diameter, layer thickness, raster angle, build orientation, printing speed, etc. [7,10,11]. The parameters identified as the most influential, based on previous studies, are selected, to determine the mechanical properties of FFF printed parts.

The infill density refers to the amount of material filling the internal structure of the printed part, which can vary from primarily hollow (0%) to mostly solid (100%), depending on the design and requirements. Extensive research has shown that infill density is one of the primary factors that significantly affects the printed part's strength. Studies have demonstrated that increasing the infill density enhances the strength of printed materials such as ABS, PLA, and nylon [12–15].

In addition to the infill density, the internal geometric structure of the FFF printed parts is also crucial for achieving the desired mechanical properties. The internal structure defines how the infilled filaments interact when the part is subjected to loading. Different internal structure shapes, such as triangular, gyroid, cubic, and more, can be utilized. The effect of infill patterns on mechanical properties has been investigated in studies [16,17]. It is important to note that each infill pattern may yield different results for specific mechanical properties. While a particular pattern might be effective for enhancing tensile or compressive properties, it may not perform as well for components subjected to other loads. Therefore, careful consideration and evaluation of the intended application and desired mechanical behavior are necessary when selecting the appropriate infill pattern for FFF printed parts.

The printing speed, in FFF, refers to the rate at which the nozzle and other movable parts of a 3D printer move, in relation to the stationary components. This parameter aims to strike a balance between the printing time and the quality of the printed parts. When the printing speed is set too high, it can result in a weak interaction between the extruded layer and the solidified part. Several studies [18–20] have demonstrated that increasing the printing speed negatively affects the tensile strength of the FFF parts. It is essential to consider the printing speed carefully, to ensure optimal printing outcomes. Finding the right balance between the printing speed and the desired mechanical properties of the printed part is crucial for achieving satisfactory results in FFF printing.

Therefore, as indicated in the literature mentioned earlier [12–20], the infill density, infill pattern, and printing speed play a crucial role in determining the tensile strength of 3D-printed parts. By meticulously selecting and effectively controlling these key process parameters, researchers and manufacturers can optimize the mechanical properties and overall performance of parts produced using the fused filament fabrication (FFF) technique.

As mentioned earlier, FFF technology utilizes thermoplastic materials with high strength, such as ABS and nylon. Nylon, known for its excellent impact strength, stress resistance, high tensile and flexural strength, thermal stability, and cost-effectiveness, is currently used in automotive parts [21] and medical components [22]. However, ABS is highly resistant to heat and chemicals, and is suitable for machining. Its affordability also contributes to its popularity as a material choice. ABS is applied in the automotive, household-goods, electronics, and medical-application industries, among others. In addition, FFF parts can be reinforced to enhance their mechanical and physical properties [23–25]. One commonly used reinforcing fiber is carbon fiber, which exhibits progressive deformation behavior, and is widely employed in strengthening polymers.

The literature review summary for ABS and nylon is presented in Table 1. Zhang H. [26] examined the effect of the printing orientation on the tensile properties of ABS. Tensile, compression, and three-point bending tests were conducted, using FFF with an ABS filament [27]. Lay M. et al. [28] suggested a relatively weaker FFF specimen when comparing the mechanical performance of ABS and nylon 6 fabricated through FFF, and through conventional injection molding. Shabana R. et al. [29] investigated the mechanical characteristics of 3D-printed ABS and PLA, concluding that the PLA showed a superior ultimate tensile strength. Kannan S. et al. [30] compared the mechanical properties of ABS, PC, and PC-ABS, using the same process parameters. The PLA and ABS were subjected to tensile and flexural tests [31], with the PLA exhibiting 7–9% higher strength. Panes A. et al. [32] also found that PLA outperformed ABS in terms of mechanical performance, considering the layer height, infill density, and layer orientation. Algarni M. et al. [33] applied ANOVA to evaluate the effects of various factors on the mechanical properties of PLA, ABS, PEEK, and PETG. The effects of the raster angles, layer height, and infill density on ABS were investigated [34]. The study in [35] explored the combined effects of different infill patterns, infill densities, and layer thicknesses.

The effects of the melting temperature and infill orientations on nylon and ABS were investigated in [36], revealing significant alterations in the sample attributes based on the nozzle temperature and infill line orientations. Terekhina et al. [37] studied the impact of internal filling on the strength characteristics of nylon, finding a significant increase in strength when the volume fraction of the infill structure was above 60%. The influence of various FFF process parameters on the tensile strength and the modulus of elasticity of 3D-printed nylon 12 was examined using Taguchi's L18 orthogonal array in [38]. The effects of the printing speed, layer height, and infill density on the nylon printed parts were explored in [13]. Moradi M. et al. [39] investigated the effects of the infill percentage, layer thickness, and number of contours, and their interactions, on the mechanical properties of nylon 12, using a design-of-experiment method. The study in [40] focused on the influence of the air gap, raster angle, and build orientation on the flexural strength of FFF-manufactured nylon 12 parts, highlighting the significant impact of the air gap and raster angle. In [41], the process parameters were optimized for the FFF process, using Taguchi's L9 orthogonal array for ABS and nylon.

Table 1. The literature review summary.

Material	Printed Parameters	Results	Ref
ABS	- nozzle temperature: 220–230 °C - nozzle diameter: 0.5 mm - printing speed: 30 mm/s - layer height: 0.1 mm - infill density: 100%	Young's modulus for 0, 45, and 90 deg of printing orientation are 1.81 GPa, 1.80 GPa, and 1.78 GPa, respectively, while the ultimate strength values are 22.4, 20.7, and 19.0 MPa, respectively.	[26]
ABS	- melting temperature: 220 °C - nozzle diameter: 0.5 mm - printing speed: 50 mm/s - layer height: 0.15 mm - infill density: 50%	Elasticity modulus: 0.65 GPa, strength: 21.68 MPa.	[27]
ABS Nylon	- melt. temp.: 230 °C (ABS), 250 °C (nylon) - nozzle diameter: 0.4 mm - printing speed: 60 mm/s - layer height: 0.15 mm - infill density: 100%	ABS: elasticity modulus: 1.3 GPa, ultimate strength: 43 MPa. Nylon: elasticity modulus: 1.45 GPa, ultimate strength: 49 MPa.	[28]
ABS	- melting temperature: 250 °C - printing speed: 35–55 mm/s - layer height: 0.1 mm - infill density: 100% - raster orientation: ±45 deg	Elasticity modulus: 1.3 GPa, ultimate strength: 34.5 MPa.	[30]
ABS	- raster angle: 0–90 deg - layer thickness: 0.1–0.3 mm - infill percentage: 60–100% - printing speed: 20–40 mm/s	The infill density is the key process variable that affects ABS strength and elastic modulus, the layer thickness is the second, the raster angle is the third, and the printing speed is the last.	[33]
ABS	- layer height: 0.35, 0.4, and 0.5 mm - infill density: 40, 60, and 80% - raster angles: 45, 55,and 65 deg	The optimum parameters are an 80% infill percentage, 0.5 mm layer thickness, and 65° raster angle. Tensile strength 31.57 MPa, elastic modulus 0.77 GPa, yield strength 19.95 MPa.	[34]
ABS	- layer height: 0.1, 0.2, and 0.3 mm - infill density: 75, 80, and 85% - infill patterns: line, triang., and concent.	The best results were obtained with a concentric infill pattern, along with an 80% infill density and 100 μm layer thickness. Tensile strength 38.95 MPa, yield strength 30.07 MPa.	[35]
Nylon	- melting temperature: 240 °C - nozzle diameter: 0.3 mm - printing speed: 40 mm/s - layer height: 0.15 mm - infill density: 20, 40, 60, 80, 100%	At the infill density of 20–40%, neighboring tracks of the same layer do not touch each other. When it is increased to 60%, the parallel tracks make contact, which leads to the formation of a continuous layer, and increases the strength of the entire sample.	[37]
Nylon	- infill density: 10–100% - layer thickness: 0.178–0.33 mm - no. of contours: 1–5 - raster pattern: sparse, double - raster width: 0.457–0.71 mm - no. of shells: 2–4 - raster orientation: 0–45 deg	The infill density is the most significant parameter. The number of contours is the second. The layer thickness ranks third. The number of shells is the fourth. The raster pattern has the lowest significance. The first four parameters control about 80% of the response value.	[38]
Nylon	- printing speed: 60–70 mm/s - layer height: 0.1–0.3 mm - infill density: 50–100%	The infill density has the highest contributing factor to mechanical characteristics, the layer thickness is second, and the printing speed is the last.	[13]
Nylon	- layer thickness: 0.15–0.35 mm - infill percentage: 15–55% - number of contours: 2–6	The layer thickness is the significant primary variable for all responses.	[39]
ABS Nylon	- layer thickness: 0.1–0.3 mm - orientation angle: 0–30 deg - shell thickness: 0.4–1.2 mm	For ABS, the tensile strength is max. for 0.2 mm layer thick, a 150 orient. angle, and 1.2 mm shell thickness. For nylon, the tensile strength is max. for 0.1 mm layer thick, a 300 orient. angle, and 1.2 mm shell thickness.	[40]

Although numerous studies have examined the mechanical behavior of FFF-printed parts, most of them concentrate on modifying one or two process parameters. Additionally, a significant portion of the research focuses on the commonly used plastics in FFF; namely, ABS and PLA. On the other hand, nylon, which is less frequently utilized, is applied in the production of gears and friction pairs that undergo cyclic loads, meaning that it is more often subjected to fatigue testing. Consequently, there is a relative scarcity of data in the literature regarding the mechanical properties of nylon, with only a limited number of reports available on its tensile strength. However, the prospects for utilizing nylon as a matrix for composite materials are growing, because of its high compatibility with biodegradable natural fibers. This can significantly expand the scope of the material [35,37,41].

The previous research of the authors focused on the fatigue performance of FDM printed parts made of ABS and nylon [42]. This current study focused on an advanced parametric analysis of these materials, using experimental testing. The objective was to identify the optimal parameter configurations that could enhance the mechanical properties of FFF-printed ABS and nylon. Tensile testing was employed, to characterize the mechanical properties of the printed parts. The study investigated various factors that influence the performance of 3D-printed parts, including the internal geometric structure (infill pattern), printing speed, and infill density.

2. Materials and Methods

Figure 2 shows the methodological steps that were used during this study. It starts with the selection of filament types, and continues with defining the parameters for the tensile test. Both ABS and PA6 (nylon) filaments, manufactured by UltiMaker (Utrecht, Netherlands), were selected to be examined in an experimental tensile test. For nylon, the impact strength was 14 kJ/m^2, the thermal resistance was 89 °C, and the filament diameter was 2.85 mm. For ABS, the impact strength was 14 kJ/m^2, the thermal resistance was 87 °C, and the filament diameter was 2.85 mm. Additional information can be found on the website of the manufacturer. To study the tensile strength of the FFF-printed materials, three parameters were chosen: the infill density, infill pattern, and printing speed.

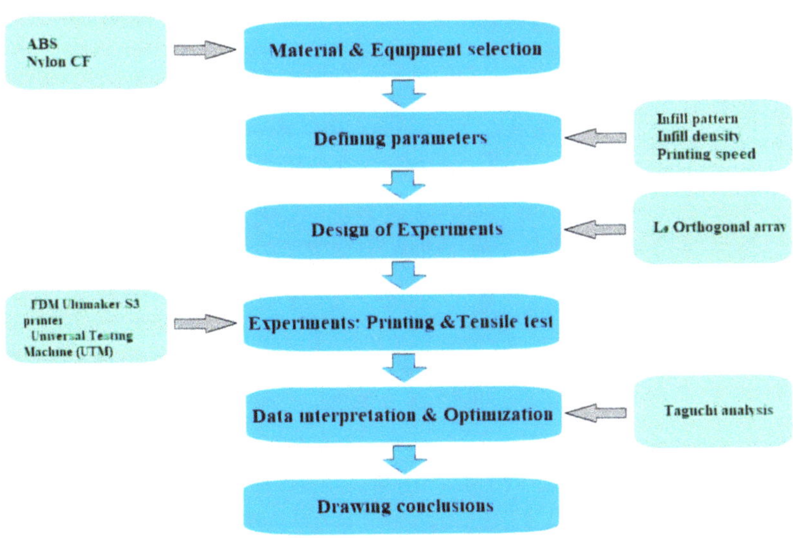

Figure 2. Methodology steps.

The L9 orthogonal array for Taguchi optimization was implemented to build the design of experiments (DOE). During the experimental part of the work, samples were printed according to the DOE. Then, tensile testing was conducted with the samples, to study their Young's modulus, ultimate tensile strength, and yield strength at different parameter configurations. The test results were analyzed using Taguchi analysis. After that, conclusions were drawn on the best configurations of the printing parameters.

2.1. Specimen Design

For the tensile test, the specimen was designed according to the ASTM D638 Type I standard (165.0 mm × 19.0 mm × 3.2 mm), as shown in Figure 3. The model was fabricated in SolidWorks software, and exported as an STL file.

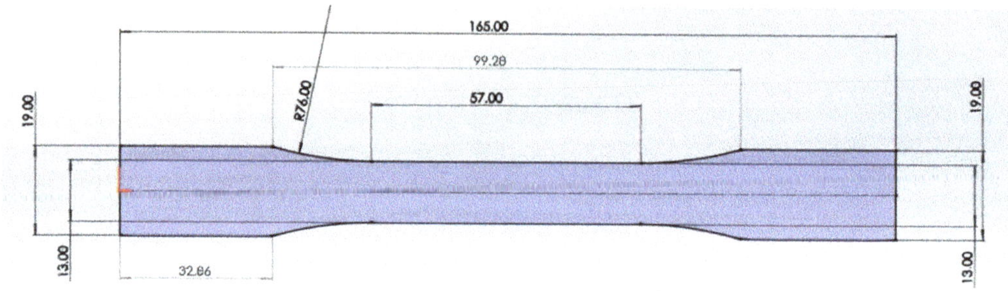

Figure 3. Specimen for the tensile test.

2.2. Design of Experiment

The Taguchi method, developed by Genichi Taguchi, is a statistical approach that aims to design experiments, and optimize processes, to enhance product quality, reduce variability, and minimize the influence of noise factors on product performance [43]. Utilizing orthogonal arrays as the experimental design ensures a minimal number of tests, while encompassing all relevant factors and their levels. The method primarily focuses on determining the optimal parameter settings that mitigate the impact of noise factors on product performance. Additionally, it emphasizes the design of processes that can accommodate variations in product performance, thus promoting consistent and reliable results. With its systematic and efficient approach to experiment design and parameter optimization, the Taguchi method enables better control and understanding of the factors that affect product performance.

The design of the experiment was based on the Taguchi L9 orthogonal array, using three parameters with three levels each, as shown in Table 2. This design allowed for a total of only nine distinct experiments per material, in contrast to the full factorial experiment, which would have required 27 tests (3 to the power of 3).

Table 2. Selected independent variables for the tensile test.

Parameters	Level 1	Level 2	Level 3
A: Infill pattern	Tri-hexagonal	Triangular	Octet
B: Infill density	10%	50%	100%
C: Printing speed	60 mm/s	65 mm/s	70 mm/s

The infill patterns of specimens are shown in Figure 4. Table 2 represents the parameters and their levels, and Table 3 shows the L9 orthogonal array. After the specimens were printed, they were weighed, and the density was measured by dividing their mass by their volume. It varied from 0.94 to 1.07 g/cm^3 for nylon, and from 0.92 to 1.03 g/cm^3 for ABS.

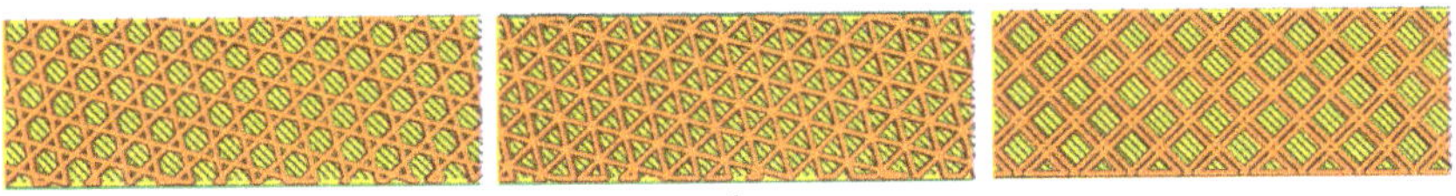

Figure 4. The profiles of the infill patterns: (**a**) tri-hexagonal, (**b**) triangular, and (**c**) octet.

Table 3. L9 Orthogonal array for the design of experiments (DOE).

#	A: Infill Pattern	B: Infill Density	C: Printing Speed
1	Tri-hexagonal	10%	60 mm/s
2	Tri-hexagonal	50%	65 mm/s
3	Tri-hexagonal	100%	70 mm/s
4	Triangular	10%	70 mm/s
5	Triangular	50%	65 mm/s
6	Triangular	100%	60 mm/s
7	Octet	10%	70 mm/s
8	Octet	50%	60 mm/s
9	Octet	100%	65 mm/s

2.3. Experimental Procedure

The experimental procedure involved two main stages: 3D printing and tensile testing. The 3D printing was performed using Ultimaker S3 and Ultimaker S5. The tensile testing was conducted utilizing MTS Electromechanical Universal Test Systems, with a maximum rated force capacity of 25 kN.

The parameters for printing were modified using the Ultimaker Cura software, and an STL input file with a designated 3D part. Apart from the parameters that were examined, there were constant variables, that are summarized in Table 4.

Table 4. Constant variables for printing.

Parameter	ABS	Nylon
Layer height	0.15 mm	0.15 mm
Orientation	horizontal	horizontal
Wall thickness	1.3 mm	1.3 mm
Wall line Count	4	4
Horizontal expansion	0 mm	0 mm
Top/bottom thickness	1.2 mm	1.2 mm
Top layers	8	8
Bottom layers	8	8
Nozzle diameter	0.4 mm	0.4 mm
Fan speed	2%	40%
Printing temperature	240 °C	245 °C

After the printing process, the dimensions of a part were measured using a caliper, because they could be modified after being printed. Then, the FFF-printed parts were tested. The tensile testing was performed using an automated material-testing system, with crosshead speeds of 2 mm/s.

2.4. Data Interpretation and Optimization

The tensile test results were analyzed firstly by their means and delta values, to perform sensitivity analysis, and rank the parameters by significance. The Taguchi method offers the calculation of the signal-to-noise ratio number, to clearly see the effect of every one changing parameter on the experimental results, by reducing the effect of the other two. The larger-the-better condition was applied, as we wanted to investigate to what extent each parameter affected the results. This condition is defined with the application of the following Equation (1):

$$S/N = -10 \cdot \log\left(\frac{1}{n}\sum_{i}^{n}\frac{1}{y_i^2}\right) \quad (1)$$

where n is the number of observations per test, and y is the observed data.

3. Results and Discussion

3.1. Experimental Results

The printed ABS and nylon parts, and the result of the tensile testing, are presented in this section. Overall, 18 samples were printed, with two repetitions of each of nine experimental runs, described in the DoE. Table 5 presents the ultimate tensile strength, yield strength, and Young's modulus calculated. The test code was the parameter-level configuration corresponding to each experimental run, presented in Table 3. Additionally, the coefficients of variations of the determined parameters were calculated:

$$k = 100\%\ SD/x_m \quad (2)$$

where SD is the standard deviation, and x_m is the mean value. Thus, the k values of the ultimate tensile strength, yield strength, and Young's modulus were equal to 4.7%, 3.9%, and 5.0% for ABS, and 2.0%, 3.3%, and 5.1% for nylon, respectively.

Table 5. Calculations for each experimental test code in tensile testing.

Test Code	Ultimate Tensile Strength, MPa		Young's Modulus, GPa		Yield Strength, MPa	
	ABS	Nylon	ABS	Nylon	ABS	Nylon
111	26.1	47.2	1.61	1.50	23.6	34.4
122	30.6	51.0	1.63	1.52	28.2	37.3
133	31.3	49.6	1.67	1.63	29.0	36.1
213	23.3	46.1	1.55	1.44	20.8	33.5
222	28.3	49.4	1.58	1.57	25.2	36.1
231	30.3	56.2	1.72	1.75	27.0	41.0
313	24.9	47.9	1.52	1.58	23.3	35.0
321	27.6	50.4	1.60	1.63	24.6	36.9
332	30.6	58.0	1.61	1.68	27.3	42.1

A graphical representation that illustrates the relationship between stress and strain for the weakest and strongest printed parts of nylon and ABS is shown in Figure 5. The 3D-printed specimens after tensile tests are presented in Figure 6. As one can see, nylon and ABS initially display linear elastic behavior, where stress and strain are directly proportional. As the load increases, the materials transition into the plastic-deformation region, and eventually reach their ultimate tensile strength.

At the point of failure, the nylon specimen undergoes necking (as depicted in Figure 6), which refers to a localized reduction in the cross-sectional area. This necking phenomenon arises due to the non-uniform stress–strain distribution within the material. As the material elongates, it becomes thinner at a specific location, resulting in a narrower region. The failure of nylon predominantly takes place at the necked region. The fracture occurs in a ductile manner.

During the tensile testing of ABS parts, it is common to observe the occurrence of multiple cracks (as indicated by the rectangular regions in Figure 6). Eventually, one of these cracks propagates, and leads to failure. ABS exhibits a combination of both ductile and brittle behavior. However, compared to nylon, the fracture of ABS occurs in a more brittle manner. ABS shows less plastic deformation before reaching the fracture point when compared to nylon, indicating a lower ability to sustain plastic deformation.

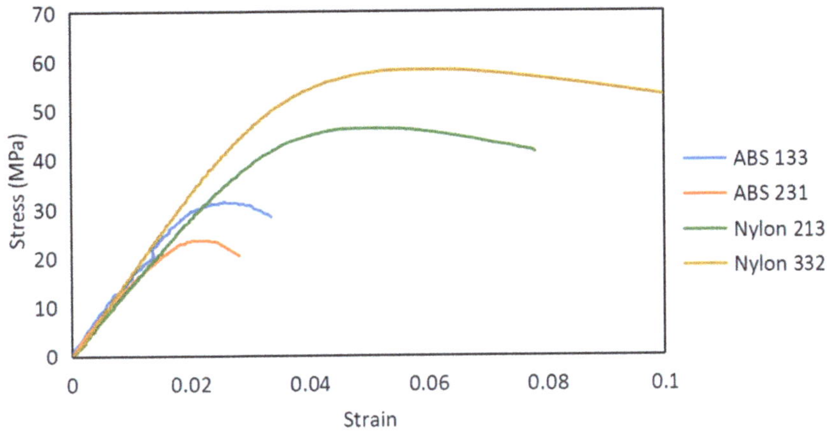

Figure 5. Stress–strain curves of ABS 133, ABS 231, nylon 213, and nylon 332 3D-printed parts.

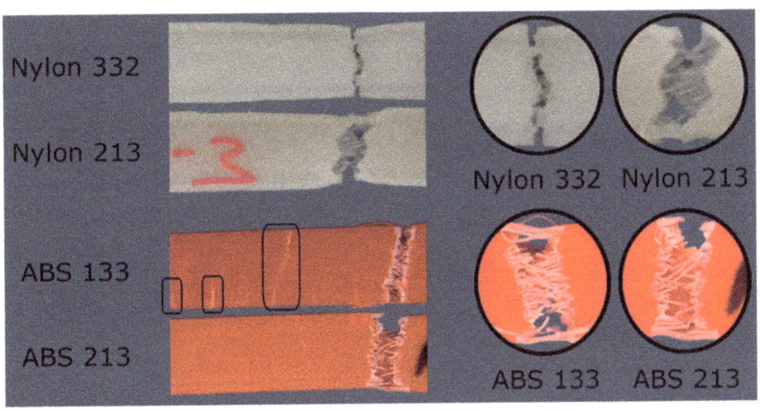

Figure 6. Tested 3D-printed specimens for ABS and nylon.

The experimental results are shown in Table 5. The ultimate tensile strength values for ABS and nylon exhibit variations, within the ranges of 23.3–31.3 MPa and 46.1–58 MPa, respectively. The yield-strength values range between 20.8–29 MPa for ABS, and 33.5–42.1 MPa for nylon. The Young's modulus values range from 1.52–1.72 GPa for ABS, and 1.44–1.75 GPa for nylon.

Table 6 shows the calculated mean value of each result, consisting of the ultimate tensile strength, yield strength, and Young's modulus, in terms of signal-to-noise (S/N) ratios. The delta is the difference between the corresponding category's maximum and minimum mean values. According to this delta value, the ranking was made, to conclude the sensitivity analysis.

The main plots of effects for ABS as a visual representation of the effect of each parameter on the results were depicted according to the mean value comparison, as shown in Figure 7.

According to our comparison and ranking, the infill density parameter had the greatest influence on the results, and the printing speed had the second highest. In contrast, the infill pattern of the specimen accompanied the least-sensitive effect. In the case of the infill pattern analysis, the ultimate tensile strength and yield strength had the highest larger-the-better results with the tri-hexagon design for ABS. The Young's modulus was also noted to be higher with this structure. Unlike the tri-hexagon geometry, the triangular showed the weakest results for ultimate tensile strength and yield strength; and the octet, for Young's modulus.

Table 6. The comparison of the mean values of the data, according to their categories and ranking of parameters, for ABS material.

S/N Ratios	Parameter Level	Infill Pattern	Infill Density	Printing Speed
Ultimate tensile strength	1	29.32	27.87	28.93
	2	28.67	29.19	29.49
	3	28.82	29.75	28.39
	Delta	0.65	1.88	1.09
	Rank	3	1	2
Young's modulus	1	4.28	3.85	4.30
	2	4.16	4.11	4.12
	3	3.94	4.42	3.96
	Delta	0.34	0.57	0.34
	Rank	2	1	2
Yield strength	1	28.56	27.04	27.96
	2	27.67	28.29	28.59
	3	27.96	28.86	27.64
	Delta	0.89	1.82	0.95
	Rank	3	1	2

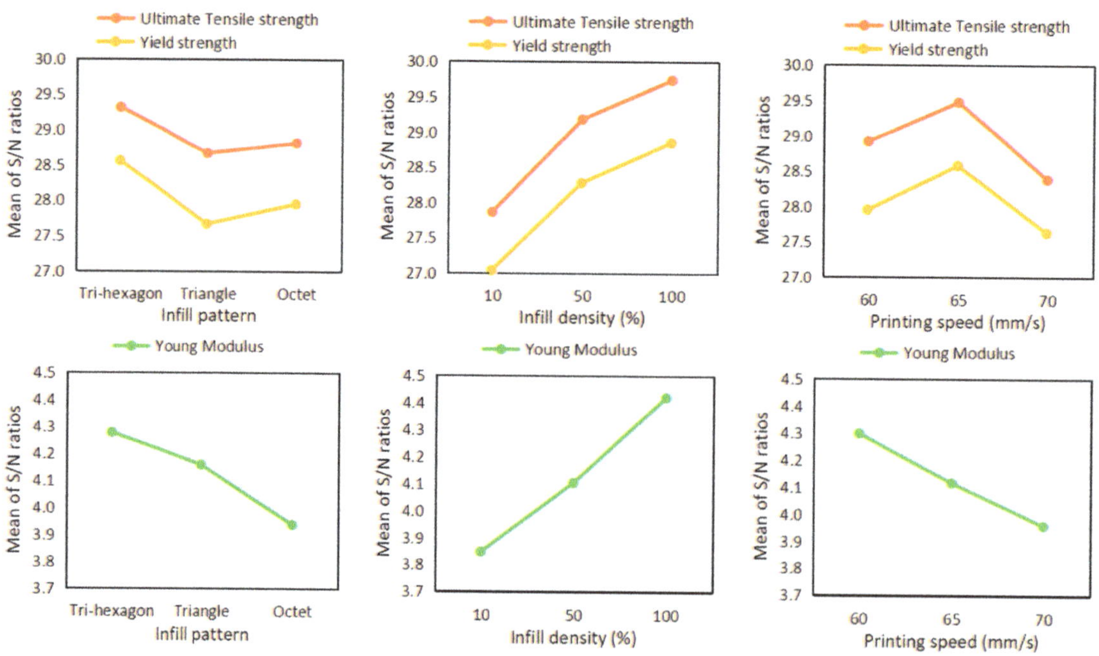

Figure 7. Sensitivity analysis for ABS parts in tensile testing.

The infill pattern of ABS influences filament interaction inside the part. Therefore, the tri-hexagon structure resulted in higher mechanical properties. It enables more dense, crisscrossed supporting offsets that help layers to stick to each other and form a support. In the case of triangular and octet structures, they have fewer dense, crisscrossing offset numbers, and thus a weaker support, wherein octets showed slightly higher results than triangular structures.

Regarding the infill density for ABS material, the "higher infill density—better specimen characteristics" trend was observed. This is because the infill density is associated with the amount of material inside a part. Therefore, increasing the infill pattern causes an increase in the load-bearing capacity of a part, and a higher mechanical property. However, in this case, the sensitivity factor was more interesting than the ranking; that is, how far the results would be scattered along the value scale when the infill density was set to 10%, 50%, and 100%. It is important to note that the infill density percentage is not solely related to the extent to which the cross-sectional area will change as a result of printing. This is because various printing parameters remain constant, such as the wall thickness, number of bottom and top layers, nozzle diameter, printing orientation, and layer thickness, and the nozzle diameter ratio to the printing part's XYZ dimensions.

The best printing speed for ABS material came up as being 65 mm/s. A higher or lower speed than 65 mm/s leads to a decrease in strength. The overall trend that could be observed for the results associated with printing speed was: at a lower printing speed, the interaction is stronger between the extruded layer and the solidified part. Thus, in most cases, the interaction becomes lower at higher speeds, and layers are loosely attracted to each other. In this work, such a trend was observed only for Young's modulus, whereas one cannot distinguish a clear pattern or trends associated with printing speed for the ultimate tensile and yield strength. One possible reason for such results might be the small ranges selected for the printing speed [44]. It varied only from 60 to 70 mm/s. The tensile properties may be more sensitive for a larger speed range, which warrants further investigation.

A similar calculation methodology was also applied to the nylon material, to perform a sensitivity analysis of how different parameters affected the tensile testing results, as shown in Table 7. The results of the analysis are shown in Figure 8.

Table 7. The comparison of the mean values of the data according to their categories, and ranking of parameters for nylon material.

S/N Ratios	Parameter Level	Infill Pattern	Infill Density	Printing Speed
Ultimate tensile strength	1	33.85	33.45	34.17
	2	34.05	34.02	34.43
	3	34.31	34.72	33.60
	Delta	0.46	1.27	0.83
	Rank	3	1	2
Young's modulus	1	3.82	3.56	4.21
	2	3.98	3.93	4.03
	3	4.24	4.55	3.80
	Delta	0.42	0.99	0.42
	Rank	2	1	2
Yield strength	1	31.10	30.70	31.44
	2	31.30	31.30	31.68
	3	31.56	31.97	30.84
	Delta	0.46	1.27	0.84
	Rank	3	1	2

Next, in the same analysis for nylon material, the octet geometry resulted in the highest values for the material tensile and yield strength, and Young's modulus. Moreover, 65 mm/s of printing speed was the optimal choice for nylon in terms of the tensile and yield strength values. Otherwise, nylon shows similar behavior patterns to ABS.

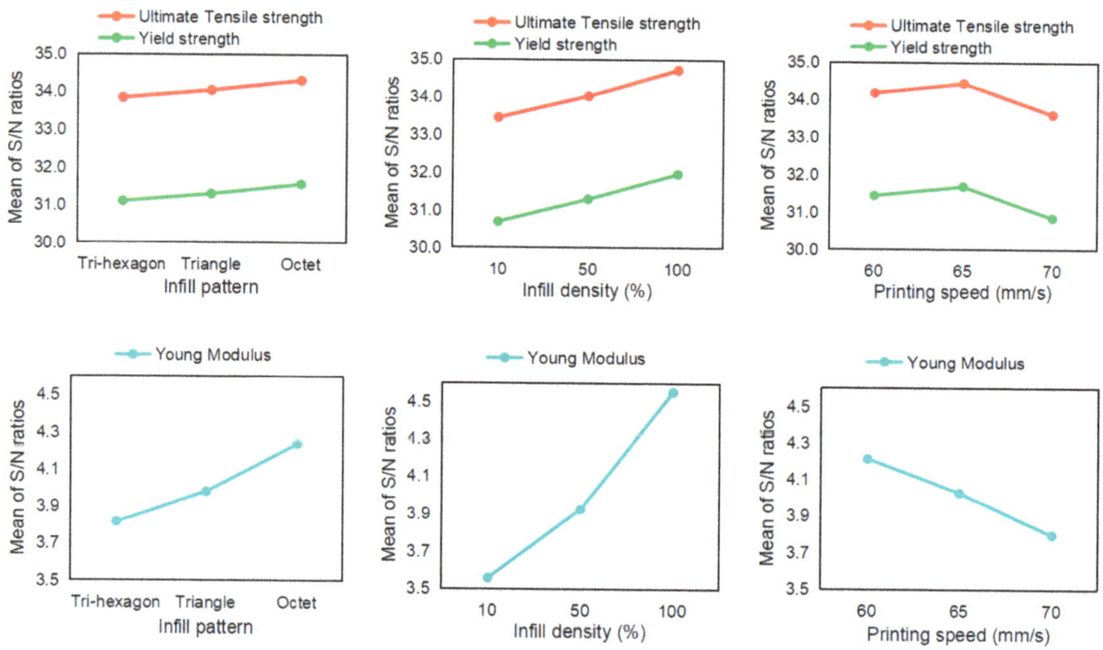

Figure 8. Sensitivity analysis for nylon parts in tensile testing.

3.2. *Validation of Results*

After running the experiments and identifying the best parameters with the highest effect on the material performance under tensile tests, the hypothesis provided by the sensitivity analysis was checked. In order to do that, the new specimens were printed, following the best configuration parameters for each material.

The specimens then underwent the corresponding test, and the results were calculated. As a result, the corresponding specimens resulted in the highest or the second-highest strength values. Obtaining the second-highest value might be related to the fact that the Taguchi method does not consider interaction effects in contrast to, for example, the full factorial DoE. The latter could give more precise results. The optimum parameter configurations for ABS and nylon parts are shown in Table 8.

Table 8. Best configuration parameter specimens, and the corresponding comparison with the set.

Material	Best Configuration Parameters			Comparison to Other Configuration Results
	Parameter 1	Parameter 2	Parameter 3	
ABS	Tri-hexagon geometric structure	100% infill density	65 mm/s printing speed	Highest
Nylon	Octet geometric structure	100% infill density	65 mm/s printing speed	The same, with the second-highest configuration

4. Conclusions

A parametric study of the effect of printing parameters on the tensile mechanical properties of FFF-printed ABS and nylon is presented in this paper. Taguchi analysis was applied, to study the effect of the infill pattern, infill density, and printing speed on the ultimate tensile strength, yield strength, and Young's modulus of the materials during tensile tests. The range of the parameter levels in the study was based on data available in the literature.

During tensile testing, the infill density had the greatest influence on both ABS and nylon, as expected. The mechanical properties, including the ultimate tensile strength, yield strength, and Young's modulus improved as the material infill density increased. These relationships were almost linear for all parameters, except for ABS's ultimate tensile and yield strength. The slope was higher when the infill density changed from 10 to 50%, compared to when it changed from 50 to 100%. The printing-speed value of 65 mm/s was optimal for both materials. The infill pattern showed a limited effect on the tensile strength. ABS had a slightly higher load capacity with the tri-hexagon pattern, while nylon had a slightly higher load capacity with the octet structure.

Thus, an experimental investigation of the tensile features of 3D-printed ABS and nylon parts was conducted. The findings can serve as reference data for future parametric studies, and in the search for an optimal printing configuration for different materials. This work helps to fill the research gap associated with the difficulty in finding experimental data on a comparative tensile test of nylon and ABS.

Author Contributions: Conceptualization, A.P.; methodology, A.P., Y.A., S.D., G.S., A.T. and A.Y.; validation, S.D., G.S. and A.Y.; formal analysis, Y.A., S.D., G.S., A.T. and A.Y.; investigation, Y.A., S.D., G.S., A.T. and A.Y.; resources, A.P. and D.T.; data curation, Y.A., S.D., G.S., A.T. and A.Y.; writing—original draft preparation, A.Y. and A.P.; writing—review and editing, A.P. and D.T.; visualization, A.P. and D.T.; supervision, A.P. and D.T.; project administration, D.T. and A.P.; funding acquisition, D.T. All authors have read and agreed to the published version of the manuscript.

Funding: This research was funded under the target program No OR07665556 for the project entitled "Additive Manufacturing Systems and Metal Powders for the Kazakhstani Industry", by the Ministry of Industry and Infrastructure Development of the Republic of Kazakhstan.

Institutional Review Board Statement: Not applicable.

Informed Consent Statement: Not applicable.

Data Availability Statement: The data presented in this study are available on request from the corresponding author.

Conflicts of Interest: The authors declare no conflict of interest.

References

1. Godec, D.; Gonzalez-Gutierrez, J.; Nordin, A.; Pei, E.; Ureña Alcázar, J. *A Guide to Additive Manufacturing*; Springer: Cham, Switzerland, 2022. [CrossRef]
2. Formlabs. Guide to 3D Printing Materials: Types, Applications, and Properties. Available online: https://formlabs.com/blog/3d-printing-materials/ (accessed on 14 November 2021).
3. Rouf, S.; Malik, A.; Singh, N.; Raina, A.; Naveed, N.; Siddiqui, M.I.H.; Haq, M.I.U. Additive manufacturing technologies: Industrial and medical applications. *Sustain. Oper. Comput.* **2022**, *3*, 258–274. [CrossRef]
4. Swetham, T.; Reddy, K.M.; Huggi, A.; Kumar, M.N. A Critical Review on of 3D Printing Materials and Details of Materials used in FDM. *Int. J. Sci. Res. Sci. Eng. Technol.* **2017**, *3*, 353–361.
5. Salem Bala, A.; bin Wahab, S.; binti Ahmad, M. Elements and Materials Improve the FDM Products: A Review. *Adv. Eng. Forum* **2016**, *16*, 33–51. [CrossRef]
6. Popescu, D.; Zapciu, A.; Amza, C.; Baciu, F.; Marinescu, R. FDM process parameters influence over the mechanical properties of polymer specimens: A review. *Polym. Test.* **2018**, *69*, 157–166. [CrossRef]
7. Dey, A.; Yodo, N. A Systematic Survey of FDM Process Parameter Optimization and Their Influence on Part Characteristics. *J. Manuf. Mater. Process* **2019**, *3*, 64. [CrossRef]
8. Li, L.; Sun, Q.; Bellehumeur, C.; Gu, P. Composite modeling and analysis for fabrication of FDM prototypes with locally controlled properties. *J. Manuf. Process.* **2002**, *4*, 129–141. [CrossRef]
9. Tao, Y.; Kong, F.; Li, Z.; Zhang, J.; Zhao, X.; Yin, Q.; Xing, D.; Li, P. A review on voids of 3D printed parts by fused filament fabrication. *J. Mater. Res. Technol.* **2021**, *15*, 4860–4879. [CrossRef]
10. Solomon, I.J.; Sevvel, P.; Gunasekaran, J. A review on the various processing parameters in FDM. *Mater. Today Proc.* **2020**, *37*, 509–514. [CrossRef]
11. Syrlybayev, D.; Zharylkassyn, B.; Seisekulova, A.; Akhmetov, M.; Perveen, A.; Talamona, D. Optimisation of Strength Properties of FDM Printed Parts—A Critical Review. *Polymers* **2021**, *13*, 1587. [CrossRef] [PubMed]
12. Vicente, C.M.S.; Martins, T.S.; Leite, M.; Ribeiro, A.; Reis, L. Influence of fused deposition modeling parameters on the mechanical properties of ABS parts. *Polym. Adv. Technol.* **2020**, *31*, 501–507. [CrossRef]

13. Ramesh, M.; Panneerselvam, K. Mechanical investigation and optimization of parameter selection for Nylon material processed by FDM. *Mater. Today Proc.* **2020**, *46*, 9303–9307. [CrossRef]
14. Alafaghani, A.; Qattawi, A.; Alrawi, B.; Guzman, A. Experimental Optimization of Fused Deposition Modelling Processing Parameters: A Design-for-Manufacturing Approach. *Procedia Manuf.* **2017**, *10*, 791–803. [CrossRef]
15. de Toro, E.V.; Sobrino, J.C.; Martínez, A.M.; Eguía, V.M. Analysis of the influence of the variables of the Fused Deposition Modeling (FDM) process on the mechanical properties of a carbon fiber-reinforced polyamide. *Procedia Manuf.* **2019**, *41*, 731–738. [CrossRef]
16. Akhoundi, B.; Behravesh, A.H. Effect of filling pattern on the tensile and flexural mechanical properties of FDM 3D printed products. *Exp. Mech.* **2019**, *59*, 883–897. [CrossRef]
17. Fernandez-Vicente, M.; Calle, W.; Ferrandiz, S.; Conejero, A. Effect of Infill Parameters on Tensile Mechanical Behavior in Desktop 3D Printing. *3D Print. Addit. Manuf.* **2016**, *3*, 183–192. [CrossRef]
18. Muhamedagic, K.; Berus, L.; Potočnik, D.; Cekic, A.; Begic-Hajdarevic, D.; Cohodar Husic, M.; Ficko, M. Effect of Process Parameters on Tensile Strength of FDM Printed Carbon Fiber Reinforced Polyamide Parts. *Appl. Sci.* **2022**, *12*, 6028. [CrossRef]
19. Rezaeian, P.; Ayatollahi, M.R.; Nabavi-Kivi, A.; Razavi, S.M.J. Effect of printing speed on tensile and fracture behavior of ABS specimens produced by fused deposition modeling. *Eng. Fract. Mech.* **2022**, *266*, 108393. [CrossRef]
20. Yu, Z.; Gao, Y.; Jiang, J.; Gu, H.; Lv, S.; Ni, H.; Wang, X.; Jia, C. Study on Effects of FDM 3D Printing Parameters on Mechanical Properties of Polylactic Acid. *IOP Conf. Ser. Mater. Sci. Eng.* **2019**, *688*, 033026. [CrossRef]
21. Jia, N.; Fraenkel, H.A.; Kagan, V.A. Effects of moisture conditioning methods on mechanical properties of injection molded nylon 6. *J. Reinf. Plast. Compos.* **2004**, *23*, 729–737. [CrossRef]
22. Rahim, T.N.A.T.; Abdullah, A.M.; Akil, H.M.; Mohamad, D.; Rajion, Z.A. The improvement of mechanical and thermal properties of polyamide 12 3D printed parts by fused deposition modelling. *Express Polym. Lett.* **2017**, *11*, 963–982. [CrossRef]
23. Saharudin, M.S.; Hajnys, J.; Kozior, T.; Gogolewski, D.; Zmarzły, P. Quality of Surface Texture and Mechanical Properties of PLA and PA-Based Material Reinforced with Carbon Fibers Manufactured by FDM and CFF 3D Printing Technologies. *Polymers* **2021**, *13*, 1671. [CrossRef]
24. Zárybnická, L.; Machotová, J.; Pagáč, M.; Rychlý, J.; Vykydalová, A. The effect of filling density on flammability and mechanical properties of 3D-printed carbon fiber-reinforced nylon. *Polym. Test.* **2023**, *120*, 107944. [CrossRef]
25. Calignano, F.; Lorusso, M.; Roppolo, I.; Minetola, P. Investigation of the Mechanical Properties of a Carbon Fibre-Reinforced Nylon Filament for 3D Printing. *Machines* **2020**, *8*, 52. [CrossRef]
26. Zhang, H. Characterization of Tensile, Creep, and Fatigue Properties of 3D Printed Acrylonitrile Butadiene Styrene. Master's Thesis, Purdue University, West Lafayette, IN, USA, 2016. [CrossRef]
27. Özsoy, K.; Ercetin, A.; Çevik, Z.A. Comparison of mechanical properties of PLA and ABS based structures produced by fused deposition modelling additive manufacturing. *Eur. J. Sci. Technol.* **2021**, *27*, 802–809. [CrossRef]
28. Lay, M.; Thajudin, N.L.; Hamid, Z.A.; Rusli, A.; Abdullah, M.K.; Shuib, R.K. Comparison of physical and mechanical properties of PLA, ABS and nylon 6 fabricated using fused deposition modeling and injection molding. *Compos. B Eng.* **2019**, *176*, 107341. [CrossRef]
29. Shabana, R.; Sarojini, J.; Vikram, K.A.; Lakshmi, V. Evaluating the mechanical properties of commonly used 3D printed ABS and PLA polymers with multi layered polymers. *Int. J. Eng. Adv. Technol.* **2019**, *8*, 2351–2356. [CrossRef]
30. Kannan, S.; Ramamoorthy, M. Mechanical characterization and experimental modal analysis of 3D Printed ABS, PC and PC-ABS materials. *Mater. Res. Express* **2020**, *7*, 015341. [CrossRef]
31. Salim, M.A.; Termiti, Z.H.; Saad, A.; Mekanikal, F.K.; Teknikal, U. Mechanical properties on ABS/PLA materials for geospatial imaging printed product using 3D printer technology. *Ref. Modul. Mater. Sci. Mater. Eng.* **2019**. [CrossRef]
32. Rodríguez-Panes, A.; Claver, J.; Camacho, A.M. The influence of manufacturing parameters on the mechanical behaviour of PLA and ABS pieces manufactured by FDM: A comparative analysis. *Materials* **2018**, *11*, 1333. [CrossRef]
33. Algarni, M.; Ghazali, S. Comparative study of the sensitivity of PLA, ABS, PEEK, and PETG's mechanical properties to FDM printing process parameters. *Crystals* **2021**, *11*, 995. [CrossRef]
34. Samykano, M.; Selvamani, S.K.; Kadirgama, K.; Ngui, W.K.; Kanagaraj, G. Mechanical property of FDM printed ABS: Influence of printing parameters. *Int. J. Adv. Manuf. Technol.* **2019**, *102*, 2779–2796. [CrossRef]
35. Agrawal, A.P.; Kumar, V.; Kumar, J.; Paramasivam, P.; Dhanasekaran, S.; Prasad, L. An investigation of combined effect of infill pattern, density, and layer thickness on mechanical properties of 3D printed ABS by fused filament fabrication. *Heliyon* **2023**, *9*, 6. [CrossRef] [PubMed]
36. Venkatesh, R.; Britto, J.J.J.; Amudhan, K.; Anbumalar, V.; Prabhakaran, R.; Sakthi, R.T. Experimental investigation of mechanical properties on CF reinforced PLA, ABS and Nylon composite part. *Mater. Today Proc.* **2023**, *76*, 4. [CrossRef]
37. Terekhina, S.; Skornyakov, I.; Tarasova, T.; Egorov, S. Effects of the infill density on the mechanical properties of nylon specimens made by filament fused fabrication. *Technologies* **2019**, *7*, 57. [CrossRef]
38. Mostafa, K.G.; Montemagno, C.; Qureshi, A.J. Strength to cost ratio analysis of FDM Polyamide 12 3D Printed Parts Costing models capacity optimization in Industry between used capacity and operational efficiency. *Procedia Manuf.* **2018**, *26*, 753–762. [CrossRef]

39. Moradi, M.; Aminzadeh, A.; Rahmatabadi, D.; Rasouli, S.A. Statistical and Experimental Analysis of Process Parameters of 3D Nylon Printed Parts by Fused Deposition Modeling: Response Surface Modeling and Optimization. *J. Mater. Eng. Perform.* **2021**, *30*, 5441–5454. [CrossRef]
40. Vishwas, M.; Basavaraj, C.K.; Vinyas, M. Experimental Investigation using Taguchi Method to Optimize Process Parameters of Fused Deposition Modeling for ABS and Nylon Materials. *Mater. Today Proc.* **2018**, *5*, 7106–7114. [CrossRef]
41. Vidakis, N.; Petousis, M.; Tzounis, L.; Maniadi, A.; Velidakis, E.; Mountakis, N.; Kechagias, J.D. Sustainable Additive Manufacturing: Mechanical Response of Polyamide 12 over Multiple Recycling Processes. *Materials* **2021**, *14*, 466. [CrossRef]
42. Yankin, A.; Serik, G.; Danenova, S.; Alipov, Y.; Temirgali, A.; Talamona, D.; Perveen, A. Optimization of Fatigue Performance of FDM ABS and Nylon Printed Parts. *Micromachines* **2023**, *14*, 304. [CrossRef]
43. Roy, R.K. *A Primer on the Taguchi Method*; Society of Manufacturing Engineers: Southfield, MI, USA, 2010; 290p.
44. Verbeeten, W.M.; Arnold-Bik, R.J.; Lorenzo-Bañuelos, M. Print velocity effects on strain-rate sensitivity of acrylonitrile-butadiene-styrene using material extrusion additive manufacturing. *Polymers* **2021**, *13*, 149. [CrossRef]

polymers

MDPI

Article

Evaluation of Organofunctionalized Polydimethylsiloxane Films for the Extraction of Furanic Compounds

Yamile Pérez-Padilla [1,*], Manuel Aguilar-Vega [2], Erbin Guillermo Uc-Cayetano [1], Adriana Esparza-Ruiz [1], Marcial Alfredo Yam-Cervantes [3] and David Muñoz-Rodríguez [1]

1 Facultad de Ingeniería Química, Universidad Autónoma de Yucatán, Periférico Norte Km 33.5, Tablaje Catastral 1361, Chuburná de Hidalgo Inn, Mérida C.P. 97203, Yucatán, Mexico
2 Unidad de Materiales, Centro de Investigación Científica de Yucatán, A.C., Calle 43 No. 130 Por 32 y 34, Chuburná de Hidalgo, Mérida C.P. 97205, Yucatán, Mexico
3 Departamento de Proyectos en Ingeniería, Universidad Internacional Iberoamericana, Calle 15, No. 36, Entre 10 y 12, IMI III, Campeche C.P. 24040, Campeche, Mexico
* Correspondence: yamile.perez@correo.uady.mx; Tel.: +52-9991186536

Abstract: Hybrid membranes with three different thicknesses, PMDS_C1, PMDS_C2, and PMDS_C3 (0.21 ± 0.03 mm, 0.31 ± 0.05 mm, and 0.48 ± 0.07 mm), were synthesized by the sol–gel method using polydimethylsiloxane, hydroxy-terminated, and cyanopropyltriethoxysilane. The presence of cyano, methyl, and silicon-methyl groups was confirmed by FTIR analysis. Contact angle analysis revealed the membranes' hydrophilic nature. Solvent resistance tests conducted under vortex and ultrasonic treatments (45 and 60 min) demonstrated a preference order of acetonitrile > methanol > water. Furthermore, the membranes exhibited stability over 48 h when exposed to different pH conditions (1, 3, 6, and 9), with negligible mass losses below 1%. The thermogravimetric analysis showed that the material was stable until 400 °C. Finally, the sorption analysis showed its capacity to detect furfural, 2-furylmethylketone, 5-methylfurfural, and 2-methyl 2-furoate. The thicker membrane was able to adsorb and slightly desorb a higher concentration of furanic compounds due to its high polarity provided by the addition of the cyano groups. The results indicated that the membranes may be suitable for sorbent materials in extracting and enriching organic compounds.

Keywords: polydimethylsiloxane; hybrid polymer; sorbent; furanic compound; characterization

Citation: Pérez-Padilla, Y.; Aguilar-Vega, M.; Uc-Cayetano, E.G.; Esparza-Ruiz, A.; Yam-Cervantes, M.A.; Muñoz-Rodríguez, D. Evaluation of Organofunctionalized Polydimethylsiloxane Films for the Extraction of Furanic Compounds. *Polymers* **2023**, *15*, 2851. https://doi.org/10.3390/polym15132851

Academic Editors: Ying Yan and Yiqi Wang

Received: 26 May 2023
Revised: 23 June 2023
Accepted: 26 June 2023
Published: 28 June 2023

1. Introduction

Furanic compounds are a broad class of heterocyclic molecules containing a five-membered aromatic ring with four carbon atoms and one oxygen atom with one or more substitutes. The latter include aldehydes, ketones, esters, alcohols, acids, thiols, and sulfides, in combination with pyrazines and pyrroles. Among furan, 2–FAL and HMF have been reported to be toxic and probably carcinogenic to humans. Furthermore, furan has generated significant concern due to its widespread presence in foods, which has led to increasing efforts to keep furan and their compounds' levels as low as possible [1–6]. Furanic compounds are formed from the catalytic dehydration of glucose or fructose to 5-hydroxymethylfurfural (5–HMF) and from xylose to furfural, which can be presented as an unwanted component in transformed oil, such as oils in alcoholic beverages such as tequila [7]. The maximum furfural limit in alcoholic beverages is 5 mg/100 mL. It is also present in heated or thermally processed foods such as coffee, canned vegetables, roasted grains, fruits, and honey, among others [8–10].

Nowadays, several separation methods, including reverse osmosis, electrodeposition, coagulation, precipitation, electrocoagulation, coagulation–flocculation–sedimentation, biosorption, and adsorption have been applied for the removal of chemical analytes; among them, the absorption method has been considered adequate and widely applied due to its low cost, and high efficiency [11,12].

In recent years, research has been focused on developing new organic–inorganic materials for use in various fields due to their properties being the result of a combination of the properties of their constituent materials. The sol–gel process is an easy widely probed method that offers modification in organic and inorganic materials, variability and applicability for hybrid material elaboration, and good resistance at high temperatures. The sol–gel method is based on yielding an inorganic or hybrid polymer from metallic and/or inorganic alkoxides through a series of simple chemical reactions at relatively low temperatures (20 to 150 °C) [13–16].

Sol–gel is a process in which a liquid solution, "sol," is converted into a solid phase, "gel" [15]. Sol–gel technology offers efficient and high-purity formation of nanopowder, fibers, solid structures, and thin film coatings. Potential applications of sol–gel technology are found in a wide range of sectors, such as pharmaceuticals, medicine, construction, aerospace, transportation, the food industry, optics, agriculture, semiconductor devices, catalysis, and biotechnology [16]. Organic–inorganic hybrid materials have attracted considerable interest in recent years because they can enhance physicochemical, thermal, and mechanical properties. Organic functional silanes can generally improve thermal stability, hardness, mechanical properties, and adhesion to specific organics other than the sol–gel, ensuring an optimal hybridization in the process where it will be applied [15–17].

Different extraction methods for organic compounds from aqueous matrices based on polydimethylsiloxane (PDMS) or other polymeric supports have been developed, mainly aiming at reducing the solvents and handling faster and easier sample preparation procedures. One of the first extraction techniques employing PDMS is solid-phase microextraction (SPME), proposed by Belardi and Pawliszyn in 1989 [18,19]. The PDMS is a linear polymer. In the PDMS/TEOS network, the tetraethoxysilane (TEOS) generates a silica cluster, which crosslinks the PDMS linear chains, generating a three-dimensional structure. PDMS and composite membranes (PDMS/TEOS) have been used as supporting membranes in the latter case.

The hybrid membranes consist of hydrophobic PDMS chains crosslinked by polar organic clusters. Their design is an alternative approach to producing multifunctional materials with polar and non-polar nanodomains, leading to different intrinsic properties and thus to many applications [20,21]. Inorganic–organic hybrid polymers are increasingly crucial in materials science because they combine the properties of the materials that make them up. A critical challenge in designing these inorganic–organic hybrid systems is controlling the blend between the two different phases [22,23].

Hybrid polymers in the form of a membrane have a great variety of applications in different areas such as analytical chemistry, food, industry, and pharmaceutical industry, among others. In particular, in the extraction of analytes, several analysis techniques, both classical and instrumental, require having the sample in dissolution to separate the analytes of interest from the matrix. Sample pretreatments transform it into an appropriate form for further analysis; however, this stage requires more time in the analytical process and the consumption of considerable amounts of solvents, which increases costs and pollutes the environment [24]. Several techniques have been studied to help counteract the sample treatment problem, including methods such as solid phase extraction (SPE) and solid phase microextraction (SPME), which can adsorb and concentrate organic analytes with a minimum amount of material, seeking to reduce working times and the number of solvents at the sample treatment stage [25–28]

Previously, a no-porous hybrid-polymeric coating based on PDMS-OH and cy-anopro pyltriethoxysilane (CPTEOS) or aminopropyltriethoxysilane (APTES) or triethoxyphenylsilane (TEPS) with sorbent properties was studied using the stir bar sorptive extraction (SBSE) technique. In all cases, furanic derivatives were moderately extracted from water, but when extracted from isooctane, they showed adsorption capacity in the order of CPTEOS > APTES > TEPS [29,30].

These functional groups were selected to increase the polarity of the sorbent phase and thus improve its sorption capacity towards polar compounds as the furanic ones. It was found that the contact area in the bars did not favor the interaction with the furanic compounds due to the fact that most functional groups were inside the bar and were not available to interact with the analytes. Thus, new techniques are needed, such as membrane sorption, which offers a larger contact area and possibly more functional groups on the surface to interact with the analytes of interest with the CPTEOS. This study aimed to evaluate the performance of a hybrid membrane synthesized from PDMS and CPTEOS as a sorbent in the analytical pretreatment for the determination of furan compounds. At the same time, the effect of membrane thickness and the evolution of their sorbent material properties was assessed. A description of the determination of the appropriate conditions for PDMS/CPTEOS hybrid membrane preparation by the sol–gel technique, as well as its characterization, was performed, followed by the studies of the sorption behavior of furan compounds.

2. Materials and Methods

2.1. Reagents

Polydimethylsiloxane, hydroxy-terminated (PM 550, PDMS-OH), cyanopropyltriethoxysilane (CPTEOS), and dichloromethane (DCM) were all obtained from Sigma-Aldrich Co. Trifluoracetic acid (TFA), HPLC grade methanol, HPLC grade acetonitrile (ACN), and HPLC grade water, were provided by J.T. Baker. Sodium hydroxide (NaOH) from Fluka Analytical and hydrochloric acid (HCl) from Meyer were also used. All reagents were at least 98% pure.

2.2. Synthesis of the Hybrid Membranes

Membranes with different thicknesses were synthesized by varying the volume of sol–gel solutions CPTEOS:PDMS (3:1 molar ratio): (PDMS_C1), (PDMS_C2), and (PDMS_C3). To synthesize the hybrid polymeric membrane (PDMS_C1), we proceeded as follows: A solution of PDMS-C1 in a 50 mL Corning tube was prepared by mixing 631.5 µL of PDMS-OH and 900 µL of DCM. The mixture was stirred for 30 s in a Genie 2.0 vortex. Next, 783 µL of CPTEOS was added and stirred for 1 min. Subsequently, 150 µL of trifluoroacetic acid (TFA 95%) was added and stirred for 2 min.

The solution was placed in an ultrasonic bath for 5 min. Then, the solution was transferred in a 6.5 cm diameter tetrafluoroethylene terephthalate mold to obtain the membrane and left to stand for ten days on a smooth leveled surface at 25 °C. During this time, a glass funnel was placed over the mold to control solvent removal and to keep impurities out. Final drying was completed in a vacuum oven using a heating protocol of 60 °C for 24 h, then increased to 120 °C for another 24 h. Finally, the hybrid membrane was left at 180 °C for 4 h. The same procedure was followed to prepare the 4 mL (PDMS_C2) and 6 mL (PDMS_C3) polymeric membrane adjusting the volume relation of the monomer, alkoxysilane agent, and solvents.

2.3. FTIR

Fourier transform infrared spectroscopy (FTIR) analysis was made on a Fourier transform infrared spectroscope model Thermo 380 FT-IR using the total attenuated reflectance method (ATR) on a spectral interval of 4000–650 cm^{-1} with a resolution of 4 cm^{-1}, scan number 60, and scan target 32.

2.4. Thermogravimetric Analysis

Thermogravimetric analysis was performed to determine the thermal stability of the membrane on a Netzsch equipment model STA 449 F3 Jupiter. It was carried out in an ultra-high purity nitrogen atmosphere with a 20 mL/min flow, in a temperature range from 50 °C to 800 °C, and with a heating speed of 10 °C/min.

2.5. Contact Angle

The sessile drop method was used to determine the contact angle value in the hybrid membranes. Dataphysics OCA 15EC contact angle equipment was used for this measurement. The film strip was placed on a smooth and utterly horizontal base. Each measurement was performed with distilled water (5 µL) on the film. Once the drop was deposited, the contact angle was measured by making a digital capture using the equipment's software. Ten measurements were made for each membrane and the average and standard deviation of the contact angle was reported.

2.6. Thickness Measurement

A Mitutoyo micrometer (H-2781) was used for the thickness measurement of the polymer membranes. The measurement was taken at 9 points distributed in the polymer membrane and the average and standard deviation of the results obtained by the equipment were reported.

2.7. pH Stability

Stability tests for different pH solutions (1, 3, 6, and 9) were carried out. Hybrid membranes were cut into 0.5 cm × 3 cm and immersed in the different solutions for 3, 6, 24, and 48 h. To ensure the membrane had no moisture, they were dried in a convection oven at 80 °C for 3 h before testing. The variation of the weight of the membranes between the beginning and the end of the immersion was determined on an analytical balance (Sartorius Model 54). The tests were repeated three times. The weight variations and the standard deviations were reported.

2.8. Stability to Solvents

The solvents evaluated were methanol (MeOH), water (grade HPLC), and acetonitrile (ACN). The membranes were immersed in solvents and the weight variation of the membranes was determined when two different tests were carried out; the membranes were immersed into the solvents and vortex stirring at 15 and 45 min and an ultrasonic bath for 30 and 60 min, respectively. The tests were carried out in triplicate.

2.9. Analyte Extraction

The extraction test was performed by depositing a polymeric membrane (1 cm × 1 cm trimmed into an Eppendorf tube with 20 µL of a polar organic compound solution: 5-(hydroxymethyl) furfural (5–HMF), furfural (2–FAL), 2-furilmethylKetone (2–FMC), 5-methyl furfural (5–MFA), methyl 2-furoate (FEMA) at a known concentration (1.0 µg mL^{-1}), and 1980 µL of a 30% NaCl saline solution (*w*/*v*). The solution was shaken in a vortex for 3 h. Then, the polymer membrane was removed from the solution and placed in an Eppendorf tube with 500 µL of a methanol:water mixture 1:1. Again, the membrane was stirred in a vortex for 20 min. The membrane was removed and the solution was filtered with a 1 mL syringe. The extraction was performed by taking a 20 µL aliquot and it was injected into a high-pressure liquid chromatograph (HPLC model 1100, Agilent Technologies) with a diode array detector (G1315B), degasser (G1379A), and quaternary pump (G1311A). The specific conditions used in the chromatographic analysis are described in Table 1.

Table 1. Chromatographic and detection conditions in the experimental section.

Column	Mobile Phase	Circumvention: Gradient	Injection Volume	Analytical Wavelength
C18 5 µm, 4.6 mm × 50 mm	A: Acetonitrile B: Water	0.4 min 95% B 4–5 min 80% B 4–5 min 80% B 7 min 95% B	20 µL	278 nm, 265 nm, 248 nm

3. Results

Synthesis of PDMS_CX Hybrid Membranes by Sol–Gel Solution

Three hybrid membranes with different thicknesses, based on polydimethylsiloxane with terminal hydroxides (PDMS-OH) and cyanopropyltriethoxysilane (CPTEOS), were synthesized; translucent membranes, free of pores and cracks, were obtained with the different synthesized thicknesses. According to the literature, these characteristics corresponded to an adequate sol–gel reaction [14,24]. The following reaction scheme (Scheme 1) showed the sol–gel synthesis route. First, catalytic hydrolysis of CPTEOS by TFA occurred, followed by condensation via a reaction between the hydrolyzed CPTEOS and PDMS-OH to form an evolving sol–gel network. The curing conditions of the material were responsible for the homogeneous non-porous structure of the obtained membranes.

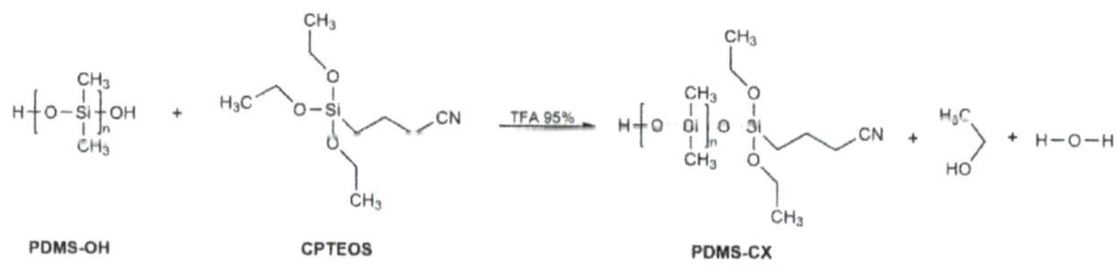

Scheme 1. General reaction scheme for the synthesis of hybrid polymers by sol–gel.

The thickness measurement on the PDMS_CX membranes was carried out by dividing them into nine sections; an image of which can be seen in Figure 1.

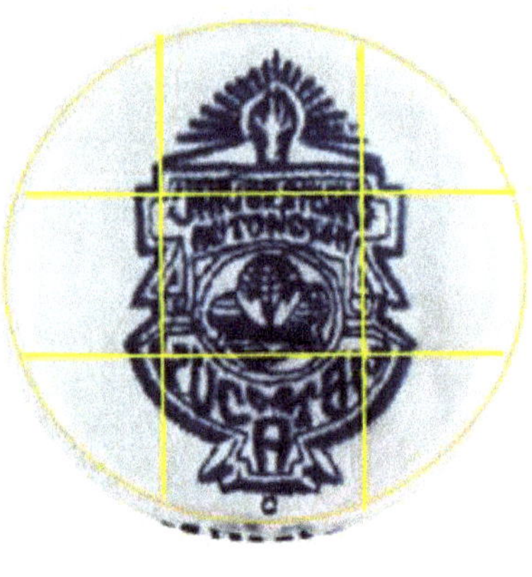

Figure 1. Image of synthesized membrane PDMS-C1 with thickness measurement marks.

Figure 2 shows membrane thickness results and the standard deviation (SD). Membrane PDMS_C1 showed 0.22 +/− 0.05 mm thickness SD. PDMS_C2 had a 0.33 +/− 0.05 mm and PDMS_C3 showed a 0.55 +/− 0.05 mm thickness. Thickness homogeneity for PDMS_CX membranes was achieved by controlling the amount of solution deposited in a properly leveled mold, followed by the elimination of solvent during the forming process. We were able to identify the synthesis conditions and could corroborate the reproducibility of the process.

The FTIR spectra of hybrid polymeric membranes prepared with PDMS-OH and CPTEOS are depicted in Figure 3. Based on previous studies from our research group and a comprehensive literature review, the overlapping bands observed at 1005 and 1090 cm^{-1} were attributed to the Si-O-Si groups present in both PDMS-OH and the hybrid polymer PDMS_CX. Additionally, the band at 1095 cm^{-1} signified the network structure of the synthesized hybrid polymer. The band corresponding to the stretching methyl -CH groups of PDMS were observed at 2962 cm^{-1}. The band corresponding to the aliphatic Si-CH_3 groups was located at 1261 cm^{-1}. The band corresponding to the cyano group -CN was observed at 2250 cm^{-1}. These results were similar to those obtained for a stir bar coating prepared with PDMS previously reported by our research group and others in the literature [29–32]. Notably, no discernible differences were observed in the spectra concerning the variation in membrane thickness.

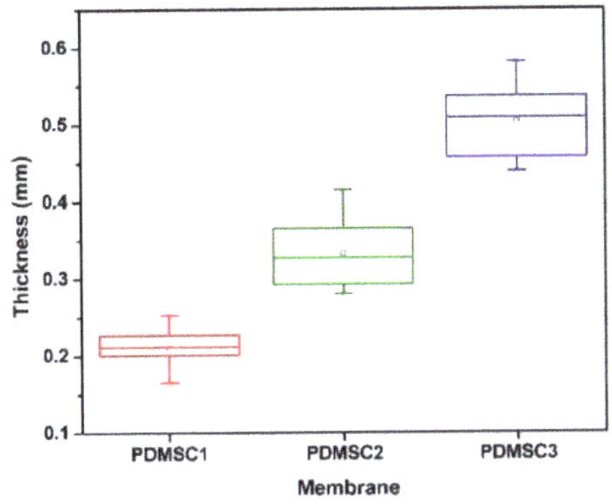

Figure 2. The standard deviation of the thicknesses obtained in the membranes.

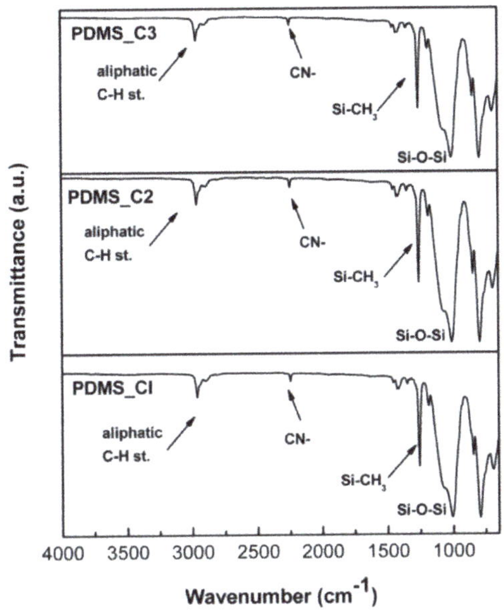

Figure 3. FTIR–ATR spectrums of PDMD_CX hybrid polymeric membranes.

Figure 4 shows the thermal decomposition of the hybrid membranes with PDMS and cyanopropyltriethoxysilane. The thermal degradation curves showed a very high thermal resistance with an onset of decomposition temperature, T_d, at 400 °C due to the decomposition of the organic material contained in the hybrid membrane. Several studies have reported that hybrid membranes showed an initial weight loss at 300 °C that was related to the decomposition of the organic hybrid membrane; these results were reported by Ballistreri et al. [33] and other studies conducted by our working group when an active coating on a stir bar was prepared with these monomers. It is essential to observe that the material did not show weight losses related to residual solvent or humidity as others have reported, indicating a good curing process.

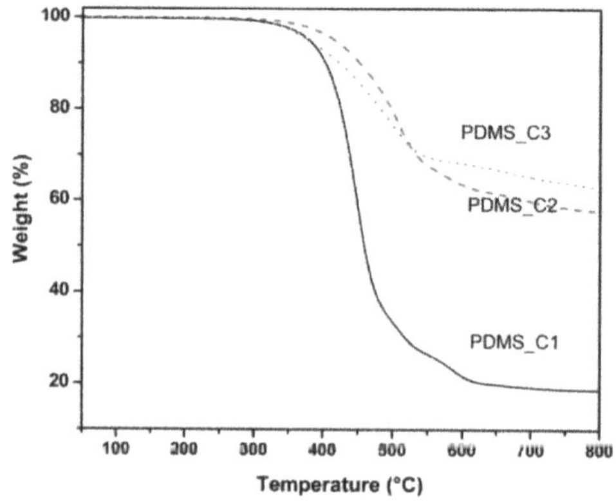

Figure 4. PDMS_Cx thermogram for hybrid polymeric membranes.

A comparison of the thermal behavior of the different membranes is also shown in Figure 4. The membrane thickness variation did not cause differences in stability as expected. This indicated that the methodology followed for curing the hybrid membrane was adequate. The thickness variations reflected changes in the percentages of residual mass attributed to the inorganic phase, demonstrating that crosslinking was greater, as this variation increased and agreed with the number of monomers added.

Contact angle in PDMS_CX hybrid membranes was conducted to determine the synthesized polymers' wettability and hydrophilicity. According to the literature, the contact angle value for PDMS is 100° [31]. The results in Table 2 show that adding the alkoxylate agent reduced the contact angle values when the membranes were evaluated with distilled water.

Table 2. Contact angle average of PDMS_CX hybrid membranes.

PDMS_CX Membranes	Contact Angle (°)
PDMS_C1	68.45 ± 2.14
PDMS_C2	68.29 ± 2.44
PDMS_C3	68.04 ± 2.46

Contact angle values for PDMS_CX hybrid membranes with different thicknesses did not show any significant difference between them. Additionally, it could be confirmed that the thickness did not influence this property. The values obtained in this study were lower than those reported by Burgos et al. [29], 74 ± 1.78°, and 77.4 ± 1.2° by Avila et al. [34], respectively, for the evaluation of a stir bar coating, using the same precursors. This fact was attributed to the cyanopropyltriethoxysilane group´s high polarity, which increased the hydrophilic capacity of the membranes and caused a contact angle to decrease value on PDMS when a hybrid polymer was formed with the alkoxylate agent.

The hybrid membranes were synthesized (Figure 5). PDMS_C1, PDMS_C2, and PDMS_C3 membranes presented good stability at the submitted pH (1, 3, 6, and 9) for up to 48 h of exposure with a mass loss of less than 1% w/w. It was also observed that the membrane structure was unaffected and did not suffer fractures, indicating that under these conditions, it could be evaluated without affecting the efficiency in the application corresponding to the extraction and continued to maintain its characteristics.

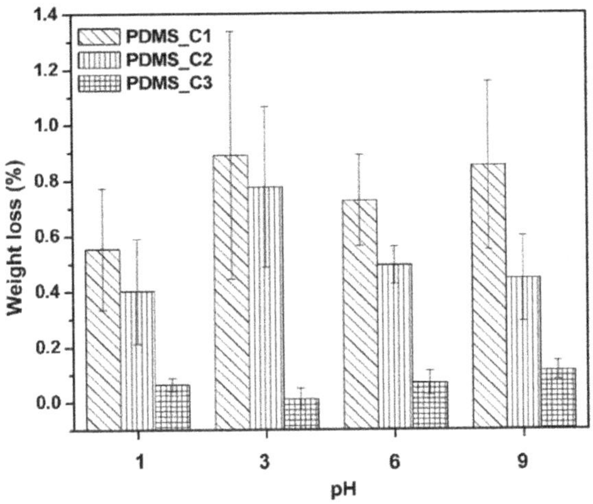

Figure 5. Stability of PDMS_C1, PDMS_C2, and PDMS_C3 membranes subjected to various pHs after 48 h of treatment.

When the tests were carried out, we observed that PDMS_C3 membranes suffered minor damage compared with the thinner membranes. Likewise, the membranes with lower thickness, PDMS_C1 and PDMS_C2, were more vulnerable to acidic and alkaline pH effects, while the PDMS_C3 membranes were very stable regarding mass loss.

The importance of sorbent materials lies in the stability to which they can be exposed and to the several pHs (acidic or alkaline media) during the extraction of ionizable compounds; based on this, these PDMS_C1, PDMS_C2, and PDMS_C3 membranes may be suitable as an alternative as a sorbent material for analyte extraction.

The results of the PDMS_CX membrane immersion in different solvents after indicated times are shown in Figure 6, as well as the immersion bath treatments in both vortex and ultrasonic stirrings. The membranes were more stable when immersed in solvents and vortexes than those subjected to ultrasonic baths for the different thicknesses tested. In the vortexes' agitation of solutions, the movement was transmitted to the liquid, generating a vortex in the sample; this allowed vigorous mixing of the product. Stirrers aimed to force liquids or gases into motion through mechanical force. In the ultrasonic bath, an electric current transmitted its energy to a mechanical system that converted it into high-intensity vibrations that generated ultrasound waves. The ultrasound, in turn, generated vibrations in the target material. If it contained liquids, millions of microscopic bubbles would be generated, which underwent rapid expansion and collapse processes that could transmit their energy to other materials, causing erosion on their surfaces.

In addition, the results showed that the highest losses occurred in both treatments when the membranes were immersed in acetonitrile. For PDMS_C1, a total mass loss of 17.9% was obtained, higher than PDMS_C2 with 11.22% and PDMS_C1 with 5.34% w/w of total mass loss, respectively, on vortex treatment. PDMS_C1 membrane immersed in an ultrasonic bath of acetonitrile solvent showed 49% w/w mass loss, reflected in a fracture in the polymeric material with a higher mass variation. However, according to Figure 6 and the standard deviation, the membranes of different thicknesses evaluated immersed with acetonitrile in the ultrasonic bath did not show differences in weight loss. The solubility of polymers in solvents was more complex due to the high molecular weights; the solvent molecules penetrated the polymer, sometimes swelling it and forming a gel until dissolution was achieved.

In addition, protic solvents are generally very polar molecules that contain acidic protons (H+) and, therefore, can form hydrogen bonds with solutes. On the other hand, aprotic solvents are solvents that do not contain acidic hydrogens and, therefore, cannot form hydrogen bonds. Acetonitrile is an aprotic polar solvent; methanol and water are polar solvents. According to the results, the hybrid polymer interacted more with acetonitrile than with methanol and water. The trend of the mass variation for the solvent under test in the vortex agitation treatment and the ultrasonic bath was ACN > MeOH > H_2O. In general, the PDMS_C1 membrane was more affected than expected for both treatments since it had a lower thickness and resistance to these solvents (Figure 5). The results suggested that when the thickness of the membrane was lower, it could suffer damage that compromised its structure in both treatments.

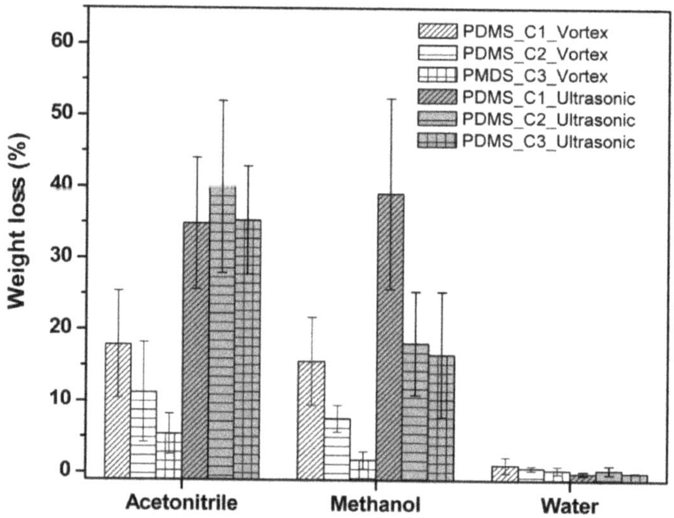

Figure 6. Weight loss percentage in PDMS_C1, PDMS_C2, and PDMS_C3 membranes; evaluation after immersion in different solvents and treatment: vortex 60 min; ultrasonic bath 90 min.

Table 3 and Figure 7 present the results of the analysis of furanic compounds in the extracts after applying the membrane for the analytical pretreatment. The goal was to achieve an enrichment factor of four because the extraction step involved 2 mL of furanic (1 µg/mL) and the desorption step involved 0.5 mL of methanol: water (1:1). However, enrichment factors higher than two were obtained only for FEMA and 5–MFA, regardless of the thickness. The RSD of the extraction experiments ranged from 0.3 to 7%.

Table 3. Furanic compound concentration after extraction from aqueous solution with PDMS_C1, PDMS_C2, and PDMS_C3 polymeric membrane and subsequent desorption with 0.50 mL MeOH: H_2O (1:1).

Furanic Compound	$\lambda_{Analytical}$ (nm)	t_R (min)	Average Concentration (ppm)		
			PDMS_C1 (0.22 mm)	PDMS_C2 (0.33 mm)	PDMS_C3 (0.55 mm)
5–HMF	278	1.040	nd	nd	nd-
2–FAL	278	1.381	0.58 ± 0.03	0.58 ± 0.06	1.08 ± 0.02
2–FMC	278	2.231	1.01 ± 0.06	1.02 ± 0.11	1.66 ± 0.03
5–MFA	278	2.659	1.45 ± 0.06	1.51 ± 0.15	2.30 ± 0.03
FEMA	248	3.734	2.79 ± 0.01	3.06 ± 0.21	3.78 ± 0.05

t_R—retention time; λ—analytical wavelength, nd—not detected.

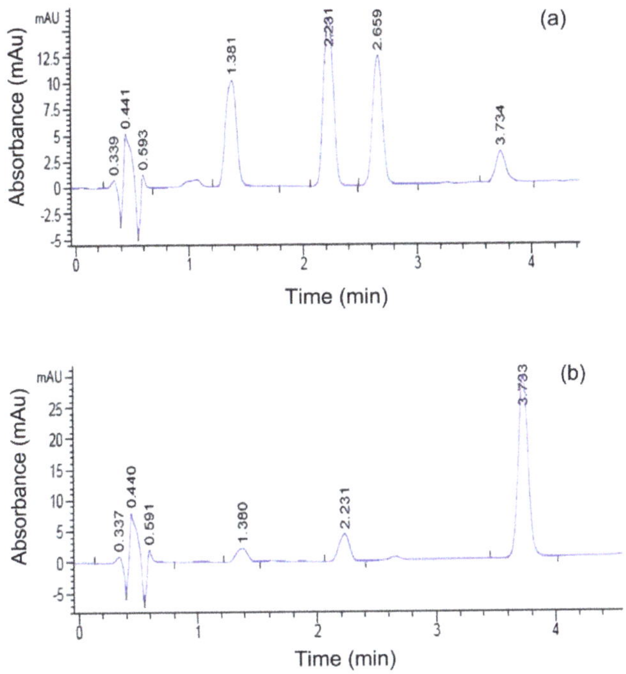

Figure 7. Typical chromatogram of the furanic compounds after their extraction from aqueous solution with PDMS_C3 membrane (0.48 ± 0.07 mm) (**a**) and their desorption with methanol: water (1:1), (**b**). Above, 278 nm; below, 248 nm.

The enrichment factors for FEMA and 5–MFA were ascribed to their high affinity toward the membrane based on PDMS. They had log *p* values of 1.0 and 0.67, respectively, and were relatively prone to interact with lipophilic materials. The dispersion forces between PDMS and cyanopropyl groups with furanic compounds could drive the affinity. In addition, the thickness was directly proportional to the concentration of FEMA and 5–MFA in the extracts. In addition, 2–FAL and 2–FMC showed the highest concentration with PDMS_C3 (0.48 ± 0.07 mm). This could be related to the increment of cyanopropyl groups as the thickness increased; as a result, there were more sites for interacting with the compounds. In contrast, 5–HMF was not detected in the final extracts because it was the most hydrophilic compound (log $p = -0.09$), regardless of the thickness. The interactions between 5–HMF and PDMS membranes were not strong enough to break the hydrogen bonding between the water molecule and the -OH group of 5–HMF.

4. Conclusions

Three different hybrid membranes with different thicknesses (0.21 ± 0.03 mm, 0.31 ± 0.05 mm, and 0.48 ± 0.07 mm) were obtained by the sol–gel method using polydimethylsiloxane, hydroxy-terminated, and cyanopropyltriethoxysilane. The presence of cyano, methyl, and silicon–methyl groups was corroborated by FTIR analysis. The contact angle results on PMDS_C1, PMDS_C2, and PMDS_C3 hybrid membranes corroborated the hydrophilic character. The solvent resistance test, conducted with 45 and 60 min exposures to vortex and ultrasonic treatments, respectively, revealed a distinct preference order for solvents: ACN > MeOH > H_2O. Additionally, when subjected to various pH conditions (1, 3, 6, and 9) over 48 h, the membranes demonstrated exceptional stability, exhibiting minimal mass losses of less than 1% *w*/*w*. These findings highlighted the robust performance of the membranes under different solvent and pH environments, further validating their suitability for practical applications. The thermogravimetric analysis showed that the material tended to lose 90% of its initial mass at 600 °C. Finally, the sorption analysis showed its capacity to adsorb furfural, 2-furylmethylketone, 5-methylfurfural, and 2-methyl 2-furoate. The thicker membrane could adsorb and slightly desorb a higher concentration of furanic compounds due to its high polarity provided by the addition of the cyano groups. These results confirmed that the synthesized membranes have the potential to be applied in the extraction of polar compounds as sorbent materials.

Author Contributions: Conceptualization, Y.P.-P.; methodology, Y.P.-P., D.M.-R., M.A.-V. and M.A.Y.-C.; validation, D.M.-R.; investigation, M.A.-V., D.M.-R., E.G.U.-C. and A.E.-R.; writing—original draft preparation, Y.P.-P. and M.A.Y.-C.; writing—review and editing, M.A.-V., E.G.U.-C. and A.E.-R.; supervision Y.P.-P.; project administration, Y.P.-P.; funding acquisition, Y.P.-P. All authors have read and agreed to the published version of the manuscript.

Funding: This research received no external funding.

Institutional Review Board Statement: Not applicable.

Data Availability Statement: The data presented in this study are available on request from the corresponding author.

Acknowledgments: The support provided by National Council of Humanities, Science and Technology, CONAHCYT (Project CB-A1-S-38216) is acknowledged. The authors are grateful to Maria Isabel de los Dolores Loría-Bastarrachea for her technical support.

Conflicts of Interest: The authors declare no conflict of interest.

References

1. Buffo, C.; Candelli-Freire, C. Coffee flavor: An overview. *Flavor Fragr. J.* **2004**, *19*, 99–104. [CrossRef]
2. ARC Working Group on the Evaluation of Carcinogenic Risks to Humans. Dry Cleaning, Some Chlorinated Solvents and Other Industrial Chemicals. Lyon (FR): International Agency for Research on Cancer. *IARC Monogr. Eval. Carcinog. Risks Hum.* **1995**, *63*, 394–406. Available online: https://www.ncbi.nlm.nih.gov/books/NBK464353/ (accessed on 21 April 2023).
3. Glatt, H.; Sommer, Y. Health risks of 5-hydorxymethylfurfural (HMF) and related compounds. In *Acrylamide and Other Hazardous Compounds in Head-Treated Foods*; Skog, K., Alexander, J.J., Eds.; Wood Heads Publ. Ltd.: Boca Raton, FL, USA, 2006; pp. 328–357. [CrossRef]
4. Anese, M. Chapter 65—Furan and Other Furanic Compounds in Coffee: Occurrence, Mitigation Strategies, and Importance of Processing. In *Processing and Impact on Active Components in Food*; Preedy, V., Ed.; Academic Press: Cambridge, MA, USA, 2015; pp. 541–547, ISBN 9780124046993. [CrossRef]
5. Tsao, W.-X.; Chen, B.-H.; Lin, P.; You, S.-H.; Kao, T.-H. Analysis of Furan and Its Derivatives in Food Matrices Using Solid Phase Extraction Coupled with Gas Chromatography-Tandem Mass Spectrometry. *Molecules* **2023**, *28*, 1639. [CrossRef] [PubMed]
6. Nie, S.; Huang, G.; Zhang, Y.; Hua, J.; Wang, S.; Shen, M.; Li, C.; Mancorte, M.; Xie, M. Analysis of furan in heat-processed foods in China by automated headspace gas chromatography-mass spectrometry (HS-GC-MS). *Food Control.* **2012**, *30*, 62–68. [CrossRef]
7. Mancilla-Margalli, N.A.; López, M.G. Generation of Maillard Compounds from Inulin during the Thermal Processing of Agave tequilana Weber Var. azul. *J. Agric. Food Chem.* **2002**, *50*, 806–812. [CrossRef] [PubMed]
8. Chaichi, M.; Mohammadi, A.; Hashemi, M. Optimization and application of headspace liquid-phase microextraction coupled with gas chromatography-mass spectrometry for determination of furanic compounds in coffee using response surface methodology. *J. Microc.* **2013**, *108*, 46–52. [CrossRef]
9. Nunkululeko, S.M.; Somandla, N.; Fanyana, M.M.; Lawrence, M.M.; Vusumzi, E.P. Determination of furanic compounds in Mopane worms, corn, and peanuts using headspace solid-phase microextraction with gas chromatography-flame ionization detector. *Food Chem.* **2022**, *369*, 130944. [CrossRef]
10. Wang, X.; Jiang, X.; Zhang, H.; Wei, J.; Huo, Z.; Ji, W. Simultaneous detection of furfural, 5-methylfurfural and 5-hydroxymethylfurfural in tsamba, roasted highland barley flour, by UPLC-MS/MS. *J. Food Compost. Anal.* **2023**, *116*, 105095. [CrossRef]
11. Fan, H.T.; Tang, Q.; Sun, Y.; Zhang, Z.G.; Li, W.X. Selective removal of antimony(III) from aqueous solution using antimony(III)-imprinted organic–inorganic hybrid sorbents by combination of surface imprinting technique with sol–gel process. *Chem. Eng. J.* **2014**, *258*, 146–156. [CrossRef]
12. Naseer, D.; Ha, H.; Lee, J.; Lee, H.J.; Song, I.-H. High-Performance γ-Al2O3 Multichannel Tube-Type Tight Ultrafiltration Membrane Using a Modified Sol-Gel Method. *Membranes* **2023**, *13*, 405. [CrossRef]
13. Dimitriev, Y.; Iordanova, R. History of sol-gel science and technology. *J. Chem. Technol. Metall.* **2008**, *43*, 181–192.

14. Wan Ibrahim, W.I.; Abdul Keyon, A.S.; Prastomo, N.; Matsuda, A. Synthesis and characterization of polydimethylsiloxane, cyanopropyltriethoxysilane-derived hybrid coating for stir bar sorptive extraction. *J. Sol-Gel. Sci. Technol.* **2011**, *59*, 128–134. [CrossRef]
15. Azmami, O.; Sajid, L.; Boukhriss, A.; Majid, S.; El Ahmadi, Z.; Benayada, A.; Gmouh, S. Sol-gel and polyurethane based flame retardant and water repellent coating for Palm/PES nonwovens composite. *J. Sol-Gel Sci. Technol.* **2021**, *97*, 92–105. [CrossRef]
16. Ismail, W.N.W. Sol–gel technology for innovative fabric finishing—A Review. *J. Sol-Gel Sci. Technol.* **2016**, *78*, 698–707. [CrossRef]
17. Woo-Chan, C.; Seong-Hyeop, K.; Won-Ki, L.; Saravanan, N.; Chang-Sik, H. UV-curable organic–inorganic hybrid hard coatings for metal sheets. *J. Coat. Technol. Res.* **2019**, *16*, 771–780. [CrossRef]
18. Richter, P.; Leiva, C.; Choque, C.; Giordano, A. and Sepúlveda, B. Rotating-disk sorptive extraction of nonylphenol from water simples. *J. Chromatogr. A* **2009**, *1216*, 8598–8602. [CrossRef]
19. Pliego, J.R.; Schiavon, M.A. Jacobsen's Catalyst Interaction with Polydimethylsiloxane/Tetraethoxysilane Network and Solvent Molecules: Theoretical Design of a New Polymeric Membrane. *J. Phys. Chem. C* **2008**, *112*, 14830–14834. [CrossRef]
20. Guedes, D.F.C.; Mac Leod, T.C.O.; Gotardo, M.C.A.F.; Schiavon, M.A.; Yoshida, I.V.P.; Ciuffi, K.J.; Assis, M.D. Investigation of a new oxidative catalytic system involving Jacobsen's catalyst in the absence of organic solvents. *Appl. Catal. A Gen.* **2005**, *296*, 120–127. [CrossRef]
21. Kickelbick, G. Concepts for the incorporation of inorganic building blocks into organic polymers on a nanoscale. *Prog. Polym. Sci.* **2003**, *28*, 83–114. [CrossRef]
22. Ng, N.T.; Kamaruddin, A.F.; Wan Ibrahim, W.A.; Sanagi, M.M.; Abdul Keyon, A.S. Advances in organic–inorganic hybrid sorbents for the extraction of organic and inorganic pollutants in different types of food and environmental samples. *J. Sep. Sci.* **2018**, *41*, 195–208. [CrossRef]
23. Majors, R.E. Highlights of HPLC 2002. (Column Watch). *LC-GC N. Am.* **2002**, *20*, 830.
24. Hu, Y.; Zheng, Y.; Zhu, F.; Li, G. Sol–gel coated polydimtehylsoloxane/b-cyclodextrin as novel stationary phase for stir bar sorptive extraction and its application to analysis of estrogens and bisphenol A. *J. Chromatogr. A* **2007**, *1148*, 16–22. [CrossRef] [PubMed]
25. Melo, A.M.; Lancas, F.M.; Queiroz, M.E.C. Polydimethylsiloxane/polypyrrole stir bar sorptive extraction and liquid chromatography (SBSE/LC-UV) analysis of antidepressants in plasma samples. *Anal. Chim. Acta.* **2009**, *633*, 57–64. [CrossRef] [PubMed]
26. Wan Ibrahim, W.A.; Keyon, A.; Sanagi, M.M. Preparation and characterization of a new sol–gel hybrid based tetraethoxysilane-polydimethylsiloxane as a stir bar extraction sorbent materials. *J. Sol-Gel Sci. Technol.* **2011**, *58*, 602–611. [CrossRef]
27. Bratkowska, D.; Cormack, P.A.G.; Borrulla, F.; Marcéa, R.M. Preparation of a polar monolithic stir bar based on methacrylic acid and divinylbenzene for the sorptive extraction of polar pharmaceuticals from complex water samples. *J. Chromatogr.* **2012**, *1225*, 1–7. [CrossRef]
28. Guan, W.; Xu, F.; Guan, Y. Poly(phthalazine ether sulfone ketone) as novel stationary phase for stir bar sorptive extraction of organochlorine compounds and organophosphorus pesticides. *J. Chromatogr.* **2008**, *1177*, 28–35. [CrossRef]
29. Burgos-Tan, M.J.; Pérez-Padilla, Y.; Avila-Ortega, A.; Barrón-Zambrano, J.A.; Vilchis-Néstor, A.R.; Carrera-Figueiras, C.; Muñoz-Rodríguez, D. Preparation, characterization and evaluation of a hybrid polymeric coating with sorbent properties. *Chem. Pap.* **2017**, *71*, 1205–1215. [CrossRef]
30. Muñoz-Rodriguéz, D.; Pérez-Padilla, Y.; Ávila-Ortega, A.; Barrón-Zambrano, J.A.; Carrera-Figueiras, C. Coating evaluation of polidimethylsiloxane-aminopropylsiloxane for stir bar sorptive extraction. *J. Coat. Technol. Res.* **2021**, *18*, 1143–1151. [CrossRef]
31. Liu, L.; Sheardown, H. Glucose permeable poly (dimethyl siloxane) poly (N-isopropyl acrylamide) interpenetrating networks as ophthalmic biomaterials. *Biomaterials* **2005**, *26*, 233–244. [CrossRef]
32. Padilla-Hernández, R.E.; Medina-Ramirez, A.; Avila-Ortega, A.; Barrón-Zambrano, J.A.; Muñoz-Rodríguez, D.; Carrera-Figueiras, C.; Pérez-Padilla, Y. Synthesis of hybrid polymeric fibers of different functionalized alkoxysilane coupling agents obtained via sol-gel and electrospinning technique: Effect on the morphology by addition of PVA. *J. Solgel Sci. Technol.* **2021**, *99*, 25–31. [CrossRef]
33. Ballisteri, A.; Garozzo, D.; Montaudo, G. Mass spectral characterization and thermal decomposition mechanism of poly(dimethylsiloxane). *Macromolecules* **1984**, *7*, 1312–1315. [CrossRef]
34. Ávila-Martínez, M.A.; Pérez-Padilla, Y.; Medina-Peralta, S.; Ávila-Ortega, A.; Muñoz-Rodríguez, D. Preparation and characterization of polydimethylsiloxane containing cyano/phenyl groups for potential use in sorptive extraction. *J. Coat. Technol. Res.* **2021**, *18*, 1087–1094. [CrossRef]

polymers

Article

UV Irradiation of Wood Surface: Bonding Properties

Tomislav Gržan [1], Lucianna Grieco [2], Vjekoslav Živković [1] and Goran Mihulja [1,*]

[1] Faculty of Forestry and Wood Technology, University of Zagreb, 10000 Zagreb, Croatia; tgrzan@sumfak.unizg.hr (T.G.); vzivkovic@sumfak.unizg.hr (V.Ž.)
[2] School of Agricultural, Forest, Food and Environmental Sciences, University of Basilicata, 85100 Potenza, Italy; lucianna.grieco@unibas.it
* Correspondence: gmihulja@sumfak.unizg.hr

Abstract: Various surface modification techniques have been developed to improve synthetic polymer surfaces' wetting, adhesion, and printing by adding various functional (polar) groups. UV irradiation has been proposed as a suitable procedure to achieve adequate surface modifications of such polymers, which can be of further use to bond many compounds of interest. The activation of the surface, the favourable wetting properties, and the increased micro tensile strength of the substrate after short-term UV irradiation suggest that such pretreatment can also improve the bonding of the wood-glue system. Thus, this study aims to determine the feasibility of UV irradiation for pretreatment of wood surfaces before gluing and to determine the properties of wooden glued joints prepared in this way. UV irradiation was used to modify variously machined pieces of beech wood (*Fagus sylvatica* L.) before gluing. Six sets of samples were prepared for each machining process. Samples prepared in this manner were exposed to irradiation on a UV line. Each radiation level had a certain number of passes through the UV line, the more passes, the stronger the irradiation. Thus, the radiation levels were as follows: 1, 5, 10, 20, and 50 passes. The dose (energy delivered on the wood surface) in one pass was 2.36 J/cm^2. A wetting angle test with glue, a compressive shear strength test of lap joints, and designation of main failure patterns were used to evaluate the properties of wooden glued joints. Wetting angle test was performed according to EN 828, while the compressive shear strength test samples were prepared and tested following the ISO 6238 standard. The tests were conducted using a polyvinyl acetate adhesive. The study found that UV irradiation before gluing improved the bonding properties of variously machined wood.

Keywords: UV light; UV irradiation; machining; surface modification; bonding properties

check for updates

Citation: Gržan, T.; Grieco, L.; Živković, V.; Mihulja, G. UV Irradiation of Wood Surface: Bonding Properties. *Polymers* **2023**, *15*, 2552. https://doi.org/10.3390/polym15112552

Academic Editors: Ying Yan and Yiqi Wang

Received: 8 May 2023
Revised: 23 May 2023
Accepted: 27 May 2023
Published: 31 May 2023

1. Introduction

Nowadays, almost every wooden product is glued at some point during production, and it is nearly impossible to imagine wood products, or rational uses of wood, without the bonding process. Due to its complex texture, wood is much more demanding to glue than other materials. Wood's bonding strength refers to its ability to resist separation from its bonded components. A quality bonded joint will have high strength and durability only if the bonding process considers all the factors that may affect it. So, this process is highly complex and bonding performance is significantly affected by wood specie, adhesive type, surface preparation, and gluing technique [1–3].

Surface properties are important parameters that often determine wood's potential uses [4–6]. Various surface preparations result in different surface properties (e.g., roughness, wettability, surface energy), which means that surfaces should be adequately treated to reduce their adverse impact. A good adhesive bond can be achieved when the liquid glue wets the wood surface well. Wetting can be improved by increasing the surface energy of the wood or by decreasing the surface tension of the glue state [7–10]. Using adhesives on inactivated wood substrates results in weak bonds, as the inactivation process reduces an adhesive's ability to properly wet, flow, penetrate, and cure [1]. Thus, the surface must be altered or modified, better said activated, in some way. This can be accomplished through mechanical methods, chemical activation, or irradiation [1,11,12]. In recent years, several surface modification techniques have been developed to improve the wetting, adhesion, and printing of polymer surfaces by adding various functional (polar) groups [13–16]. Relevant techniques in polymer surface modification are wet chemical deposition, organosilane monolayer treatment, ionised gas treatments (e.g., plasma, corona discharge, flame treatment), machining processes (sanding or planing), and ultraviolet (UV) irradiation. In all techniques, functionalised polymer surfaces are produced, and/or the old inactive layer is removed, revealing a fresh surface of the polymer [16,17]. In contrast to other activation methods, UV irradiation can be applied in-line during manufacturing [18]. UV irradiation has been proposed as a suitable procedure to achieve adequate polymer surface modifications by generating active sites, which can be of further use to bond many compounds of interest [14]. As a result of UV light exposure, polymer surfaces produce reactive sites, which can become functional groups when exposed to gases or can be used to initiate UV-induced graft polymerisation [19].

Many researchers have applied the UV irradiation method to enhance adhesion at the interface of dissimilar materials, including polyethylene (PE)/polyetheretherketone (PEEK) [20], copper/epoxy, copper/graphene [21] rubber/polyurethane [22], polymethyl methacrylate (PMMA)/gold, PMMA/copper [23], silicon/gold [24], and carbon fibre reinforced PEEK/phenylene-sulfide [12]. Quan et al. reported that UV irradiation is a rapid, eco-friendly, and low-cost method for activating thermoplastic composite (TPC) surfaces, resulting in the excellent structural integrity of adhesive joints [12]. As a heterogeneous and porous biopolymer, wood presents more challenges to activate by UV light than synthetic polymer surfaces (smooth, homogeneous, non-porous, etc.) [16,18,25].

Bogner (1993) reports that electromagnetic radiation can increase the energy of wood's surface layer. An increase in the amount of surface energy increases the wetting of the wood surface with glue, and thus the adhesion of the wood-glue system [26]. Wood absorbs light rapidly and intensely, especially in UV than in other areas of the spectrum [27]. The UV and visible radiation penetration depth into the wood is not great, so the changes are limited to the shallow surface layer. UV light is completely absorbed by a 75 to 80 μm-thick layer of wood's surface [28,29]. Exposure of polysaccharides to UV irradiation is reported to enhance chemical reactions due to photooxidative degradation. The quantum energy associated with this process can break many chemical bonds present in wood components (carbon–carbon, carbon–oxygen, carbon–hydrogen bonds) [13,14]. Free radicals are the primary photochemical products generated by this homolytic process. It can cause depolymerisation and the formation of chromophoric groups, including carbonyls, carboxyls, quinones, peroxides, hydroperoxides, and conjugated double bonds, either with or without oxygen and water involvement. Secondary reactions, chain formation of free radicals, depend on the temperature and increased presence of oxygen and water [30–33]. The chemical components of wood (i.e., lignin, cellulose, hemicellulose, and extractives) are sensitive to UV radiation with a consequential deterioration effect [16,19]. Hon and Shiraishi (1991) asserted that lignin contributes 80–95% to the absorption coefficient of wood, carbohydrates 5–20%, and extractive substances about 2% [27]. Lignin is easily degraded in light shorter than 350 nm, so it is therefore very susceptible to degradation when exposed to the UV part of sunlight. Light with wavelengths longer than 350 nm does not degrade but causes bleaching and lightening of lignin. The initial and most prominent absorption occurs on chromophore groups within the phenolic molecular network of lignin, resulting in stronger decomposition than cellulose [34]. The presence of lignin protects cellulose from photodegradation to a certain extent because lignin exhibits pronounced absorption properties and a propensity for autoxidation [35].

On the other hand, the primary process in the photochemical degradation of cellulose is thought to be the breaking of its chains and this process is not dependent on the presence of oxygen [34,36]. It is believed that the organised regions of cellulose are impermeable to light due to their high degree of crystallinity, so depolymerisation occurs only in the light-accessible, amorphous regions [30]. However, the crystal surface is saturated with hydroxyl groups that are also capable of reacting with light [31]. According to the above, UV irradiation can clean the wood surface, causing pits to open, altering surface morphology, and to some extent, altering the chemical composition of the wood surface. The deterioration of connections between cell elements was related to the degradation of the middle lamella (S1 layer), which also confirmed the degradation of lignin [36,37]. Consequently, when hydrophobic lignin degrades, surface hydrophilicity increases due to abundant cellulose [18,38]. So, the effects of UV activation are most likely seen in surface morphology changes and surface free energy (wetting angle). UV irradiation of wood initiates surface oxidation (an increase in acid/base or polar components), resulting in functional group formation (carboxyl groups), which increases surface free energy, wettability (decreased contact angles), surface roughness, and adhesion [39–42]. The destruction of pits allows coatings and adhesives to penetrate deeper into wood surfaces, enhancing mechanical anchoring [18]. UV irradiation also increases the strength of the surface layer of wet wood [33,35].

The activation of the surface, the favourable wetting properties, and the increased micro-tensile strength of the substrate after short-term UV irradiation suggest that such pretreatment can also improve the bonding of the wood-glue system. Thus, this study aims to determine the feasibility of UV radiation for the pretreatment of wood surfaces before gluing and to determine the properties of wooden glued joints prepared in this way. Therefore, UV irradiation was used to activate face-milled, planed, and sanded surfaces of beech wood (*Fagus sylvatica* L.) before gluing.

2. Materials and Methods

2.1. Wood

The study was performed with slightly steamed beech wood (*Fagus sylvatica*), obtained from a local supplier. Kiln-dried boards of approximately 1000 × 150 × 30 mm^3 (L × R × T) were formatted into 1000 × 130 × 13 mm^3 (L × R × T) boards with a circular saw followed by face milling to ensure the flatness and uniformness for further mechanical processing. Six high-quality boards were chosen: without bio-damages, knots, or other inhomogeneities; with radial texture—the angle between the surface and annual rings was 60°–90°. This is particularly important since it has been shown that a flat growth ring angle is related to poor bonding performance [43,44]. Prior to other technological operations, boards were conditioned at a temperature of 23 ± 2 °C and a relative air humidity of 50 ± 5%, achieving an equilibrium moisture content of 9 ± 2%.

2.2. Wood Surface Machining

Wood surfaces were prepared first mechanically and then by UV irradiation. The surfaces were mechanically prepared through three surface machining methods: planing, sanding, and face milling. Face milling was performed on Homag Venture 316M CNC machining centre (Zagreb, Croatia), while planing was performed on the combined planer and thicknesser Rojek, MSP 415. Planing cutter head had four sharpened knives, and the face milling cutter head had three newly installed insertion knives (cutting edges). Sanding was carried out with a manual oscillating sander-belt Makita provided with a new belt of grit P120. Additional machine characteristics and process parameters are provided in Table 1. Therefore, 1000 × 130 × 13 mm^3 (L × R × T) boards were surface machined differently and then sized to produce three sets of 24 boards measuring 200 × 130 × 10 mm^3 (L × R × T).

Table 1. Surfacing machines and process parameters.

Surface Method	Sanding	Planing	Face Milling
Surfacing Machine	Manual oscillating sander Makita	Combined planer and thicknesser Rojek, MSP 415	CNC machining centre Homag Venture 316M
Characterisation of Knives/Sanding Belt	P120	VHW	HW, insertion knives with four-sided blades and rake angle 35 *
Tool Diameter (mm)	n/a	95	80
Number of Cutting Edges, z	n/a	4	3
Cutting Speed, v_c (m/s)	n/a	20.89	83.78
Feed Speed, v_f (m/min)	5	5	5
Feed, f_z (mm)	n/a	0.3	0.08
Cutting Depth (mm)	0.5	0.8	0.8
Rake Angle (°)	n/a	25	−10 *
Clearance Angle (°)	n/a	27	47 *
Cutter (rpm)	n/a	4200	20,000

* As the blade is highly specific, a detailed view of the tool is presented in Figure 1.

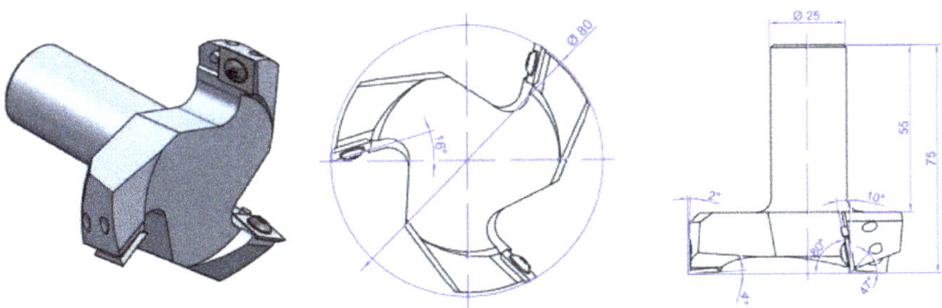

Figure 1. Face milling tool shown in isometry and orthogonal projection.

2.3. UV Irradiation

After mechanical preparation, surfaces were additionally modified by UV irradiation. For each method of surface treatment, six sets were prepared, five sets for different levels (more in Section 2.3.2.) of irradiation and one set as a control. There were four boards in each set measuring 200 × 130 × 10 mm^3 (L × R × T). Furthermore, the boards were cut into two parts before UV irradiation. One part was used for measuring adhesive wetting angles (27 × 125 × 10 mm^3, L × R × T), and the other one for preparing shear test specimens (170 × 130 × 10 mm^3, L × R × T). Once the boards had been mechanically treated and cut, they were placed in a darkened plastic box without the presence of light until they were exposed to UV radiation. The box was stored in the standard laboratory environment. UV irradiation was performed within 12 h of mechanical treatment.

2.3.1. Work Principles of the UV Line

Each radiation level had a certain number of passes through the UV line, the more passes, the stronger the radiation was. Irradiation of the surface of the samples was carried out on a device for UV drying and hardening of varnishes from brand Trivec (Figure 2). The device consists of horizontal, powered input and output tables, electric folding panels (integrated into the lower part of the case), UV exposure units radiation, exhaust fan, and transport engine. The device works on the principle of passing the boards under UV lamps where it changes the distance from the lamp, the belt speed (%), and the intensity of the lamp irradiance (W/cm) to achieve a certain dose of radiation (J/cm^2). The device contains mercury (Hg) and gallium (Ga) lamps.

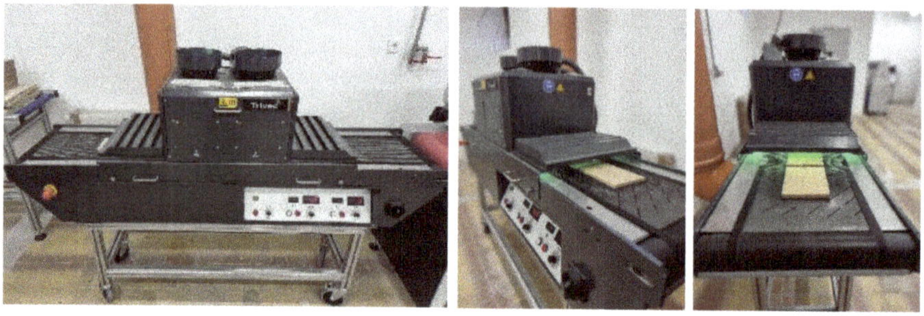

Figure 2. UV irradiation process: UV line Trivec (**a**), input (**b**), and output (**c**) of specimens.

2.3.2. Determination of the Irradiation Dose

During irradiation, the samples were passed under the UV lamp which was positioned 130 mm from the surface of the specimen, at the lowest belt speed, and at the highest intensity. For this research, a medium-pressure Hg-lamp (Heraeus brand) with a UV emission spectrum according to Figure 3, was used. Medium-pressure UV lamps have a broadband emission in the UV and visible light spectral range, therefore spectral lines from 200 nm to 600 nm have high radiation output in the UVC spectrum. Their strong UV radiation flux results in high penetration depth and efficient disinfection. Their high UV intensity also makes them particularly suitable for photochemical processes (Heraeus brochure). Further characteristics of the lamp are shown in Table 2.

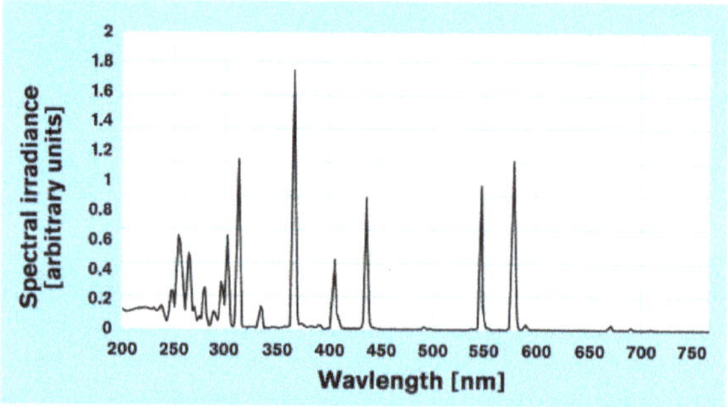

Figure 3. UV emission spectrum of the medium-pressure Hg-lamp (Heraeus) used (Heraeus brochure).

Table 2. Characteristics of the medium-pressure Hg-lamp (Heraeus) used.

Medium-Pressure Hg-Lamp Properties	
Effective Spectral Range (nm)	200–300
Specific Electrical Power (W/cm)	80–120
Specific Radiation Flux UVC (W/cm)	12–18
Power Range (kW)	0.5–15
Surface Temperature (°C)	600–900

Due to the fact that the purpose of this work is to demonstrate the general ability of UV light to activate different wooden surfaces before gluing, working at a constant distance makes more sense, as it minimises measurement variation. One of the possible optimisation steps after confirmation of the activation of wood surfaces by UV light is adjusting the distance between the UV lamp and the wood surface. Under previously mentioned conditions, it was measured with a radiometer that the irradiation dose during one pass of the specimen under the lamp is 2.36 J/cm^2. Irradiation levels were performed in 1, 5, 10, and 50 passes. The total intensity of radiation was determined by multiplying the radiation dose during one pass (2.36 J/cm^2) and the number of passes of the sample under the lamp (Table 3).

Table 3. Irradiation levels.

Irradiation Level (Number of Passes)	0 (Control)	1	5	10	20	50
Dose (J/cm^2)	0	2.36	11.80	23.60	45.20	118.00

2.4. Adhesive and Wood Bonding

The adhesive used was polyvinyl acetate adhesive (PVAc, Kleiberit 300.0, Weingarten, Germany), whose properties are listed in Table 4. As indicated in the literature, polyvinyl acetate is a thermoplastic adhesive and belongs to the prepolymerized adhesives category [45].

Table 4. Selected bonding parameters and adhesive properties for PVAc adhesive.

Bonding Parameters and Adhesive Properties	
Base	PVAc dispersion
pH value	3 ± 0.5
Viscosity at 20 °C (mPas)	12,000 ± 3000
Open time at 20 °C (min)	6–10
Water Resistance (EN 204)	Stress Group D3

The bonding material was 72 pieces of flat, differently machined, and UV-irradiated boards 170 mm × 130 mm × 10 mm (L × R × T). The adhesive was applied one-sided with a spread of 200 g/m^2 (A = 0.0221 m^2; spread = 4.42 g per joint), manually, using a metal-toothed spatula. The bonding process was performed in the press for 60 min at a pressure of 1 N/cm^2. The bonding operations were all performed within the specifications of the adhesive manufacturers. Bonding was performed immediately after UV irradiation, i.e., within 12 h after machining. Upon the hardening of the adhesive, two glued boards (170 mm × 130 mm × 20 mm) were obtained for each combination of mechanical treatment and UV irradiation level, from which 32 shear-test specimens were made. Therefore, from each glued board, 16 lap-shear specimens 35 × 25 × 20 mm^3 (L × R × T) were prepared according to the instructions of ISO 6238 standard [46] (more in Section 2.6). All steps of the bonding process and seven-day specimen conditioning before shear testing were carried out under standard laboratory conditions.

2.5. Wetting Angle

On the radial textured specimens, dimensions of 30 × 130 × 10 mm^3 (L × R × T), the adhesive (PVAc) wetting angle was measured using the sessile drop method with instruction of EN 828 standard [47]. The device for measuring the wetting angle consisted of a Dinolite 2.0 digital camera, lighting fixture, a specimen and needle stand, and a control device to adjust the volume and dosage of drops (Figure 4a). A total of 24 specimens were tested for each surface treatment with different levels of irradiation: therefore, 4 specimens for each of the 6 levels of irradiation (5 levels and control). Five drops were recorded on each specimen, and the volume of each drop was 4 μL. Thus, a total of 20 measurements were performed on each type of prepared sample. The test was performed under standard laboratory conditions immediately after UV irradiation, i.e., within 12 h after machining. Each adhesive drop was recorded for about a minute, and it was found that the droplet stopped changing and assumed its final shape after about 10 s. Thus, the wetting angle between a liquid drop and a wood surface was determined to be 10 s after applying the adhesive drop. The recordings were displayed, the final figures of drops were captured, and the wetting angles were evaluated and measured in the Dinolite 2.0 capture program (AnMo Electronics Corporation, New Taipei City, Taiwan) (Figure 4b).

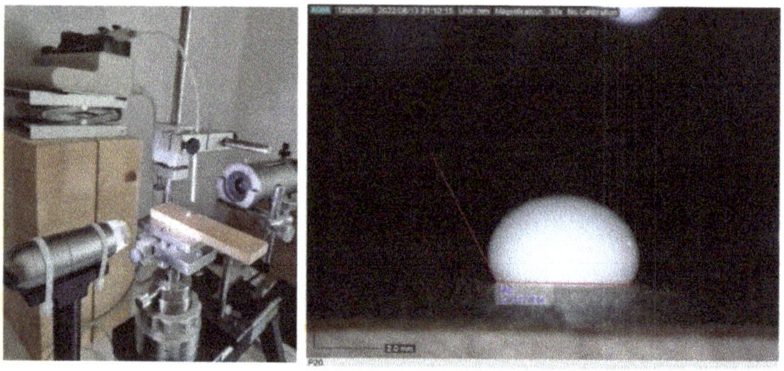

(a) (b)

Figure 4. Device for measuring the wetting angles—goniometer OCA20, Data Physics (**a**) and wetting angle measurement and evaluation in the Dinolite 2.0 capture program (**b**).

2.6. Bonding Quality

Bonding quality was expressed through the shear strength test and designation of main failure patterns. The bonded beech wood adhesion strength test—the compressive shear strength of lap joints was performed and evaluated by the standard ISO 6238 [46].

The shear test was performed using a 100 kN universal mechanical testing machine Shimadzu (Kyoto, Japan). A total of 572 specimens were tested and evaluated for three surface treatments with six levels of irradiation: therefore, 32 specimens for each combination. The specimens were tested at a constant testing speed until the separation of the substrates in the force-controlled mode indicated by the standards. Using the maximum load observed at the breakpoint (F_{max} in Newtons) and the bonding surface of the sample (mm^2) (a is glue face width, and b is glue face length—shown in Figure 5), shear strength was calculated according to the following equation:

$$S = \frac{F_{max}}{a \times b}, \left[\frac{N}{mm^2}\right]. \quad (1)$$

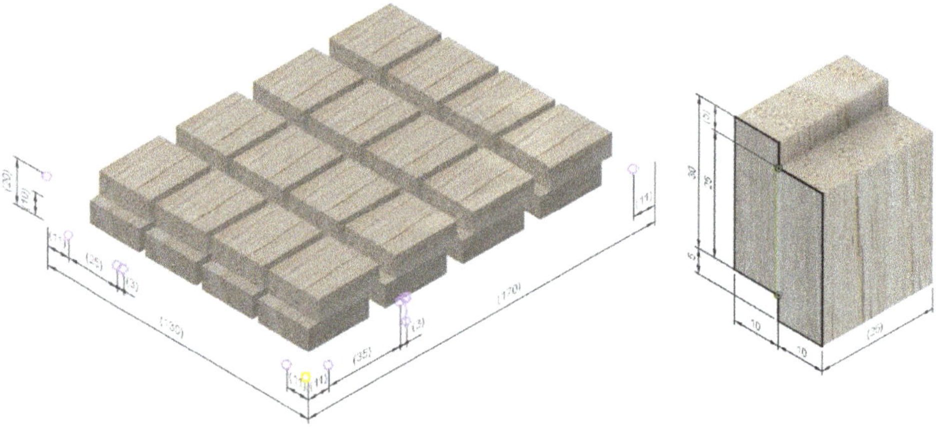

Figure 5. Compression shear strength test specimen according to ISO 6238: Layout of the specimen (**right**) and schematic representation of its production (**left**). Shear loaded bonding surface is shown in green (modified according to [46]).

After the shear test, the specimen's fracture surfaces were evaluated for the following main failure patterns: wood failure percentage (WFP), glue failure percentage (GFP), interface failure percentage (IFP), and adhesion failure percentage (AFP). Both sides of the failure surfaces of the specimen were examined visually, using a light source to illuminate the bonding surface at a specific angle, as recommended by the standard [46]. The failure percentage was evaluated at an approximation of 5%. To facilitate identification and evaluation, surfaces were contrasted with Lugol's indicator solution (1% iodine) [46].

2.7. Statistical Analysis

Statistical analysis and data visualisation was performed using the R (ver. 4.2.0) and R Studio software with the help of 'ggstatsplot' [48] and 'ggplot2' [49] packages. Analysis of variance (ANOVA) was used to evaluate the variation of adhesive wetting angles and shear strength of wood joints grouped by type of surface machining. The source of variation was the UV irradiation dose. The Shapiro–Wilk test confirmed the normality of the data, and Levene's test checked the homogeneity of the variances. When the assumption of homogeneity was met, a one-way ANOVA test was used; if not, Welch's ANOVA test was used. A *p*-value, ω^2, and a 95% confidence interval were used to estimate the effect size in omnibus tests. Effect size (ω^2) of omnibus test was interpretated as per Fields's (2013) conventions: $\omega^2 < 0.01$—very small, $0.01 \leq \omega^2 < 0.06$—small, $0.06 \leq \omega^2 < 0.14$—medium, ≥ 0.14—large [50]. An appropriate pairwise comparison (post hoc) test was performed if the omnibus test proved statistical significance. Thus, the Student's pairwise test (equal variances) was used after the one-way ANOVA, while the Games Howell pairwise test (non-equal variances) was used after Welch's ANOVA. Correlation and strength of relationships between the two variables were indicated with the Pearson correlation coefficient. The effect sizes of the pairwise test and Pearson correlations coefficient were estimated using holm-adjusted *p*-values. Descriptive statistics were also used to display the results: arithmetic means (red points), standard deviation, median, and 25th and 75th percentiles. Graphic displays were made using box whiskers and bar plots. The results are shown with a 95% family-wise confidence level.

3. Results and Discussion

3.1. Wetting Angles

The UV irradiation influenced the wetting of the differently machined beech wood surfaces, which was determined by adhesive wetting angle measurements. The results of the measurement of adhesive wetting angle on untreated and UV-irradiated wood surfaces are shown in Figure 6. One-way ANOVA was performed for each type of surface machining to compare the influence of UV irradiation on the adhesive wetting angle of differently treated surfaces. There was a statistically significant difference in mean wetting angle values between at least two levels of UV irradiation at the $p < 0.05$ for face-milled (F_F (5, 114) = 11.74, $p = 0.000$), planed (F_W (5, 48.9) = 4.03, $p = 0.004$), and sanded (F_F (5, 114) = 34.81, $p = 0.000$) surfaces. Additionally, the effect sizes indicate that the effect of UV irradiation in all three methods of surface machining, planing (${\omega_p}^2 = 0.22$), sanding (${\omega_p}^2 = 0.58$), and face milling (${\omega_p}^2 = 0.31$) were large (${\omega_p}^2 \geq 0.14$—large).

Also, based on the Pearson correlation test, all mechanical treatments tended to decrease adhesive wetting angles (e.g., better wettability) with increasing UV irradiation dose. There was a weak negative correlation between adhesive wetting angle and UV irradiation dose in face milling ($r_{Pearson} = -0.31$) and planing ($r_{Pearson} = -0.34$), but a strong negative correlation in sanding ($r_{Pearson} = -0.66$) (Figure 7).

Furthermore, multiple comparisons test found that the mean values of adhesive wetting angles were significantly different between some irradiation levels compared with control. This difference does not have the same trend for all surface preparations.

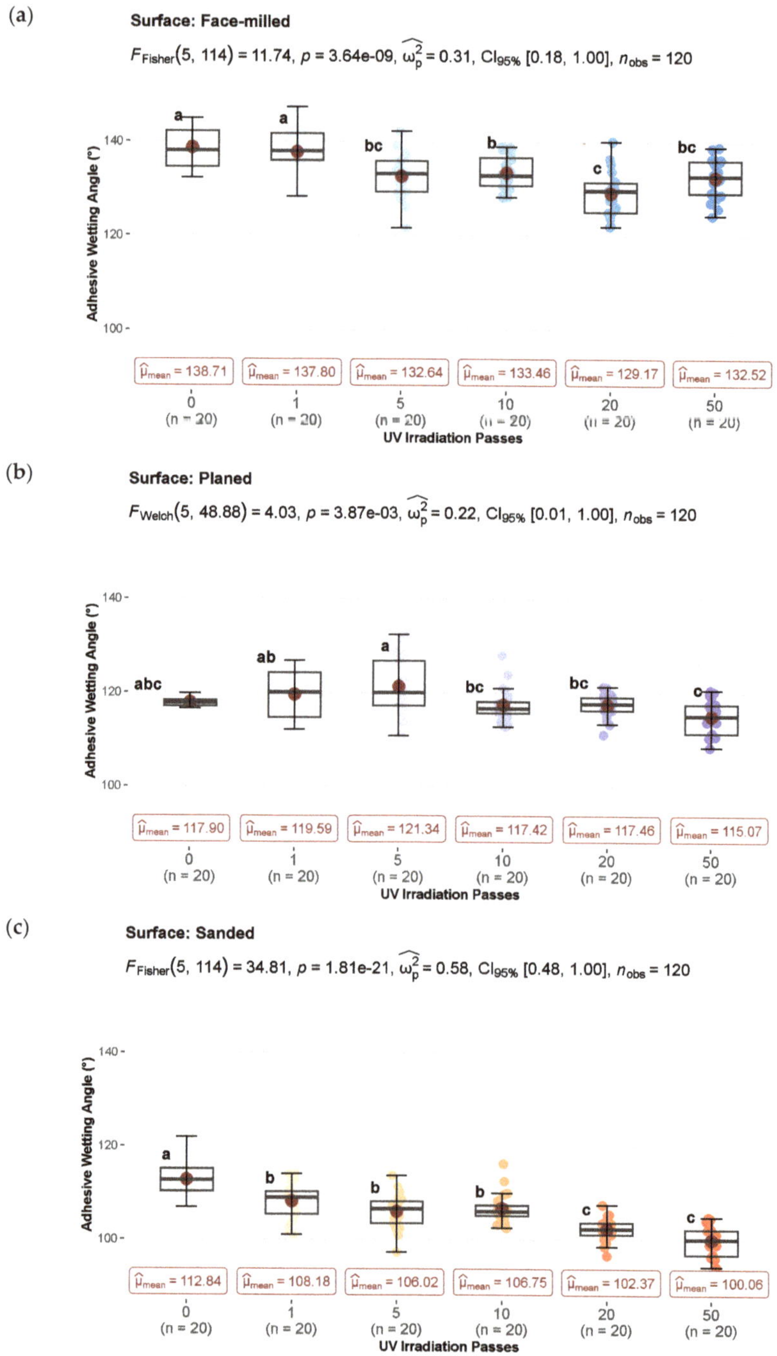

Figure 6. The effect of UV-irradiation before gluing on the adhesive wetting angles of face-milled (**a**), planed (**b**), and sanded (**c**) surface: whiskers show 5th–95th percentile, red point in box displays arithmetic mean value, and horizontal line shows median (means within the UV irradiation level followed by the same letter are not significantly different at 5% level of significance).

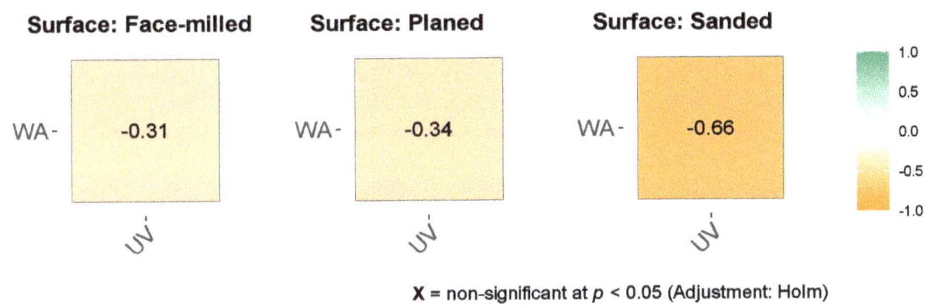

Figure 7. The bivariate correlation between UV irradiation (UV) and adhesive wetting angle (WA) for the face-milled, planed, and sanded joints are represented using a color-coding scheme. Dark green is associated with Pearson Correlation Coefficient, $r_{Pearson}$, equal to 1 and dark orange is associated to $r_{Pearson} = -1$. The Pearson correlation coefficients are indicated on plot.

Wetting angles of the face-milled surfaces indicate that control and short-term irradiated samples (1 pass) belong to group a, whereas the other samples belong to groups b, bc, and c. The adhesive wetting angle was significantly lower for all radiation doses except 1 pass (2.36 J/cm^2) compared to the control. Although the maximum wetting was achieved after 20 passes (45.20 J/cm^2), a statistically significant improvement in wetting already occurred after 5 passes, and after that, there was no statistically significant difference compared to other irradiation doses. There was no significant difference between 20 (c), 5 (bc), and 50 passes (bc), but only between 10 (b) and 20 passes (c) (Figure 6a). In the case of planed surfaces, there was no statistically significant difference between the control sample (abc) and irradiated samples (a, ab, b, or c) in adhesive wetting angle. If the wetting angle is compared only between doses of irradiation, then there were statistically significant differences in some cases. Irradiation dose of 10 (bc), 20 (bc), and 50 (c) passes had a significantly lower adhesive wetting angle than 5 passes (a). In comparison, 50 passes (c) had a lower wetting angle than 1 (ab) and 5 passes (a) (Figure 6b). The effect of UV irradiation on the wettability of planed surfaces has not been statistically proven.

The statistically significant wetting improvement of sanded surfaces was achieved after one pass (b) of UV irradiation. A further significant improvement occurred after 20 passes (c) (Figure 6c). In general, sanded surfaces showed the lowest wetting angles in all investigated samples, followed by planed and face-milled surfaces. This is expected because sanded surfaces show good wettability and high process roughness [51]. A thin layer of wood flour combined with a high level of fibrillation created by sanding provides the best conditions for water spreading [52]. Further, more hydrophilic active groups (hydroxyl groups) are exposed on the surface, reducing the angle of contact between water-based glues or coatings and the wood [53,54].

3.2. Bonding Quality

The results of the joints shear strength for untreated and UV-irradiated wood surfaces divided according to surface machining are shown in Figures 8–10. One-way ANOVA was performed for each type of surface machining to compare the effect of UV irradiation on joint shear strength results. There was a statistically significant difference in mean shear strength values between at least two levels of UV irradiation at the $p < 0.05$ for face-milled (F_W (5, 78.7) = 90.01, p = 0.000), planed (F_F (5, 185) = 19.65, p = 0.000), and sanded (F_W (5, 85.2) = 22.60, p = 0.000) surfaces. As well, the effect sizes indicate that the effect of UV irradiation in all three methods of surface machining, planing ($\omega_p^2 = 0.33$), sanding ($\omega_p^2 = 0.54$), and face milling ($\omega_p^2 = 0.33$) were large ($\omega_p^2 \geq 0.14$—large). Furthermore, multiple comparisons test found that the mean values of shear strength was significantly different between some irradiation levels compared with control. It indicates that increases in shear strength depend on the UV irradiation dose and the surface machining used.

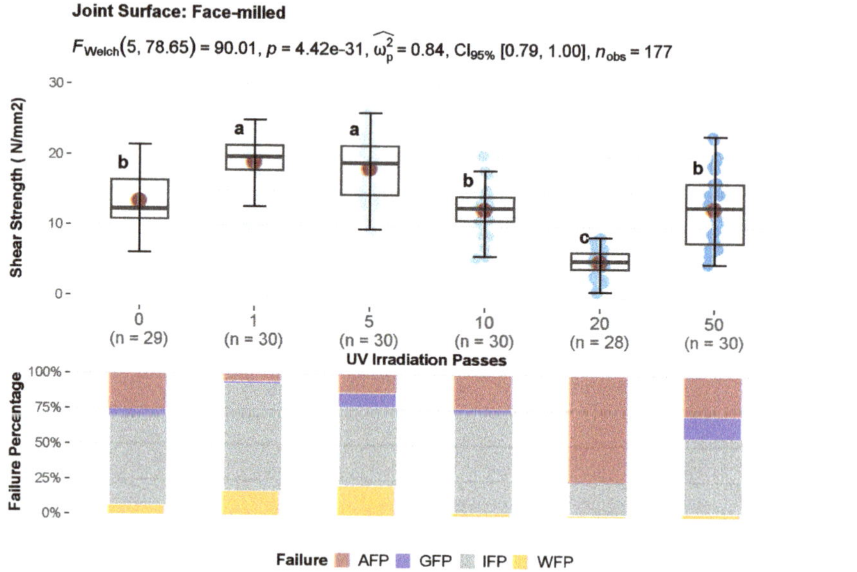

Figure 8. The effect of UV irradiation before gluing on the compressive shear strength (**top**) and average percentages of main failure patterns (**bottom**) of face-milled joints: whiskers show 5th–95th percentile, red point in box displays arithmetic mean value, and the horizontal line shows median (means within the UV irradiation level followed by the same letter are not significantly different at 5% level of significance).

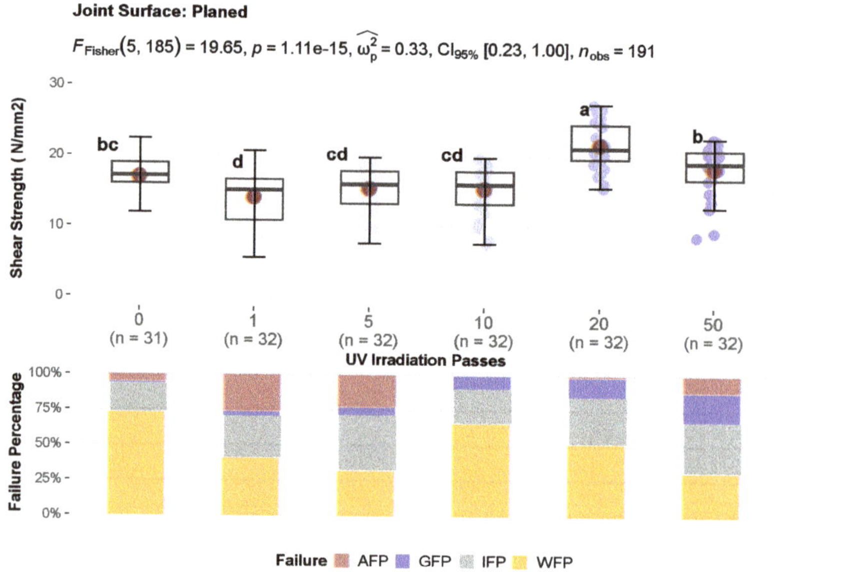

Figure 9. The effect of UV irradiation before gluing on the compressive shear strength (**top**) and average percentages of main failure patterns (**bottom**) of planed joints: whiskers show 5th–95th percentile, red point in box displays arithmetic mean value, and horizontal line shows median (means within the UV irradiation level followed by the same letter are not significantly different at 5% level of significance).

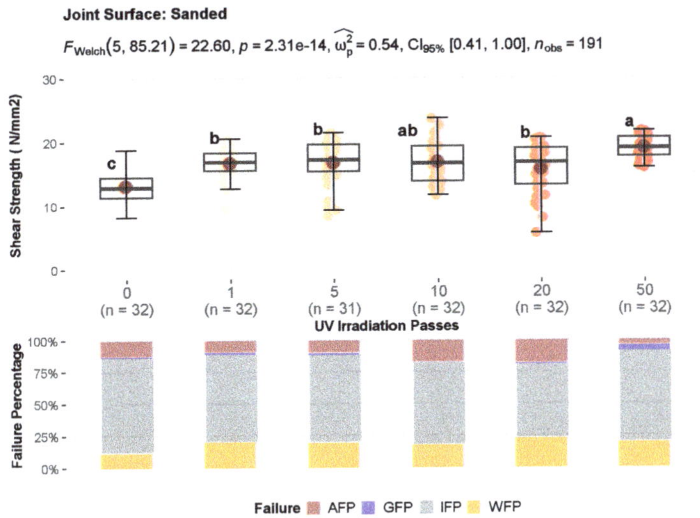

Figure 10. The effect of UV irradiation before gluing on the compressive shear strength (**top**) and average percentages of main failure patterns (**bottom**) of sanded joints: whiskers show 5th–95th percentile, red point in box displays arithmetic mean value, and the horizontal line shows median (means within the UV irradiation level followed by the same letter are not significantly different at 5% level of significance).

Also, the correlation coefficients between UV radiation dose and shear strength showed a weak negative correlation for face-milled joints ($r_{Pearson} = -0.3$), while sanded ($r_{Pearson} = 0.35$) and planed ($r_{Pearson} = 0.3$) joints showed a weak positive correlation (Figure 11). Generally, when increasing the dose of UV irradiation, the strength of face-milled joints decreases, and the strength of sanded and planed joints increases. However, with increasing the UV irradiation dose, the trend of shear strength did not follow the same pattern for each mechanical surface treatment. Therefore, a detailed analysis is given in the following text.

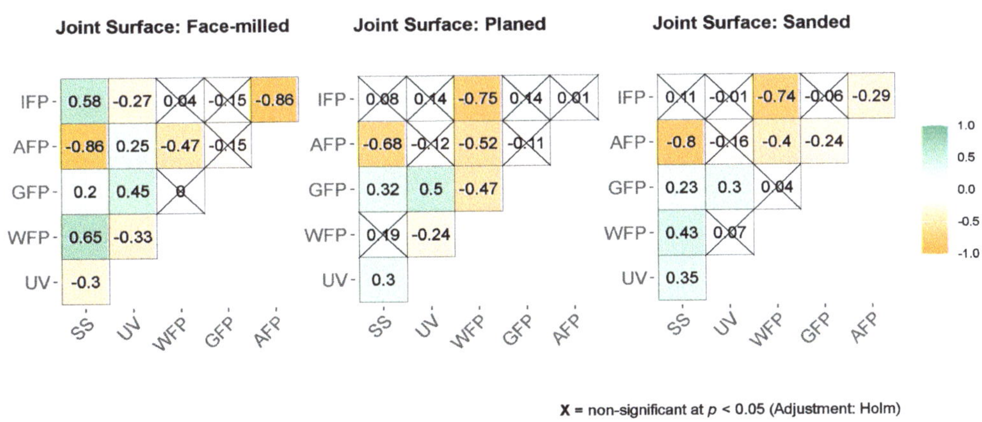

Figure 11. The bivariate correlations between UV irradiation (UV), shear strength (SS), wood (WFP), glue (GFP), interface (IFP) and adhesion (AFP) failure percentage for the face-milled, planed, and sanded joints are indicated using a color-coding scheme. Dark green is associated with Pearson Correlation Coefficient, $r_{Pearson}$, equal to 1 and dark orange is associated to $r_{Pearson} = -1$. The Pearson correlation coefficients are represented on plot.

Joints with face-milled and sanded surfaces showed a statistically significant increase in shear strength after 1 pass of UV irradiation (2.36 J/cm^2). That increase of 27% in sanded joints was approximately constant up to the dose of 20 passes (group b) and then at the dose of 50 passes (group a), it additionally increased by 45% regarding control (Figure 10). Contrarily, in face-milled joints, the increase of 42% was constant up to 5 passes and then sharply decreased with increasing UV irradiation (Figure 8). So, doses of one and five passes had statistically significant higher shear strength than control sample and other sample doses. The control, doses of 10 and 50 passes, belong to group b and there was no statistically significant difference in shear strength between them. The dose of 20 passes had statistically significant lower shear strength than other samples. At the dose of 10 passes, the shear strength is equal to the control, then decreases sharply and shows only 40% of the control shear strength at the 20 passes (45.2 J/cm^2). Interestingly, with higher doses of 20 passes, the shear strength does not decrease further, at 50 passes it shows the same values as the control. In the case of joints with planed surfaces, a statistically significant improvement of 22% in shear strength was determined after 20 passes of UV irradiation (45.2 J/cm^2) (Figure 9). However, joints with planed surfaces had a decrease in shear strength by 17% after 1 pass of UV irradiation.

The measured strength in some cases may or may not be a suitable criterion for determining bonding quality. It is very important to support the shear strength results with the type of failure in the joint. Surface fracture analysis and estimation of main failure patterns give more information about joint bonding quality than pure shear strength. Wood failure is common with quality glued joints and indicates that the strength of the joint (wood impregnated with glue) is greater than the cohesive strength of the wooden substrate [1]. However, with such failure, the actual shear strength of the wood-glue composite layer remains unknown. It is assumed to be equal to or greater than the one obtained. Therefore, the interface failure indicates the real shear strength of the wooden joint. The most preferred types of failure are cohesive wood failure and interface failure. Adhesion and glue failure are unfavourable patterns of surface fracture, although adhesive failure often shows high shear strengths due to the high cohesive strength of the adhesive. Therefore, stacked bar plots in Figures 8–10 show the average percentages of main failure patterns by the radiation doses for each surface machining. In Figure 11, the main failure patterns of the joints are contrasted on all fracture surfaces of the samples.

It is evident that with all doses of irradiation, except for 20 passes, the failure was primarily in the interface layer of face-milled joints (Figures 8, 11 and 12a–b). This confirms the joint's shear strength since the fracture was initiated correctly and shows the shear strength of the glue-impregnated wood layer. As well, it is obvious that the change in shear strength depended on the percentage of adhesion failure. The adhesion failure was dominant at the dose of 20 passes, hence the low shear strength. So, on one surface of the joints, pure glue layer was visible (red to red-brown surface), and on the other, no traces of glue were visible (the surface is the color of wood); there were no wood fibers on the glue layer or traces of individual surface fibers were barely noticeable (Figure 6b, middle two rows). Based on the Pearson correlation test (Figure 11), the decrease in the shear strength of the joint is strongly related to the increase in adhesion failure in the face-milled ($r_{Pearson} = -0.86$), planed ($r_{Pearson} = -0.68$), and sanded ($r_{Pearson} = -0.80$) joint surface. However, a weak positive correlation between adhesion failure percentage and UV irradiation dose was only observed in the face-milled joints ($r_{Pearson} = 0.25$).

In the case of planed joints, the failure was primarily in wood, but in some doses, it was equally in the wood and interface layer. By increasing the dose of UV irradiation, the wood failure slightly decreased ($r_{Pearson} = -0.24$), but glue failure increased ($r_{Pearson} = 0.5$), and it became more pronounced after a dose of 10 passes (Figures 9, 11 and 12c,d). In lower doses of irradiation (1–5 passes) adhesion failure was present. Compared to the control, at doses of 10 and 20 passes, no adhesion failure occurs. Therefore, the satisfying failure model and the significant increase in strength make the dose of 20 passes the most favourable for the UV modification of planed surfaces before gluing. In planing, high pressure is exerted on the wood cells, whereby they are crushed, and the surface becomes plasticised [1]. Therefore, a higher irradiation dose was required to activate such a compressed and plasticised surface.

(a) (b) (c) (d) (e) (f)

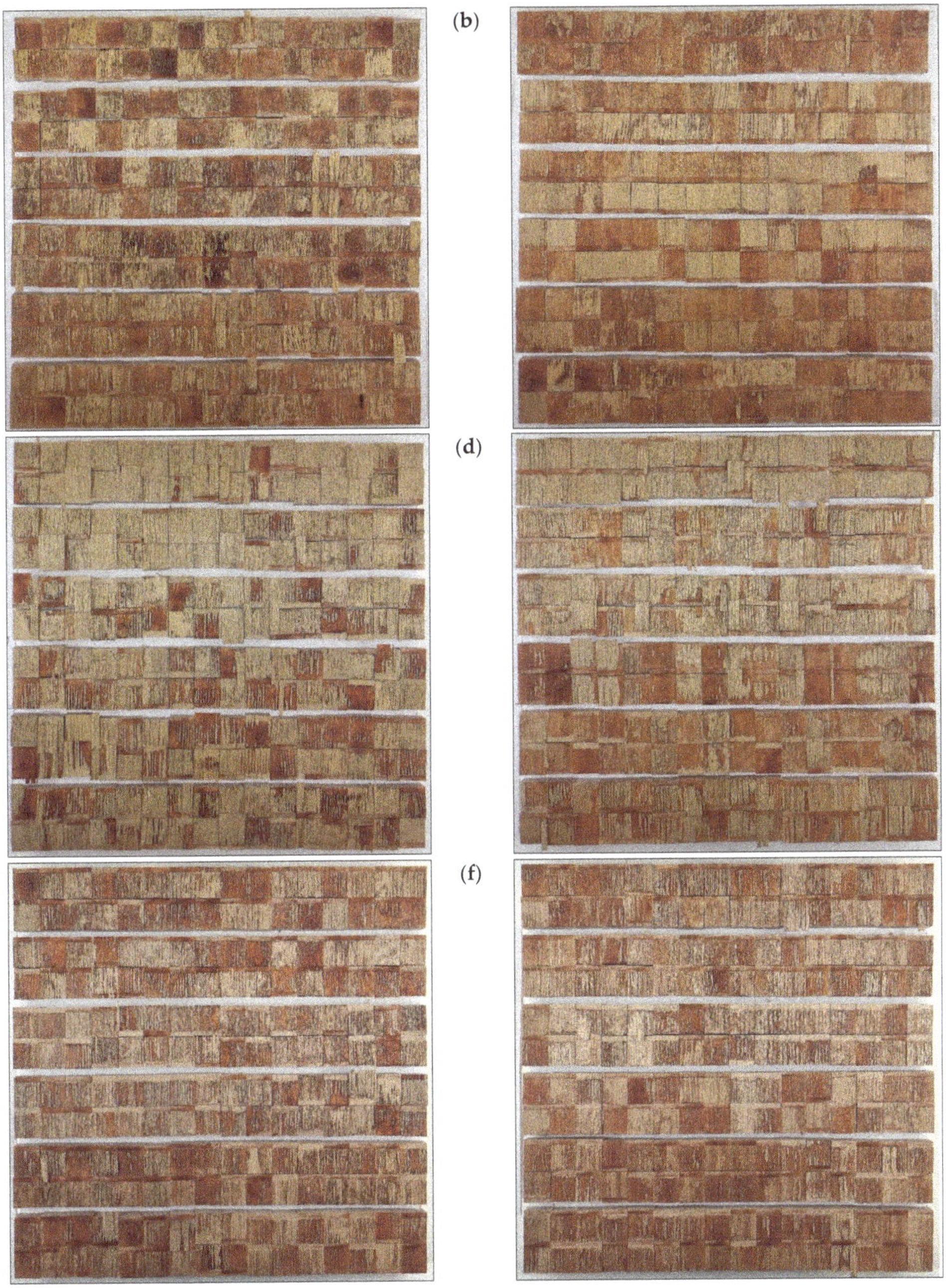

Figure 12. Main failure patterns of joints with: face-milled (**a**,**b**), planed (**c**,**d**), and sanded (**e**,**f**) surfaces before (**a**,**c**,**e**) 0, 1, 5 passes, and (**b**,**d**,**f**) 10, 20, 50 passes of UV irradiation. Every two rows of specimens represent a different dose of UV irradiation.

The failure of sanded joints dominantly occurred in the interface layer (Figure 12e–f). As shown in Figures 10 and 11, increasing UV irradiation did not result in significant changes in failure patterns.

Therefore, after only one pass of UV irradiation (2.36 J/cm^2), the shear strength of joints with face-milled and sanded surfaces improved statistically significantly, but for joints with planed surfaces, it improved after 20 passes (45.2 J/cm^2). Several factors may contribute to the difference in shear strength between surface machining methods including surface roughness, fibrillation, and cell damage under the surface [55]. The results could be explained by the fact that sanding and face milling commonly cause cell wall fibrillation (surface fibrillation) [56], whereas planing crushes rather than cuts the cells, which results in the plasticisation of the surface. As a result of fibrillation and a proper longitudinal cell cut, cells are exposed on the surface, allowing UV light to intensively reach the pits and middle lamella. In contrast, light cannot reach the middle lamella as intensely on plasticised surfaces, where cells are crushed rather than cut. A direct result of UV light's effect on wood cells exposed in this manner is the degradation of lignin. Additionally, the deterioration of connections between cellular elements is associated with the degradation of the middle lamella (layer S1), which also confirms the degradation of lignin [36,37]. When hydrophobic lignin degrades, surface hydrophilicity is increased because cellulose becomes more abundant on the wood surface [18]. A sharp increase in shear strength was therefore the result of the glue's better mechanical and chemical adhesion. More surface area is available for mechanical anchorage with an adhesive, potentially contributing to improved adhesion strength. Moreover, UV light also causes pits to open, allowing coatings and adhesives to penetrate deeper into the wood, enhancing mechanical anchoring [16,18,57]. Beechwood is a diffuse-porous hardwood specie with small, uniformly distributed vessel element cells (pores) in each growth ring, making their anatomical structure more predictable compared to ring-porous hardwoods. As a result, this wood's structure was not having an impact on the process of UV irradiation, gluing, and fracture initiation, even though it is an anisotropic biopolymer.

In addition, machining results in considerable deformation of the wood surface, resulting in a weak boundary layer. A chemically weak boundary layer on the molecular level and/or a mechanically weak boundary layer on the macro level may prevent adhesives and coatings from penetrating and anchoring to intact wood [16,57]. So, according to the conclusions stated earlier, UV light successfully activates and modifies the weak boundary layer by cleaning the wood surface, altering surface morphology, and to some extent, altering the chemical composition of the wood surface.

However, after the sharp increase in shear strength of face-milled and sanded joints with short-term intense UV radiation, the shear strength trend did not follow the same pattern. In both cases, the shear strength was initially relatively constant, and then, at higher doses of irradiation, it increased in the sanded joints, while it decreased in the face-milled joints. The values of the wetting angles of face-milled surfaces do not correlate directly with the shear strength, but it is one of the components that should be considered. The surface energy of the sanded surface has a greater share in the total adhesion than that of face-milled surfaces, which is also visible at lower wetting angles (Figure 6). UV exposure caused a slight decrease in the contact angle (increase in wettability) [18,57].

It has been suggested by Mihulja et al. (1999) that the decrease in shear strength of face-milled joints is the result of excessive drying of the surface by ultraviolet radiation, as well as too fast drying of the glue applied to the surface [42]. Besides the weak relationship between wetting and shear strength at high doses of radiation, high proportion of adhesion failure can also support this conclusion (Figures 8 and 11b). When the glue is applied to such a surface, it dries rapidly and then cannot adhere to another wood piece adhesively, especially when it is one-side applied. For example, extending the adhesive's open time in such cases would be desirable. The above is not the case with sanding and planing because the wetting angles decreased (wetting increased), and the shear strength increased when the radiation dose was increased. Moreover, the surface morphology after machining prevented undesirable glue and surface factors. It is known that a layer of crushed cells characterises sanded wood surface and subsurface, lumens clogged by fine dust, scratches, and packets of microfibrils torn out from cell walls [58–60]. Furthermore, sanding produces smooth, homogeneous surfaces for applying glues and defect-free surfaces for absorption of coatings [58]. Conversely, the face milling process enhanced the penetration of adhesives, especially for hardwood species, and produced significantly greater tensile shear strength than planing [55,60].

4. Conclusions

In this research, it was found that UV light, with its intense effect on the wood surface, improves the bonding properties (shear strength, wetting angle) of variously machined beech wood (*Fagus sylvatica* L.). The reason for this is that UV irradiation probably cleans wood surfaces, alters surface morphology, and to some extent, modifies chemical composition, thus activating and modifying weak boundary layers. Moreover, the dose of UV radiation affected the bonding properties differently depending on the surface machining.

The adhesive wetting angle decreased (increase in the energy of the surface layer of wood) while shear strength increased significantly for all tested surfaces with processing time and absorbed UV light. In joints with face-milled and sanded surfaces, the shear strength of the joint increased statistically significantly after UV radiation of 2.36 J/cm^2, but only after 45.2 J/cm^2 for joints with planed surfaces. There was an increase in average shear strength of 27% in sanded joints, 42% in face-milled joints, and 22% in planed joints, compared with the control samples. A sharp increase in shear strength was therefore the result of the glue's better mechanical and chemical adhesion. This is because sanding and face milling lead to fibrillation and proper longitudinal cutting of the cells; cells are exposed on the surface; UV light is allowed to reach the pits and the middle lamella very intensively. In contrast, light cannot reach the central lamella so intensely on planed (plasticised) surfaces, where cells are crushed rather than cut. When hydrophobic lignin degrades, surface hydrophilicity is increased because cellulose becomes more abundant on the wood surface. More surface area is available for mechanical anchorage with an adhesive, potentially contributing to improved adhesion strength. Moreover, UV light also causes pits to open, allowing coatings and adhesives to penetrate deeper into the wood, enhancing mechanical anchoring.

However, after the sharp increase in shear strength of face-milled and sanded joints with short-term intense UV irradiation, the trend did not follow the same pattern. In both cases, the shear strength was initially relatively constant, and then, at higher doses of irradiation, it increased in the sanded joints, while it decreased in the face-milled joints. The relationship between wetting and shear strength at high doses of radiation is connected with the proportion of adhesion failure, rapid drying of glue and wood surfaces, and surface morphology after machining.

Furthermore, several procedures must be carried out to identify all the key parameters more precisely to develop this method in greater detail. Short-term UV radiation can suddenly and intensively activate wooden surfaces, so it is necessary to define radiation regimes within this range, taking into account the change in surface energy. Additionally, it is important to solve the technological problem of drying the glue, for example, by extending the open time of the glue, in order to achieve a complete joining of the joints and to determine how this impacts the strength of the joint. Depending on the results, further research would be conducted with other wood types and the possibilities of technological applications to improve the strength of glued joints, as well as the adhesion and durability of wood coatings.

Author Contributions: Conceptualisation, G.M. and T.G.; methodology, G.M. and T.G.; software, T.G.; validation, G.M., V.Ž. and T.G.; formal analysis, G.M., V.Ž. and T.G.; investigation, T.G., L.G. and G.M.; resources, V.Ž. and G.M.; data curation, T.G.; writing—original draft preparation, T.G.; writing—review and editing, G.M. and V.Ž.; visualisation, T.G.; supervision G.M. and V.Ž. All authors have read and agreed to the published version of the manuscript.

Funding: This research received no external funding.

Institutional Review Board Statement: Not applicable.

Data Availability Statement: The data presented in this study are available upon request from the corresponding author.

Acknowledgments: We would like to express our gratitude to our colleagues from the Laboratory for Wood in Construction.

Conflicts of Interest: The authors declare no conflict of interest.

References

1. Mihulja, G.; Bogner, A. Strength and Durability of Glued Wood Part One: Factors of Glued Joint Strength. *Drv. Ind.* **2005**, *56*, 69–78.
2. Hänsel, A.; Sandak, J.; Sandak, A.; Mai, J.; Niemz, P. Selected Previous Findings on the Factors Influencing the Gluing Quality of Solid Wood Products in Timber Construction and Possible Developments: A Review. *Wood Mater. Sci. Eng.* **2021**, *17*, 230–241. [CrossRef]
3. Iždinský, J.; Reinprecht, L.; Sedliačik, J.; Kúdela, J.; Kučerová, V. Bonding of Selected Hardwoods with PVAc Adhesive. *Appl. Sci.* **2020**, *11*, 67. [CrossRef]
4. Lavisci, P.; Masson, D.; Deglise, X. Physico-Chemical Parameters of Quercus Cerris. *Holzforschung* **1991**, *45*, 415–418. [CrossRef]
5. Kilic, M.; Hiziroglu, S.; Burdurlu, E. Effect of Machining on Surface Roughness of Wood. *Build. Environ.* **2006**, *41*, 1074–1078. [CrossRef]
6. Knorz, M.; Neuhaeuser, E.; Torno, S.; Van De Kuilen, J.W. Influence of Surface Preparation Methods on Moisture-Related Performance of Structural Hardwood-Adhesive Bonds. *Int. J. Adhes. Adhes.* **2015**, *57*, 40–48. [CrossRef]
7. Stables, C.L.; Frazier, C.E.; Roman, M.; Zink-Sharp, A. Wetting and Penetration Behavior of Resin/Wood Interfaces. Ph.D. Thesis, Virginia Tech, Blacksburg, VA, USA, 2017.
8. Gardner, D.J.; Blumentritt, M.; Wang, L.; Yildirim, N. Adhesion Theories in Wood Adhesive Bonding: A Critical Review. *Rev. Adhes. Adhes.* **2014**, *2*, 127–172. [CrossRef]
9. Qin, Z.; Gao, Q.; Zhang, S.; Li, J. Surface Free Energy and Dynamic Wettability of Differently Machined Poplar Woods. *Bioresources* **2014**, *9*, 3088–3103. [CrossRef]
10. Kamke, F.A.; Lee, J.N. Adhesive Penetration in Wood-A Review. *Wood Fiber Sci.* **2007**, *39*, 205–220.
11. Aydin, I. Activation of Wood Surfaces for Glue Bonds by Mechanical Pre-Treatment and Its Effects on Some Properties of Veneer Surfaces and Plywood Panels. *Appl. Surf. Sci.* **2004**, *233*, 268–274. [CrossRef]
12. Quan, D.; Deegan, B.; Byrne, L.; Scarselli, G.; Ivanković, A.; Murphy, N. Rapid Surface Activation of Carbon Fibre Reinforced PEEK and PPS Composites by High-Power UV-Irradiation for the Adhesive Joining of Dissimilar Materials. *Compos. Part A Appl. Sci. Manuf.* **2020**, *137*, 105976. [CrossRef]
13. Rabek, J.F. *Photodegradation of Polymers: Physical Characteristics and Applications*; Springer Science & Business Media: Berlin/Heidelberg, Germany, 1996; p. 212.
14. Goddard, J.M.; Hotchkiss, J.H. Polymer Surface Modification for the Attachment of Bioactive Compounds. *Prog. Polym. Sci.* **2007**, *32*, 698–725. [CrossRef]
15. Alonso, D.; Gimeno, M.; Olayo, R.; Vázquez-Torres, H.; Sepúlveda-Sánchez, J.D.; Shirai, K. Cross-Linking Chitosan into UV-Irradiated Cellulose Fibers for the Preparation of Antimicrobial-Finished Textiles. *Carbohydr. Polym.* **2009**, *77*, 536–543. [CrossRef]
16. Gindl, M.; Sinn, G.; Stanzl-Tschegg, S.E. The Effects of Ultraviolet Light Exposure on the Wetting Properties of Wood. *J. Adhes. Sci. Technol.* **2006**, *20*, 817–828. [CrossRef]
17. Yousif, E.; Haddad, R. Photodegradation and Photostabilization of Polymers, Especially Polystyrene: Review. *Springerplus* **2013**, *2*, 398. [CrossRef]
18. Jankowska, A.; Rybak, K.; Nowacka, M.; Boruszewski, P. Insight of Weathering Processes Based on Monitoring Surface Characteristic of Tropical Wood Species. *Coatings* **2020**, *10*, 877. [CrossRef]
19. Teacă, C.A.; Roşu, D.; Bodîrlău, R.; Roşu, L. Structural Changes in Wood under Artificial UV Light Irradiation Determined by FTIR Spectroscopy and Color Measurements-a Brief Review. *Bioresources* **2013**, *8*, 1478–1507. [CrossRef]
20. Mathieson, I.; Bradley, R.H. Improved Adhesion to Polymers by UV/Ozone Surface Oxidation. *Int. J. Adhes. Adhes.* **1996**, *16*, 29–31. [CrossRef]
21. Seo, J.; Chang, W.S.; Kim, T.S. Adhesion Improvement of Graphene/Copper Interface Using UV/Ozone Treatments. *Thin. Solid Film.* **2015**, *584*, 170–175. [CrossRef]

22. Moyano, M.A.; Martín-Martínez, J.M. Surface Treatment with UV-Ozone to Improve Adhesion of Vulcanized Rubber Formulated with an Excess of Processing Oil. *Int. J. Adhes. Adhes.* **2014**, *55*, 106–113. [CrossRef]
23. Liu, J.; He, L.; Wang, L.; Man, Y.; Huang, L.; Xu, Z.; Ge, D.; Li, J.; Liu, C.; Wang, L. Significant Enhancement of the Adhesion between Metal Films and Polymer Substrates by UV-Ozone Surface Modification in Nanoscale. *ACS Appl. Mater. Interfaces* **2016**, *8*, 30576–30582. [CrossRef] [PubMed]
24. Le-The, H.; Tiggelaar, R.M.; Berenschot, E.; Van Den Berg, A.; Tas, N.; Eijkel, J.C.T. Postdeposition UV-Ozone Treatment: An Enabling Technique to Enhance the Direct Adhesion of Gold Thin Films to Oxidized Silicon. *ACS Nano* **2019**, *13*, 6782–6789. [CrossRef] [PubMed]
25. Sernek, M.; Kamke, F.A.; Glasser, W.G. Comparative Analysis of Inactivated Wood Surface. *Holzforschung* **2004**, *58*, 22–31. [CrossRef]
26. Bogner, A. Modifikacija Površine Bukovine Radi Poboljšanja Lijepljenja. Ph.D. Thesis, Faculty of Forestry, University of Zagreb, Zagreb, Croatia, 1993.
27. Hon, D.N.S.; Shiraishi, N. *Colour and Discolouration: Wood and Cellulosic Chemistry*; Marcel Dekker: New York, NY, USA; Basel, Switzerland, 1991.
28. Hon, D.N.S.; Ifju, G. Measuring Penetration of Light into Wood by Detection of Photo-Induced Free Radicals. *Wood Sci. Technol.* **1978**, *11*, 118–127.
29. Hon, D.N.S.; Chang, S.T. Surface Degradation of Wood by Ultraviolet Light. *Polym. Sci. Polym. Chem.* **1984**, *22*, 2227–2241. [CrossRef]
30. Hon, D.N.-S.; Feist, W.C. Free Radical Formation in Wood: The Role of Water. *Wood Sci.* **1981**, *14*, 41–81.
31. Hon, D.N.S.; Ifju, G.; Feist, W.C. Characteristics of Free Radicals in Wood. *Wood Fiber.* **1980**, *12*, 121–130.
32. Derbyshire, H.; Miller, E.R.; Turkulin, H. Investigations into the Photodegradation of Wood Using Microtensile Testing: Part 3: The Influence of Temperature on Photodegradation Rates. *Holz Als Roh-Und Werkst.* **1997**, *55*, 287–291. [CrossRef]
33. Turkulin, H.; Derbyshire, H.; Miller, E.R. Investigations into the Photodegradation of Wood Using Microtensile Testing—Part 5: The Influence of Moisture on Photodegradation Rates. *Holz Als Roh-Und Werkst.* **2004**, *62*, 307–312. [CrossRef]
34. Cogulet, A.; Blanchet, P.; Landry, V. Wood Degradation under UV Irradiation: A Lignin Characterization. *J. Photochem. Photobiol. B* **2016**, *158*, 184–191. [CrossRef]
35. Turkulin, H. Fotodegradacija Proizvoda Od Drva u Građevinarstvu. Ph.D. Thesis, Faculty of Forestry, University of Zagreb, Zagreb, Croatia, 1996.
36. Williams, R.S. *Handbook of Wood Chemistry and Wood Composites: Weathering of Wood*; Rowell, R.M., Ed.; Taylor and Francis: Boca Raton, FL, USA, 2005.
37. Mamoňová, M.; Reinprecht, L. The Impact of Natural and Artificial Weathering on the Anatomy of Selected Tropical Hardwoods. *IAWA J.* **2020**, *41*, 333–355. [CrossRef]
38. Jankowska, A.; Kozakiewicz, P. Ocjena Otpornosti Drva Na Utjecaj Umjetnih Atmosferskih Uvjeta Mjerenjem Tlačne Čvrstoće. *Drv. Ind.* **2016**, *67*, 3–8. [CrossRef]
39. Laskowska, A.; Dobrowolska, E.; Boruszewski, P. The Impact of Ultraviolet Radiation on the Colour and Wettability of Wood Used for Facades. *Drewno* **2016**, *59*, 99–111. [CrossRef]
40. Miklečić, J.; Lončarić, A.; Veseličić, N.; Jirouš-Rajković, V. Influence of Wood Surface Preparation on Roughness, Wettability and Coating Adhesion of Unmodified and Thermally Modified Wood. *Drv. Ind.* **2022**, *73*, 261–269. [CrossRef]
41. Khan, M.A.; Khan, R.A.; Aliya, B.S.; Nasreen, Z. Effect of the Pretreatment with UV and Gamma Radiations on the Modification of Plywood Surface by Photocuring with Epoxy Acrylate. *J. Polym. Environ.* **2006**, *14*, 111–118. [CrossRef]
42. Mihulja, G.; Bogner, A.; Turkulin, H. Modification of the Beech—Wood Surface by UV-Irradiation. *Drv. Idnustrija* **1999**, *50*, 133–140.
43. Ohnesorge, D.; Richter, K.; Becker, G. Influence of Wood Properties and Bonding Parameters on Bond Durability of European Beech (*Fagus Sylvatica* L.) Glulams. *Ann. For. Sci.* **2010**, *67*, 601. [CrossRef]
44. Hass, P.; Müller, C.; Clauss, S.; Niemz, P. Influence of Growth Ring Angle, Adhesive System and Viscosity on the Shear Strength of Adhesive Bonds. *Wood Mater. Sci. Eng.* **2009**, *4*, 140–146. [CrossRef]
45. Hunt, C.G.; Frihart, C.R.; Dunky, M.; Rohumaa, A. Understanding Wood Bonds-Going beyond What Meets the Eye: A Critical Review. *Rev. Adhes. Adhes.* **2018**, *6*, 369–463. [CrossRef]
46. *ISO 6238:2018*; Adhesives—Wood-to-Wood Adhesive Bonds—Determination of Shear Strength by Compressive Loading. ISO: Geneva, Switzerland, 2018.
47. *BS EN 828:2013*; Adhesives—Wettability—Determination by Measurement of Contact Angle and Surface Free Energy of Solid Surface. European Committee for Standardization: Brussels, Belgium, 2013.
48. Patil, I. Visualizations with Statistical Details: The "ggstatsplot" Approach. *J. Open Source Softw.* **2021**, *6*, 3167. [CrossRef]
49. Wickham, H. *Ggplot2: Elegant Graphics for Data Analysis*; Springer: New York, NY, USA, 2016; ISBN 978-3-319-24277-4.
50. Field, A. *Discovering Statistics Using IBM SPSS Statistics*, 4th ed.; Carmichael, M., Ed.; Sage: Newcastle upon Tyne, UK, 2013; ISBN 9781446273043.
51. Jankowska, A.; Zbieć, M.; Kozakiewicz, P.; Koczan, G.; Oleńska, S.; Beer, P. The Wettability and Surface Free Energy of Sawn, Sliced and Sanded European Oak Wood. *Maderas Cienc. Y Tecnol.* **2018**, *20*, 443–454. [CrossRef]

52. De Moura, L.F.; Hernández, R.E. Effects of Abrasive Mineral, Grit Size and Feed Speed on the Quality of Sanded Surfaces of Sugar Maple Wood. *Wood Sci. Technol.* **2006**, *40*, 517–530. [CrossRef]
53. Qin, Z.; Chen, H.; Gao, Q.; Zhang, S.; Li, J. Wettability of Sanded and Aged Fast-Growing Poplar Wood Surfaces: I. Surface Free Energy. *Bioresources* **2015**, *10*, 1008–1023. [CrossRef]
54. Liptáková, E.; Kúdela, J.; Bastl, Z.; Spirovová, I. Influence of Mechanical Surface Treatment of Wood on the Wetting Process. *Holzforschung* **1995**, *49*, 369–375. [CrossRef]
55. Leggate, W.; McGavin, R.L.; Outhwaite, A.; Kumar, C.; Faircloth, A.; Knackstedt, M. Influence of Mechanical Surface Preparation Methods on the Bonding of Southern Pine and Spotted Gum: Tensile Shear Strength of Lap Joints. *Bioresources* **2020**, *16*, 46–61. [CrossRef]
56. Hernández, R.E.; Cool, J. Effects of Cutting Parameters on Surface Quality of Paper Birch Wood Machined across the Grain with Two Planing Techniques. *Holz Als Roh-Und Werkst.* **2008**, *66*, 147–154. [CrossRef]
57. Gindl, M.; Reiterer, A.; Sinn, G.; Stanzl-Tschegg, S.E. Effects of Surface Ageing on Wettability, Surface Chemistry, and Adhesion of Wood. *Holz Als Roh-Und Werkst.* **2004**, *62*, 273–280. [CrossRef]
58. Cool, J.; Hernández, R.E. Improving the Sanding Process of Black Spruce Wood for Surface Quality and Water-Based Coating Adhesion. *Prod J.* **2011**, *61*, 372–380. [CrossRef]
59. Kuljich, S.; Cool, J.; Hernández, R.E. Evaluation of Two Surfacing Methods on Black Spruce Wood in Relation to Gluing Performance. *J. Wood Sci.* **2013**, *59*, 185–194. [CrossRef]
60. Fernando De Moura, L.; Hernández, R.E. Evaluation of Varnish Coating Performance for Three Surfacing Methods on Sugar Maple Wood. *Prod. J.* **2006**, 130–136.

polymers MDPI

Article

Multi-Material Additive Manufacturing of High Temperature Polyetherimide (PEI)–Based Polymer Systems for Lightweight Aerospace Applications

Ved S. Vakharia [1], Hunter Leonard [2,†], Mrityunjay Singh [3,*] and Michael C. Halbig [4,*]

1 NASA Pathway Intern, Department of Mechanical and Aerospace Engineering, University of California San Diego, La Jolla, CA 92092, USA
2 Beta Technologies, Burlington, VT 05403, USA
3 Ohio Aerospace Institute, Cleveland, OH 44142, USA
4 NASA Glenn Research Center, Cleveland, OH 44135, USA
* Correspondence: mrityunjaysingh@oai.org (M.S.); michael.c.halbig@nasa.gov (M.C.H.)
† Former NASA Intern.

Abstract: Rapid innovations in 3-D printing technology have created a demand for multifunctional composites. Advanced polymers like amorphous thermoplastic polyetherimide (PEI) can create robust, lightweight, and efficient structures while providing high-temperature stability. This work manufactured ULTEM, a PEI-based polymer, and carbon-fiber-infused ULTEM multi-material composites with varying layering patterns (e.g., AAABBB vs. ABABAB) using fused filament fabrication (FFF). The microstructure of fractured surfaces and polished cross-sections determined that the print quality of layers printed closer to the heated bed was higher than layers closer to the top surface, primarily due to the thermal insulating properties of the material itself. Mechanical properties of the multi-material parts were between those of the single-material parts: an ultimate tensile strength and elastic modulus of 59 MPa and 3.005 GPa, respectively. Multi-material parts from the same filaments but with different layering patterns showed different mechanical responses. Prints were of higher quality and demonstrated a higher elastic modulus (3.080 GPa) when consecutive layers were printed from the same filament (AAABBB) versus parts with printed layers of alternating filaments (ABABAB), which showed a higher ultimate strength (62.04 MPa). These results demonstrate the potential for creatively designing multi-material printed parts that may enhance mechanical properties.

Keywords: fused-deposition modeling; mechanical properties; polyetherimide (PEI); multi-material

Citation: Vakharia, V.S.; Leonard, H.; Singh, M.; Halbig, M.C. Multi-Material Additive Manufacturing of High Temperature Polyetherimide (PEI)–Based Polymer Systems for Lightweight Aerospace Applications. *Polymers* **2023**, *15*, 561. https://doi.org/10.3390/polym15030561

Academic Editor: Ying Yan

Received: 15 December 2022
Revised: 17 January 2023
Accepted: 19 January 2023
Published: 21 January 2023

1. Introduction

The accessibility and affordability of sizeable 3-D printing systems have opened the field of materials and manufacturing to many new possibilities. Polymer matrix composites with metal additives are being used to introduce multi-functionality to parts with complex geometries. Carbon additives are being used to enhance the durability of 3-D printed parts without adding significant weight. These, among others, are examples of 3-D printed materials being investigated for their potential to provide manufacturing materials and methods for aerospace applications. The future of aviation is limited by the weight of an aircraft's many complicated components, and additive manufacturing aims to address that challenge.

Fused filament fabrication (FFF), alternately known as fused deposition modeling (FDM), is an additive manufacturing (AM) process described as "a material extrusion process used to make thermoplastic parts through heated extrusion and deposition of materials layer by layer" [1]. FFF is one of the most widely used 3-D printing methods and has been the center of much additive manufacturing research. FFF has made great progress recently, particularly due to multi-material printing. In recent efforts, several studies have explored using FFF to 3-D print parts that integrate multiple materials [2–12]. Multi-material printing presents a huge potential for new multi-functionality and enhancement of overall mechanical properties of designs that may have been weak as a single-material part. Higher strength or functional materials can be directly deposited where they are needed based on the design.

Multi-material extrusion does, however, present limitations and challenges. The possibility of weak bond strength between layers of different materials presents a significant challenge that needs to be addressed. The bonding in a multi-material part can be influenced by a variety of variables, such as the many different printing parameters (heat, print speed, and layer thickness) or the properties of the materials themselves (thermal expansion rate and melting temperature) [13].

Traditionally, most of the multi-material AM research is limited to a few materials: acrylonitrile butadiene styrene (ABS), polylactic acid (PLA), and high-impact polystyrene (HIPS). These three materials are easily accessible, biodegradable, partially biocompatible, and have high impact resistance, heat resistance, and toughness [14–17]. There is significant research into reinforcing these polymers with metal and non-metal additions such as copper-infused PLA or ABS with carbon nanotubes [18–22].

A specific direction aviation research is currently taking is that of an electric aircraft with lighter, cooler, and more efficient motors and energy storage systems. While motors currently can reach the 95% efficiency range, motors and batteries have struggled with a heat management problem and a weight problem due to the required heavy safety systems. Potential mitigation of these problems can be addressed by replacing metals with lightweight yet strong polymers that can be 3-D printed to allow for new shapes and designs that help move heat away from sources by active or passive systems. New shapes from additive manufacturing can allow water cooling or forced air cooling of motor stators and battery cores, helping to keep systems lighter and running more efficiently at higher altitudes. PLA, ABS, and HIPS should be further investigated, but other polymers can prove more useful.

A thermoplastic material that offers the potential for applications in many engineering fields is ULTEM 9085, due to its high strength, low weight, flame retardancy, chemical resistance, and low toxicity. ULTEM 9085 can be acquired as a filament from various vendors for use in FFF. In this filament form, it exhibits a glass transition temperature (T_g) of 186 °C [23]. However, there has been limited research conducted on this material and its application as a 3-D printed material. Some studies have investigated the effects that FFF processing parameters may have on the mechanical performance of 3-D printed ULTEM 9085 parts [24–26]. ULTEM 9085 has even been used to successfully print perforated engine access doors and acoustic panels and liners [27]. However, no literature exists on the use of ULTEM 9085 in a multi-material 3-D printed part. There is potential to make ULTEM 9085 parts tougher or multi-functional with the inclusion of a second material.

In this study, single- and multi-material parts were 3-D printed from ULTEM 9085 and carbon-fiber-infused ULTEM 9085. The microstructure, tensile strength, and elastic modulus were determined as a function of the FFF layering patterns. In the multi-material parts, different layering patterns were used (i.e., AAABBB vs. ABABAB) to study any effects the patterns may have on the mechanical properties. This research aimed to determine the effectiveness of using FFF to print multi-material ULTEM 9085 parts through careful observation of microstructure evolution and mechanical testing. Polymers like ULTEM 9085 show the potential to create strong, lightweight, and efficient structures to advance innovations in aviation.

2. Materials and Methods

A HyRel Hydra 645 (HYREL 3D, Norcross, GA, USA) was used in this work to print ASTM D638 [28] tensile specimens. The materials used, in the form of continuous filaments, were ULTEM AM9085F (SABIC, Riyadh, Saudi Arabia) and ULTEM 9085 CF (3DXTech,

Grand Rapids, MI, USA) and will hereby be referred to as ULTEM 9085 and ULTEM 9085CF. The continuous filaments have a diameter of 1.75 mm and come from the supplier wound on a spool. ULTEM 9085CF is an ULTEM 9085-matrix composite with chopped carbon-fiber additions. Single-material and multi-material parts were made. The filaments used in the multi-material parts were ULTEM 9085 + ULTEM 9085CF. Each multi-material part was printed multiple times, with ULTEM 9085 + ULTEM 9085CF having different layering patterns each time. For example, one part was printed with an AAABBB pattern while another with an ABABAB pattern. The printed multi-material parts and their layering patterns are shown in Table A1.

Thermogravimetric analysis (TGA) was performed with a TGA Q500 thermogravimetric analyzer (TA Instruments, New Castle, DE, USA) for bulk filament pieces to estimate the wt.% addition of reinforcement and observe differences in impurities present in the filaments. The instrument was calibrated for temperature, heat flow, and weight according to the manufacturer's suggestions. Specimens were cut from the filament spool and tested. TGA scans were run under air from room temperature (25 °C) to 600 °C at the heating rate of 10 °C/min. The sample purge was 60 cc/min, and the balance purge was 40 cc/min under the flowing nitrogen gas of 20 psi. Samples in the weight range of 3–5 mg were placed in a high-temperature platinum pan and degraded as a function of temperature; their mass loss was recorded by the TGA Universal Software (TA Instruments, New Castle, DE, USA).

All parts were printed at 390 °C with an MK1-450 (HyRel) head. ULTEM 9085 and ULTEM 9085CF filaments were printed with a 1.0 mm nozzle. Parts were printed with three outer perimeters, 100% infill, and a layer height half the nozzle's width. Perimeters are used to create a well-defined part exterior to ensure the part is made accurate to the specific dimensions of the design, and the rest of the part is printed within this outline. The bed was preheated to 200 °C for ULTEM 9085 materials. A coating of PVA was applied to the print bed to increase bed adhesion. Tensile samples were 3 mm thick with six 0.5 mm layers. Parts were printed at alternating raster angles of 0° and 45°. The use of more print layers was investigated, but the final parts were printed with only 6 layers at 0.5 mm for two reasons: 3 mm provides a typical coupon thickness for mechanical testing, and the print quality diminishes with increased number of layers because the part is built up further from the heated bed. The final shapes of the parts were made according to the ASTM standards for tensile testing. Figure 1 shows the printing of an ULTEM 9085CF part.

Figure 1. Hyrel Hydra with a MK1-450 head printing ULTEM 9085CF on an ULTEM 9085 raft on a 200 °C glass bed with a solution of PVA glazed on top.

Samples cut from printed coupons of ULTEM 9085, ULTEM 9085CF, and ULTEM 9085 + ULTEM 9085CF were mounted in epoxy and polished for microstructural analysis of the cross-sections with an optical microscope. Measurements of % area porosity of the polished cross-sections were conducted on the optical micrographs using ImageJ (National Institutes of Health, Bethesda, MD, USA). A Keyence VR-3200 profilometer (Keyence, Itasca, IL, USA) was used to develop 3-D scans of the fracture surfaces of the tensile specimens.

For tensile testing, the strains were determined with an ARAMIS (Trilion Quality Systems, Plymouth Meeting, PA, USA) photogrammetry system using Digital Image Correlation (DIC). DIC is a non-contact optical method used to measure deformation and strain in the specimens during the tensile tests. A visualization of the test setup and an example of the DIC analysis is shown in Figures 2 and 3, respectively. Mechanical modulus and ultimate tensile strength (UTS) values were determined from the tensile data. Six coupons of pure ULTEM 9085, three of ULTEM 9085CF, and two of each layering pattern (ABBABB, ABABAB, AAABBB) underwent tensile testing. Details of the material system, print layer height, coupon dimensions, elastic modulus, and ultimate strength for each tested coupon are displayed in Table A1.

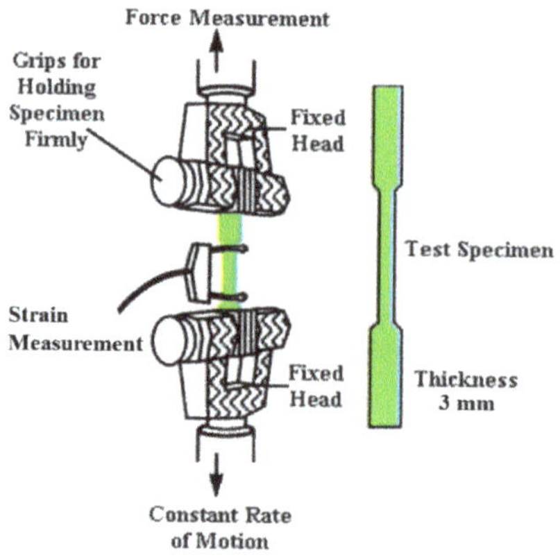

Figure 2. Visualization of tensile testing set up of specimens made in this study.

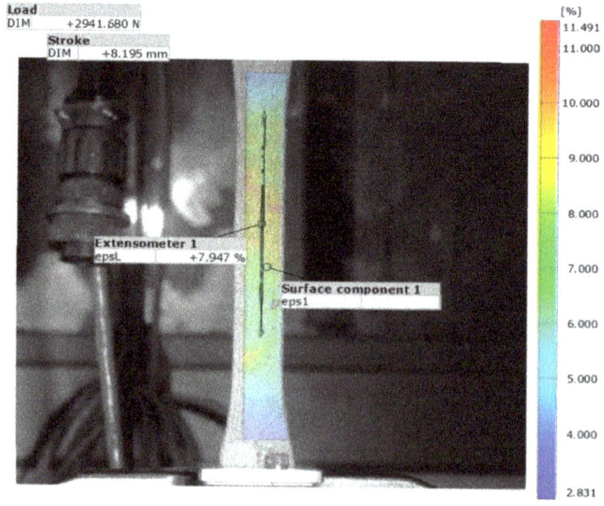

Figure 3. Example of DIC analysis of tensile test specimen.

3. Results and Discussion

3.1. Thermal Stability

The thermal stability of the filaments used in this study was investigated to characterize the high-temperature capabilities of the printed parts. Figure 4 shows the TGA (thermogravimetric analysis) results performed in air. The initial degradation temperatures (wt.% loss = 5%) for ULTEM 9085 and ULTEM 9085CF are 496 °C and 562 °C, respectively. The initial degradation temperature of ULTEM 9085 is comparable to other work, done by Padovano et al. and Kafi et al., that has investigated ULTEM 9085 parts additively manufactured with the FFF process [29,30]. Though the TGA runs in this study did not surpass 600 °C, ULTEM 9085 is expected to reach full degradation at approximately 700 °C. Compared to common FFF materials such as ABS and PLA, ULTEM materials begin degradation at higher temperatures as expected because they are known to be high-temperature polymers [22,31].

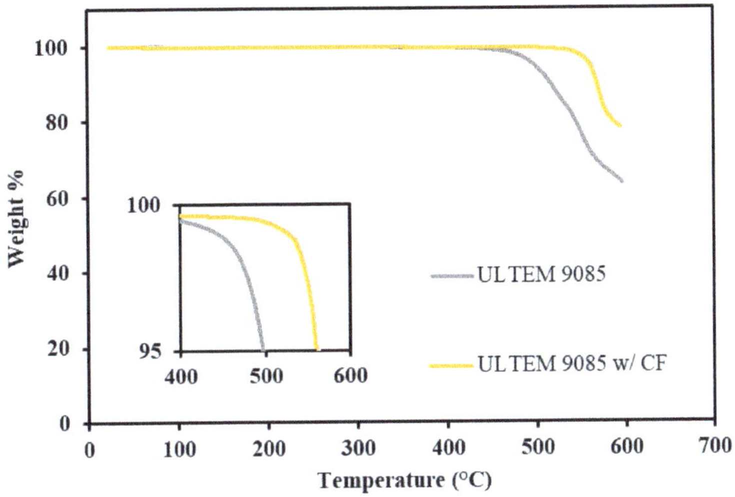

Figure 4. TGA results conducted in air with first 5% of weight loss magnified (inset).

ULTEM 9085 and ULTEM 9085CF exhibit a two-step thermal degradation which the two points of inflection can note at 570 °C and 580 °C, respectively. The two-step degradation can be explained by the compositional makeup of ULTEM 9085 materials, which is a blend of PEI and polycarbonate. According to a previous study by Feng et al., polycarbonate exhibits a single-step degradation completed at approximately 510 °C [32]. The first degradation step in ULTEM 9085 materials can, therefore, be credited to the degradation of the polycarbonate present in the material. The remaining polymer in the matrix is PEI, which will achieve complete degradation at approximately 700 °C.

ULTEM 9085CF maintains a higher remaining wt% by the end of the TGA test at 600 °C, specifically, there is 78 wt% remaining compared to 64 wt% for ULTEM 9085. The matrix of the ULTEM 9085CF materials should be functionally the same as the ULTEM 9085 material, as ULTEM 9085 is a material trademarked by SABIC. Therefore, the matrix of each material should behave the same during a TGA run. The higher 12 wt% present in the ULTEM 9085CF material can be attributed to the chopped carbon fibers added during the processing of the filament. ULTEM 9085CF, compared to ULTEM 9085, is not necessarily more stable at higher temperatures.

3.2. Microstructral Characterization

A micrograph of an entire through-thickness cross-section of an ULTEM 9085 part is shown in Figure 5. The print direction is perpendicular to the cross-section face (going into the image). As each line of the filament is printed, there is a separation of about 200 μm to account for the material expansion upon being heated during deposition. This part was intentionally printed with 14 layers to investigate the quality of each print layer in reference to its distance from the heated bed. Only the first layer exhibits enough lateral thermal expansion to account for the spacing between each pass. The second layer shows some lateral expansion, but every layer above the second shows little to no expansion. The ninth layer and above has such a slight expansion that it does not fully adhere to a second location on the previous layer and forms an empty crevice in the part that extends from the top of the eighth layer to the top surface. The poor print quality in the upper half of this part and to minimize effects of heat loss are two of the reasons why tensile specimens were made with only six print layers.

Figure 5. Micrograph of ULTEM 9085 polished cross-section showing all printed layers. Layers are labeled 1 through 14 with 1 being the bottom layer.

Micrographs of the cross-section of the multi-material ULTEM 9085 + ULTEM 9085CF parts are shown in Figure 6, with layering patterns ABABAB (Figure 6a,b) and AAABBB (Figure 6c,d). Part ABABAB exhibits inferior adhesion between layers of different materials, which in this case is every layer interface. Interestingly, the matrix of both materials is ULTEM 9085, but they are sourced from different vendors. However, it is likely that the porosity produced by the presence of chopped carbon fiber is partially causing poor adhesion between layers. The black spots in the image are pores, and the tiny white speckles are carbon fibers. Previous work has shown that the incorporation of chopped carbon fibers into a polymer-matrix led to a more porous filament which contributed to the poor adhesion between layers of a 3-D printed part, specifically in materials sourced from 3DXTech [31].

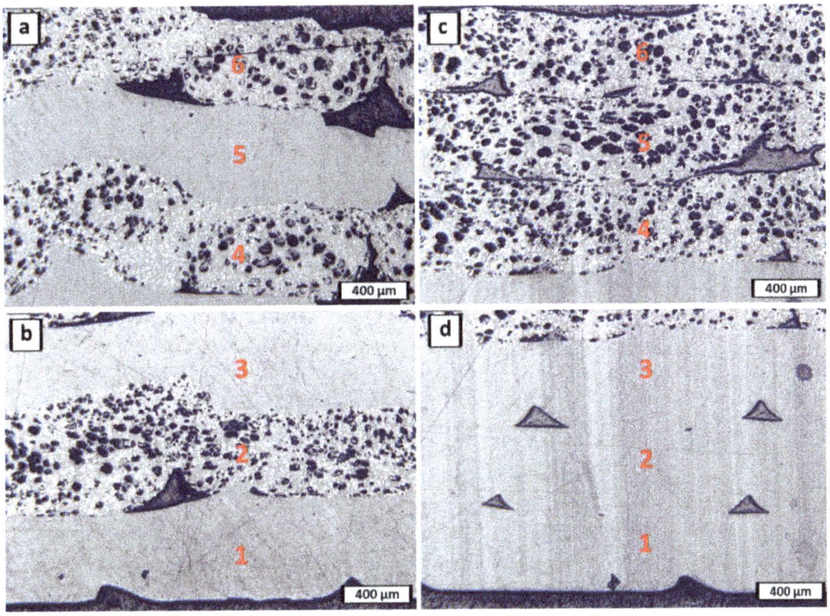

Figure 6. Micrographs of the top and bottom layers of ULTEM 9085 + ULTEM 9085 CF multi-material with layering pattern ABABAB (**a**,**b**) and AAABBB (**c**,**d**). Layers are labeled 1 through 6 with 1 being the bottom layer.

Using ImageJ, a similar porosity analysis was done, and the area % porosities of Figure 6a–d were found to be approximately 13%, 18%, 31%, and 2%, respectively. If the porosity is measured for the part's entire cross-section, then the area % porosities of ABABAB and AAABBB are approximately 16% and 29%, respectively. Interestingly, the layers of pure ULTEM 9085 in AAABBB exhibit less porosity than those in the ABABAB part. While samples printed to 14 layers, as shown in Figure 5, may experience a significant heat loss in the upper half of the part, the parts shown in Figure 6 are only six layers thick and are still well heated through the thickness. The porosity is not an effect of heat loss in the individual printed layers. However, the materials' heat transfer and conduction properties differ and can prevent proper adhesion or flow in the alternating layers. Another reason for the printed filament's poor adhesion could be the material's insufficient flow. The wetting behavior of the chopped carbon fibers in ULTEM 9085CF, and the fibers themselves, may impede proper flow of the polymer matrix if there is significant interlocking of fibers. In Part AAABBB, there is strong adhesion between the ULTEM 9085 layers. Consecutive layers are well bonded such that a layer interface cannot be identified except by the location of the few empty spaces. These patterned pores are developed between print lines of the same layer that are not printed close enough to each other and allow a gap. In Part AAABBB, the ULTEM 9085CF layers are not well-adhered to each other.

In all parts (Figures 5 and 6), the adhesion between layers worsens with each subsequent layer added. For example, the adhesion between layers 1 and 2 is much better than between layers 5 and 6. The material acts as an insulator between the print bed and the topmost layer. Thus, there is a lack of heat at the higher layers, preventing proper printing behavior of the high-temperature polymer. The print chamber was not at a sufficiently high enough temperature to prevent rapid cooling of the printed layers. Without proper heating capabilities, the surface of the top layer is not hot enough to adhere well to the subsequently printed layer. ULTEM 9085 is a material that must be printed at a high temperature to ensure proper adhesion. The high print quality of the consecutive ULTEM 9085 layers in Figure 6d show that the printing parameters are properly chosen for this polymer matrix, and the poor adhesion of ULTEM 9085CF layers are more attributed to its inherent porosity, differing heat transfer coefficients, and the wetting behavior of the chopped carbon fibers. The printed parts in this study are the best obtainable parts within the confides of the printer's capabilities. Porosity within parts containing ULTEM 9085CF is considered a design constraint, and not a function of improper printing techniques.

Some of the single and multi-material specimens after fracture are displayed in Figure 7. From a macroscopic view, the individual fractures look very similar, and differing details cannot be readily determined. However, a closer look at the fracture surfaces illuminates some key differences. 3-D scans of the fracture surfaces of the multi-material ULTEM 9085 + ULTEM 9085CF parts are shown in Figure 8. Figure 8a,b show the fracture surfaces of the ABABAB and AAABBB parts, respectively. In the multi-material parts, layers of ULTEM 9085 show a relatively flat fracture surface compared to the ULTEM 9085CF layers. In ULTEM 9085 layers, we see a mostly uniform color, showing that there is not much height change on the layer surface. The opposite is seen in the ULTEM 9085CF layers; the wide range of colors implies an uneven surface. The ULTEM 9085CF layers are jagged, regardless of the layering pattern. The difference in surface roughness can be explained by the significant porosity and the brittleness of the ULTEM 9085 CF layers. The porosity acts to deflect crack propagation but also increases the number of micro-cracks that are associated with the main crack [33]. A ductile material will deform in the presence of a crack and prevent spontaneous crack growth. In contrast, a brittle material cannot deform as much and allows spontaneous crack growth [34].

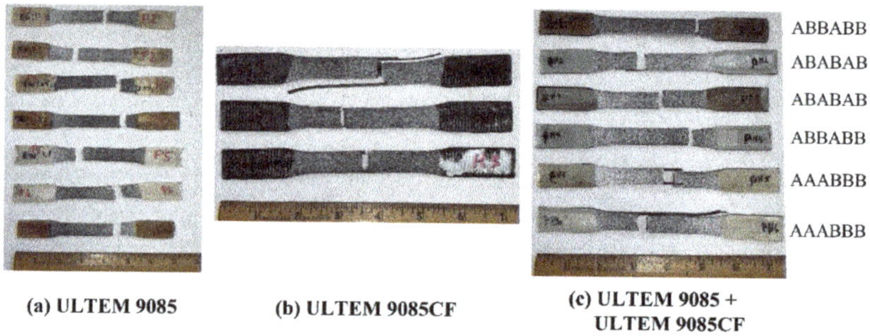

Figure 7. (**a**) ULTEM 9085, (**b**) ULTEM 9085CF, and (**c**) ULTEM 9085 + ULTEM 9085CF specimens after undergoing fracture during tensile testing.

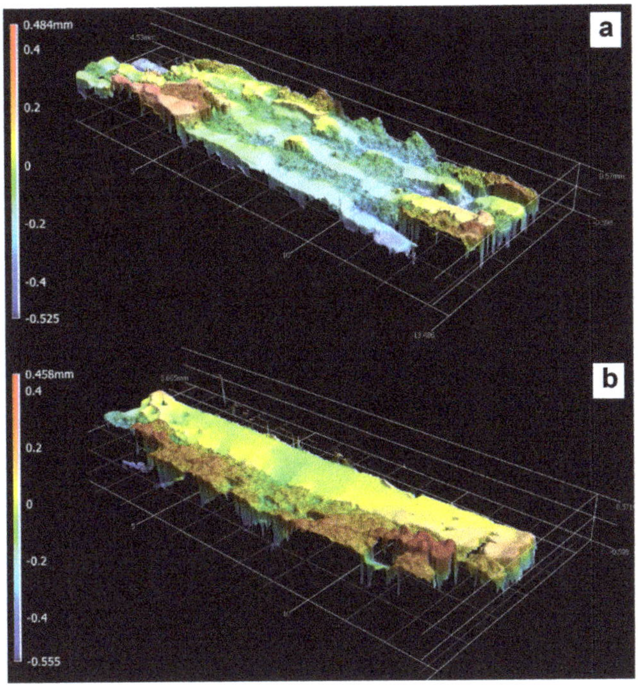

Figure 8. 3-D scans of fracture surfaces of multi-material ULTEM 9085 + ULTEM 9085 CF parts with layering patterns ABABAB (**a**) and AAABBB (**b**).

The rough fracture surface is also expected of brittle materials, whereas materials with a lower elastic modulus exhibit necking before fracture. Very ductile fractures usually form in a cone shape associated with the necking of a material, as is evident in the ULTEM 9085 part in Figure 9. The tougher ULTEM 9085 material allows for more plastic strain before failure. However, the brittle and stiff ULTEM 9085 CF layers in the multi-material parts prevented necking. The resulting fracture surfaces show smooth ULTEM 9085 layers and jagged ULTEM 9085CF layers.

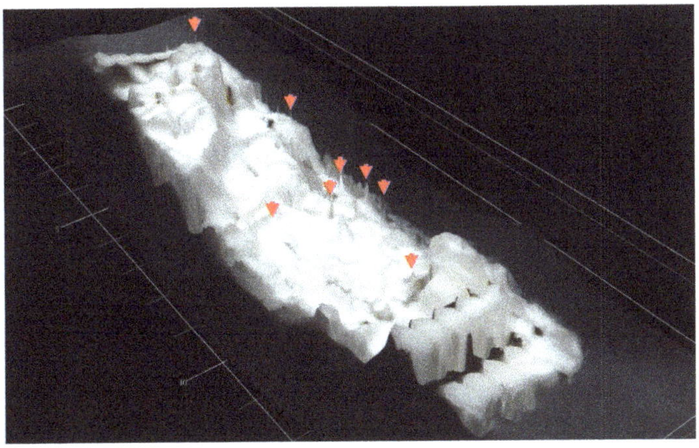

Figure 9. 3-D scan of fracture surface of single-material ULTEM 9085 part. Conical shapes formed by necking during plastic strain shown by red arrows.

3.3. Mechanical Properties

Stress-strain curves of the single material specimens are shown in Figure 10. The overall results for ULTEM 9085 parts (Figure 10a) follow expected trends for plastic materials, are reasonably consistent, and provide confidence in the reproducibility of the printed parts. Like most other polymeric materials, the strain-to-failure is slightly variable. One specimen showed significantly more necking than the others, and another achieved an abnormally high max stress. From looking at the individual dimensions and mass of each specimen, accessible in Table A1, the specimens that exhibited the large necking region and higher max stress had a significantly higher mass than the rest. The heavier specimens were 10.1 g compared to 9.7 g, the average of the other four. This difference in mass may explain what allowed such plastic strain or high max loading to occur before failure. However, the two abnormal specimens behave oppositely to each other while having similar masses. It is difficult to ascertain what caused the prolonged necking or loading, and it is possible that the specimens were incorrectly placed in the testing setup. The outliers will significantly affect the average max loading and ultimate tensile strength.

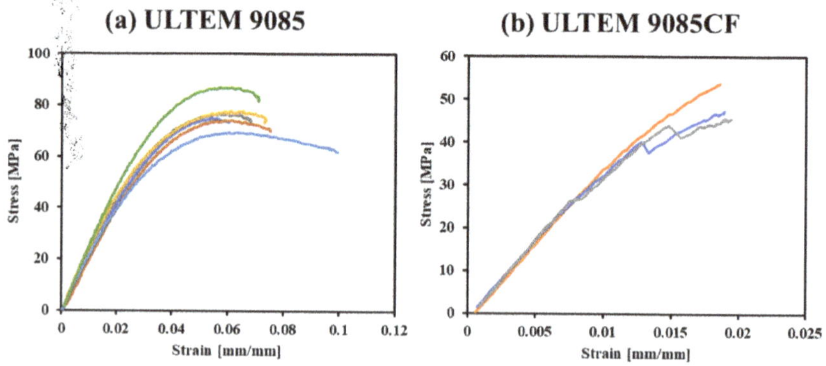

Figure 10. Tensile testing data for all single-material (**a**) ULTEM 9085 and (**b**) ULTEM 9085CF specimens.

Two specimens of ULTEM 9085CF (Figure 10b) underwent a discontinuity in the tensile response, which is common in some ductile materials. The material reaches an upper yield stress limit, undergoes strain before reaching a lower yield stress limit, and then experiences plastic strain. However, one of the three specimens shows a trend shared with brittle materials where the material undergoes little to no plastic strain before failure. The upper/lower yield stress limits and the little to no plastic strain before failure are characteristics of ductile and brittle materials, respectively, so ULTEM 9085CF can be characterized as a material type somewhere between brittle and ductile. From the micrographs of ULTEM 9085CF in Figure 6, it is clear there is significant porosity present in the material. Porosity often introduces variation in a material's mechanical behavior, which may be the reason for the variation in stress–strain trends exhibited by ULTEM 9085CF.

Stress–strain curves of the multi-material specimens are shown in Figure 11. Multiple runs of the same coupon type are displayed to show the slight variability between tests. Though there is some inconsistency between individual runs, they generally follow the same trend. The multi-material of AAABBB layering exhibits an upper and lower yield stress limit, similar to the ULTEM 9085CF parts. However, it also plastically strains past that yield stress, similar to the ULTEM 9085 parts. The remaining multi-material parts show brittle characteristics; there is no necking before failure and little plastic strain. Regardless of the layering pattern, all multi-material parts exhibit similar elastic moduli.

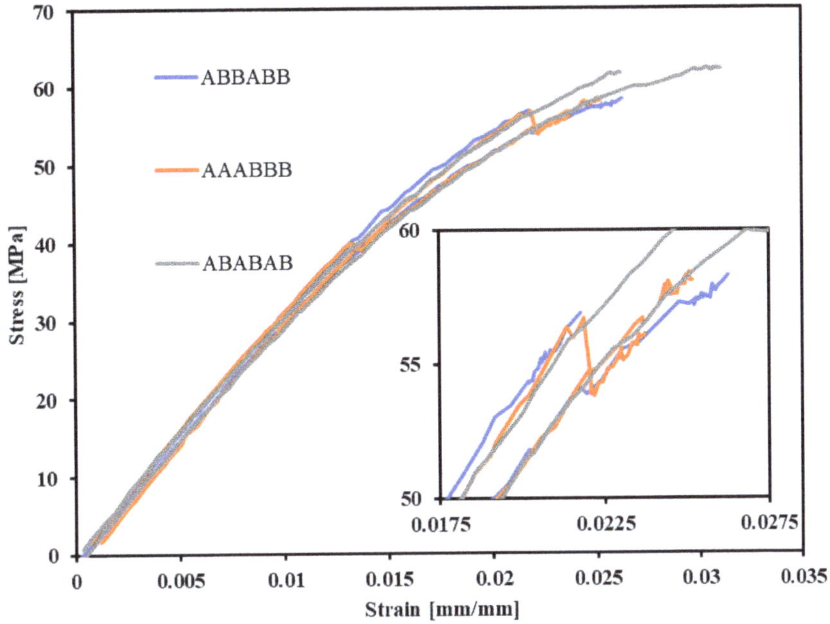

Figure 11. Tensile testing data for all multi-material A—ULTEM 9085 + B—ULTEM 9085CF specimens. Magnification of strain near fracture is shown in the inset.

The average elastic moduli and ultimate strengths of the singular-material ULTEM 9085 and the ULTEM 9085 + ULTEM 9085CF multi-material parts are shown in Figure 12 and mechanical properties of the specimens can be found in Table A1; the layering patterns differentiate the multi-material parts. The standard deviation represents how widely the values range from the mean. Specifically, it was calculated with the following formula:

$$\sqrt{\frac{\sum (x-\overline{x})^2}{n}}, \quad (1)$$

where $\overline{x}$ is the average value and n is the sample size. The large standard deviation in the mechanical properties of the pure ULTEM 9085 coupon (AAAAAA) is caused by an outlier coupon that exhibited an ultimate strength and elastic moduli of 98 MPa and 2.584 GPa, respectively. If that coupon is discounted, the average ultimate strength and elastic moduli of the pure ULTEM 9085 parts would be 75 MPa (±2.8 MPa) and 2.262 GPa (±0.103 GPa), respectively. However, the large deviation in the pure ULTEM 9085CF coupons (BBBBBB) should not be disregarded. Porosity in the as-received ULTEM 9085CF filament presents unpredictability and variability in the end print quality of the parts. The broader range of potential strength and modulus values is an important design aspect to consider when designing structural parts with this material.

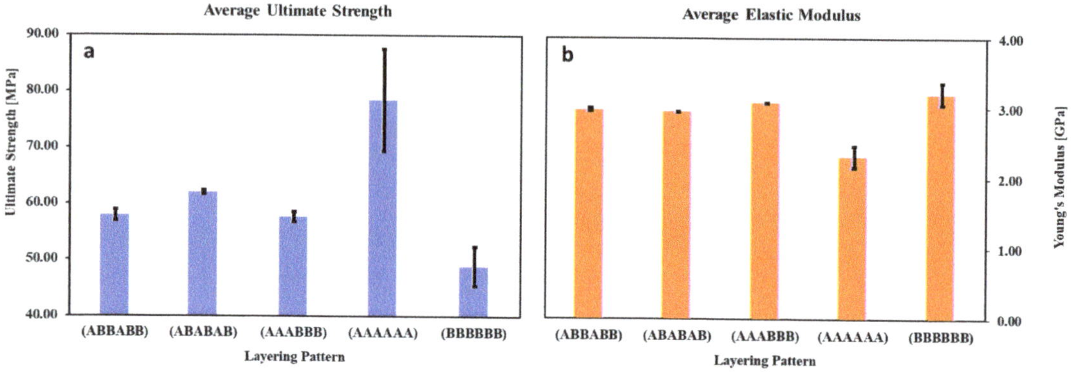

Figure 12. Average (**a**) ultimate tensile strengths and (**b**) elastic moduli of single and multi-material parts made from A—ULTEM 9085 + B—ULTEM 9085CF.

The ultimate strengths and elastic moduli observed in the ULTEM 9085 parts of this study are within the range of strengths observed in multiple studies [28,35–37]. As expected, the ULTEM 9085CF part is the stiffest (3.204 GPa), primarily due to the high modulus of chopped carbon fiber, while the ULTEM 9085 part is the most ductile. Similarly, ULTEM 9085 has the higher ultimate strength (78 MPa) and toughness, and ULTEM 9085CF has the lower (49 MPa). The low ultimate tensile strength of ULTEM 9085CF parts is partially due to the large porosity within the layers. There is potential for ULTEM 9085CF parts to be significantly more robust if the porosity can be minimized. ULTEM 9085 parts exhibited significantly more plastic strain before failure than the ULTEM 9085CF parts. All multi-material parts are in-between the two extremes in strength and stiffness (elastic modulus). The ULTEM 9085CF layers have a more significant influence on the stiffness of the multi-material part than the ULTEM 9085 layers. The multi-material parts exhibit an elastic modulus less than 300 MPa lower than the ULTEM 9085CF part but approximately 700 MPa higher than the ULTEM 9085 parts. In the case of ultimate strength, the multi-material parts exhibit strengths much closer in value to the ULTEM9085CF single-material specimens.

Interestingly, ABABAB patterned parts exhibit a lower elastic modulus but a higher ultimate strength than AAABBB patterned parts. AAABBB parts have three consecutive layers of a plastic material (ULTEM 9085) followed by three of a ductile-brittle material (ULTEM 9085CF). The three consecutive, porous ductile-brittle layers could allow crack propagation before necking begins with less resistance than the plastic ULTEM 9085 layers, which may cause the earlier failure. In comparison, the ABABAB parts show more plastic strain because the stiffer carbon-fiber-infused layers are interrupted by plastic ULTEM 9085 layers and prevent early tensile failure.

In order to compare the three material types, Figure 13 shows one stress–strain curve example of each type of specimen. ULTEM 9085, being a more plastic material, exhibits necking and a vast plastic strain region, whereas ULTEM 9085CF does not. The chopped carbon fiber and the porosity within the ULTEM 9085 matrix create a more brittle material with a higher elastic modulus. Similarly, the multi-material part does not exhibit much necking or a plastic strain region, which corroborates the conclusions made from the 3-D scans of fracture surfaces in Figure 8. The brittle carbon fiber layers of the multi-material are strong enough to prevent the necking usually exhibited by the ULTEM 9085 layers. The differences in elastic moduli and ultimate strengths are explained by a combination of the sequential order of the brittle and ductile layers and the significant presence of porosity.

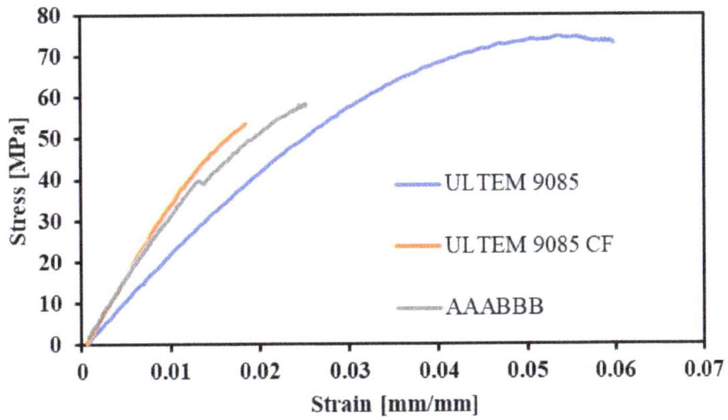

Figure 13. Stress–strain data of the single-material and multi-material (AAABBB) parts made from ULTEM 9085 and ULTEM 9085 CF.

4. Conclusions

In this study, multi-material parts were printed with ULTEM 9085 and ULTEM 9085CF. Upon studying the microstructure and the interface between consecutive layers, there is a clear difference in print quality between the first and last layers printed. ULTEM 9085 is appealing for its high thermal resistance. However, that same property is why heat does not efficiently conduct from the print bed to the highest print layers, resulting in poor application and adhesion of the material. A high-temperature material, such as ULTEM 9085, is problematic to easily 3-D print with the FFF process, and introducing a second material only complicates the optimization process.

Furthermore, ULTEM 9085CF has a large amount of apparent porosity: approximately 31% area compared to 2% area for pure ULTEM 9085 layers. This extensive porosity results in poor adhesion, independent of the consecutive layers' likeness or difference in materials, whereas successive layers of ULTEM 9085 adhere reasonably well until the printed layers are approximately 1.5 mm from the print bed. Print parameters can affect the quality of prints, but in this study, the print settings are reasonably chosen, as is shown by the high print quality of the pure ULTEM 9085 layers. However, the print chamber should be maintained at a high enough temperature to not allow significant cooling of the part during the printing process.

However, even with the print difficulties, it is evident that the order in which a multi-material part is printed, AAABBB vs. ABABAB, has a pronounced effect on the resulting mechanical properties. This work has provided insight into how carbon-fiber-infused layers prevent necking in printed ULTEM 9085 parts while also inducing a higher ultimate tensile strength. The toughness of the different materials plays a role in the fracture behavior of the tensile coupons. While a ULTEM 9085 part would experience prolonged plastic strain or necking at the fracture surface due to its high plasticity, the introduction of the brittle carbon-fiber-infused ULTEM 9085 layers prevents this necking. Instead, it exhibits behavior more consistent with a brittle fracture. The mechanical response is explained by the individual mechanical characteristics of the two different ULTEM 9085 materials. The mechanical behavior can be predicted with a strong understanding of how the different brittle and ductile materials behave under tensile stress. Multi-material parts with alternating ULTEM 9085 and ULTEM 9085 CF layers exhibit more strain before failure than those with consecutive ULTEM 9085 layers (0.033 vs. 0.025 mm/mm), most likely due to the toughness of the ULTEM 9085 layers allowing continued elongation. Less plastic strain is observed in AAABBB parts, presumably due to the brittle ULTEM 9085 CF layers causing an early brittle fracture in half of the specimen or the micro-cracks formed from the large porosity.

Further studies into multi-material printing, particularly with a high-temperature polymer such as polyetherimide, may add to the vast potential of additive manufacturing in the aeronautical and space industries. This study shows the potential for introducing multifunctionality, enhancing the already superb mechanical or thermal properties, and optimizing the printing parameters of promising part designs.

Author Contributions: Conceptualization, M.S. and M.C.H.; methodology and 3-D printing, M.S. and H.L.; mechanical testing and characterization, H.L. and V.S.V.; data analysis, V.S.V., M.C.H. and M.S.; writing—original draft preparation, V.S.V., M.C.H. and M.S.; writing—review and editing, V.S.V., M.S. and M.C.H.; supervision, M.C.H. and M.S.; funding acquisition M.C.H. All authors have read and agreed to the published version of the manuscript.

Funding: This research was supported by the Lewis' Educational and Research Collaborative Internship Project (LERCIP), the NASA Pathways Program, and the Revolutionary Vertical Lift Technology (RVLT) Project.

Institutional Review Board Statement: Not applicable.

Informed Consent Statement: Not applicable.

Data Availability Statement: Not applicable.

Acknowledgments: The authors would like to thank the NASA Internship Program, previously known as the Lewis' Educational and Research Collaborative Internship Program (LERCIP), for the support provided to Hunter Leonard and the NASA Pathways Program for the support provided to V.S.V.

Conflicts of Interest: The authors declare no conflict of interest. The funders had no role in the design of the study; in the collection, analyses, or interpretation of data; in the writing of the manuscript, or in the decision to publish the results.

Appendix A

Table A1. Details of all specimens used in this study.

Material	Layering	Width	Depth	Mass	Max Load	UTS	Elastic Modulus
		mm	mm	g	N	MPa	GPa
ULTEM 9085	AAAAAA	12.3	3.3	9.6	3012	75	2.250
ULTEM 9085	AAAAAA	12.6	3.4	9.7	3164	74	2.178
ULTEM 9085	AAAAAA	12.5	3.0	9.8	2903	77	2.360
ULTEM 9085	AAAAAA	12.7	2.9	9.6	2817	77	2.395
ULTEM 9085	AAAAAA	12.5	3.2	10.1	2974	70	2.125
ULTEM 9085	AAAAAA	12.5	2.8	10.1	3387	98	2.584
ULTEM 9085CF	BBBBBB	12.9	2.9	8.0	1795	47	3.227
ULTEM 9085CF	BBBBBB	12.8	2.9	7.9	1992	54	3.381
ULTEM 9085CF	BBBBBB	12.8	2.9	8.2	1688	46	3.005
ULTM 9085 + ULTEM 9085CF	ABBABB	13.1	3.5	9.7	2577	57	3.005
ULTM 9085 + ULTEM 9085CF	ABBABB	12.9	3.5	9.7	2631	59	2.960
ULTM 9085 + ULTEM 9085CF	ABABAB	13.0	3.3	9.8	2653	62	2.950
ULTM 9085 + ULTEM 9085CF	ABABAB	12.9	3.4	9.8	2679	62	2.956
ULTM 9085 + ULTEM 9085CF	AAABBB	13.0	3.0	8.9	2180	57	3.083
ULTM 9085 + ULTEM 9085CF	AAABBB	13.0	2.9	8.8	2164	58	3.077

References

1. ASTM International. *ASTM Committee F42 on Additive Manufacturing Technologies, Subcommittee F42. 91 on Terminology*; Standard Terminology for Additive Manufacturing Technologies; ASTM International: West Conshohocken, PA, USA, 2012.
2. Lin, W.; Shen, H.; Xu, G.; Zhang, L.; Fu, J.; Deng, X. Single-layer temperature-adjusting transition method to improve the bond strength of 3D-printed PCL/PLA parts. *Compos. Part A Appl. Sci.* **2018**, *115*, 22–30. [CrossRef]
3. Espalin, D.; Ramirez, J.A.; Medina, F.; Wicker, R. Multi-material, multi-technology FDM: Exploring build process variations. *Rapid Prototyp. J.* **2014**, *20*, 236–244. [CrossRef]
4. Baca, D.; Ahmed, R. The impact on the mechanical properties of multi-material polymers fabricated with a single mixing nozzle and multi-nozzle systems via fused deposition modeling. *Int. J. Adv. Manuf. Technol.* **2020**, *106*, 4509–4520. [CrossRef]

5. Rocha, V.G.; Saiz, E.; Tirichenko, I.S.; Garcia-Tunon, E. Direct ink writing advances in multi-material structures for a sustainable future. *J. Mater. Chem. A* **2020**, *8*, 15646. [CrossRef]
6. Rafiee, M.; Farahani, R.D.; Therriault, D. Multi-Material 3D and 4D Printing: A Survey. *Adv. Sci.* **2020**, *7*, 1902307. [CrossRef]
7. Zheng, Y.; Zhang, W.; Lopez, D.M.B.; Ahmad, R. Scientometric Analysis and Systematic Review of Multi-Material Additive Manufacturing of Polymers. *Polymers* **2021**, *13*, 1957. [CrossRef]
8. Bartlett, N.W.; Tolley, M.T.; Overvelde, J.T.B.; Weaver, J.C.; Mosadegh, B.; Bertoldi, K.; Whitesides, G.M.; Wood, R.J. A 3D-printed, functionally graded soft robot powered by combustion. *Science* **2015**, *349*, 161–165. [CrossRef]
9. Bandyopadhyay, A.; Heer, B. Additive manufacturing of multi-material structures. *Mater. Sci. Eng. R Rep.* **2018**, *129*, 1–16. [CrossRef]
10. Chen, D.; Zheng, X. Multi-material additive manufacturing of metamaterials with giant, tailorable negative Poisson's ratios. *Sci. Rep.* **2018**, *8*, 9319.
11. Bijadi, S.; De Brujin, E.; Tempelman, E.Y.; Oberdorf, J. Application of multi-material 3D printing for improved functionality and modularity of open source lowcost prosthetics—A case study. In *Frontiers in Biomedical Devices, Proceedings of the 2017 Design of Medical Devices Conference, Minneapolis, MN, USA, 10–13 April 2017*; ASME: New York, NY, USA, 2017; p. V001T10A003.
12. Singh, R.; Kumar, R.; Farina, I.; Colangelo, F.; Feo, L.; Fraternali, F. Multi-material additive manufacturing of sustainable innovative materials and structures. *Polymers* **2019**, *11*, 62. [CrossRef]
13. Vaezi, M.; Chianrabutra, S.; Mellor, B.; Yang, S. Multiple material additive manufacturing–Part 1: A review. *Virtual Phys. Prototyp.* **2013**, *8*, 19–50. [CrossRef]
14. Rutkowski, J.V.; Levin, B.C. Acrylonitrile–butadiene–styrene copolymers (ABS): Pyrolysis and combustion products and their toxicity—A review of the literature. *Fire Mater.* **1986**, *10*, 93–105. [CrossRef]
15. Singh, R.; Sandhu, G.S.; Penna, R.; Farina, I. Investigations for thermal and electrical conductivity of ABS-graphene blended prototypes. *Materials* **2017**, *10*, 881. [CrossRef] [PubMed]
16. Martin, O.; Averous, L. Poly (lactic acid): Plasticization and properties of biodegradable multiphase systems. *Polymers* **2001**, *42*, 6209–6219. [CrossRef]
17. Singh, D.; Singh, R.; Boparai, K.S.; Farina, I.; Feo, L.; Verma, A.K. In-vitro studies of SS 316 L biomedical implants prepared by FDM, vapor smoothing and investment casting. *Compos. Part B Eng.* **2018**, *132*, 107–114. [CrossRef]
18. Singh, R.; Kumar, R.; Ranjan, N. Sustainability of Recycled ABS and PA6 by Banana Fiber Reinforcement: Thermal, Mechanical and Morphological Properties. *J. Inst. Eng. (India) Ser. C* **2019**, *100*, 351–360. [CrossRef]
19. Singh, R.; Kumar, R.; Feo, L.; Fraternali, F. Friction welding of dissimilar plastic/polymer materials with metal powder reinforcement for engineering applications. *Compos. Part B Eng.* **2016**, *101*, 77–86. [CrossRef]
20. Singh, R.; Kumar, R.; Mascolo, I.; Modano, M. On the applicability of composite PA6-TiO2 filaments for the rapid prototyping of innovative materials and structures. *Compos. Part B Eng.* **2018**, *143*, 132–140. [CrossRef]
21. Singh, R.; Kumar, R.; Ranjan, N.; Penna, R.; Fraternali, F. On the recyclability of polyamide for sustainable composite structures in civil engineering. *Compos. Struct.* **2018**, *184*, 704–713. [CrossRef]
22. Vakharia, V.S.; Kuentz, L.; Salem, A.; Halbig, M.C.; Salem, J.A.; Singh, M. Additive Manufacturing and Characterization of Metal Particulate Reinforced Polylactic Acid (PLA) Polymer Composites. *Polymers* **2021**, *13*, 3545. [CrossRef]
23. Stratasys. Available online: http://www.stratasys.com/materials/fdm/ultem-9085 (accessed on 10 November 2022).
24. Gebisa, A.W.; Lemu, H.G. Investigating effects of Fused-Deposition Modeling (FDM) processing parameters on flexural properties of ULTEM 9085 using designed experiment. *Materials* **2018**, *11*, 500. [CrossRef] [PubMed]
25. Motaparti, K.P.; Taylor, G.; Leu, M.C.; Chandrashekhara, K.; Castle, J.; Matlack, M. Effects of build parameters on compression properties for ULTEM 9085 parts by Fused Deposition Modeling. In Proceedings of the 27th Annual International Solid Freeform Fabrication Symposium, Austin, TX, USA, 8–10 August 2016; pp. 964–977.
26. Motaparti, K.P.; Taylor, G.; Leu, M.C.; Chandrashekhara, K.; Castle, J.; Matlack, M. Experimental investigation of effects of build parameters on flexural properties in fused deposition modelling parts. *Virtual Phys. Prototyp.* **2017**, *12*, 207–220. [CrossRef]
27. Chuang, K.C.; Grady, J.E.; Draper, R.D.; Shin, E.-S.E.; Patternson, C.; Santelle, T.D. Additive manufacturing and characterization of ULTEM polymers and composites. In Proceedings of the CAMX—The Composites and Advanced Materials Expo, Dallas, TX, USA, 26–29 October 2015.
28. *ASTM D638-14*; Standard Test Method for Tensile Properties of Plastics. ASTM International: West Conshohocken, PA, USA, 2014.
29. Padovano, E.; Galfione, M.; Concialdi, P.; Lucco, G.; Badini, C. Mechanical and Thermal Behavior of ULTEM® 9085 Fabricated by Fused-Deposition Modeling. *Appl. Sci.* **2020**, *10*, 3170. [CrossRef]
30. Kafi, A.; Wu, H.; Langston, J.; Atak, O.; Kim, H.; Fahy, W.P.; Reber, R.; Misasi, J.; Bateman, S.; Koo, J.H. Evaluation of additively manufactured ultraperformance polymers to use as thermal protection systems for spacecraft. *J. App. Polym. Sci.* **2020**, *137*, 49117. [CrossRef]
31. Vakharia, V.S.; Singh, M.; Salem, A.; Halbig, M.C.; Salem, J.A. Effect of Reinforcements and 3-D Printing Parameters on the Microstructure and Mechanical Properties of Acrylonitrile Butadiene Styrene (ABS) Polymer Composites. *Polymers* **2022**, *14*, 2105. [CrossRef] [PubMed]
32. Feng, Y.; Wang, B.; Wang, F.; Zhao, Y.; Liu, C.; Chen, J.; Shen, C. Thermal degradation mechanism and kinetics of polycarbonate/silica nanocomposites. *Polym. Degrad. Stab.* **2014**, *107*, 129–138. [CrossRef]

33. Liu, D.; Šavija, B.; Smith, G.E.; Flewitt, P.E.J.; Lowe, T.; Schlangen, E. Towards understanding the influence of porosity on mechanical and fracture behaviour of quasi-brittle materials: Experiments and modelling. *Int. J. Fract.* **2017**, *205*, 57–72. [CrossRef] [PubMed]
34. Hayes, M.D.; Edwards, D.B.; Shah, A.R. *Fractography in Failure Analysis of Polymers*; Elsevier Inc.: Amsterdam, The Netherlands, 2015; p. 48.
35. Zaldivar, R.J.; Mclouth, T.D.; Ferrelli, G.L.; Patel, D.N.; Hopkins, A.R.; Witkin, D. Effect of initial filament moisture content on the microstructure and mechanical performance of ULTEM® 9085 3D printed parts. *Addit. Manuf.* **2018**, *24*, 457–466. [CrossRef]
36. Zaldivar, R.J.; Witkin, D.B.; Mclouth, T.; Patel, D.N.; Chmitt, K.; Nokes, J.P. Influence of processing and orientation print effects on the mechanical and thermal behavior of 3D-Printed ULTEM® 9085 Material. *Addit. Manuf.* **2017**, *13*, 71–80. [CrossRef]
37. Bagsik, A.; Schöppner, V.; Klemp, E. Long-term ageing effects on fused deposition modeling parts manufactured with ULTEM 9085. In Proceedings of the 23rd Annual International Solid Freeform Fabrication Symposium—An Additive Manufacturing Conference, Austin, TX, USA, 22 August 2012; pp. 629–640.

polymers

Article

Static Mixers Producible by Additive Manufacturing: Novel Rapid Automatic Optimisation and Practical Evaluation

Jana Sasse *, Malte Schön and Christian Hopmann

Institute for Plastics Processing (IKV) in Industry and Craft, RWTH Aachen University, Seffenter Weg 201, 52074 Aachen, Germany
* Correspondence: jana.sasse@ikv.rwth-aachen.de; Tel.: +49-241-80-27271

Abstract: In the extrusion of plastics, the thermal and material homogeneity of the plastic melt at the die entry are of high importance for the extrudate quality. While static mixers are widely used to improve the melt homogeneity, previous attempts at optimisation for reduced pressure loss and improved mixing had to be performed by hand and human experience, limiting the degrees of freedom and efficiency. A new automatic optimisation method based on the open source software OpenFOAM was developed. Using immersed boundary methods, new target functions in the pre-existing routine *adjointShapeOptimizationFoam* and an additional algorithm checking the suitability for additive manufacturing and fixing the geometry during run-time is presented. The new algorithm is used to optimise an existing static mixer based on an X-type geometry with integrated oil channels, maximising the heat exchange between oil and melt. Based on the results of these simulative optimisations, the best candidates were manufactured using selective laser melting and experimental trials were run. Experimental validation shows that with our optimisation algorithm, a pressure loss reduction of 10% could be achieved. The core melt temperature was reduced by 6 °C, improving the thermal homogenisation as well. While the main advantage of this method is the rapid optimisation taking the operating point into account, the trials also showed positive results in off-design operating points. This allows the low-cost design and manufacture of individualised static mixers.

Keywords: static mixer; extrusion; OpenFOAM; optimisation; additive manufacturing

check for updates

Citation: Sasse, J.; Schön, M.; Hopmann, C. Static Mixers Producible by Additive Manufacturing: Novel Rapid Automatic Optimisation and Practical Evaluation. *Polymers* **2022**, *14*, 4646. https://doi.org/10.3390/polym14214646

Academic Editors: Ying Yan and Yiqi Wang

Received: 26 September 2022
Accepted: 28 October 2022
Published: 1 November 2022

1. Introduction

In the extrusion of plastics, the thermal and material homogeneity of the plastic melt at the die entry are of high importance. Thermal inhomogeneities can be introduced, e.g., by high dissipation and lead to thickness variations over the outlet in the extrudate [1]. Material homogeneity on the other hand is vital in foam extrusion and other applications, where a uniform distribution of blowing agents, colourants, additives and processing agents is required. In order to avoid production scrap, static mixers are used to improve homogeneity by splitting and redistributing the melt flow. Static mixers are also used in other applications, such as the synthesis of polymers [2]. There are different types of static mixers available, and mixers of the X-type are widely used as their geometry provides numerous flow divisions and recombinations on a short length.

Most static mixers on the market are manufactured using conventional methods. On the other hand, additive manufacturing (AM), in particular selective laser melting (SLM), allows for new types of geometries for components of extrusion lines due to the added degrees of freedom [3]. In SLM, the geometry is printed in slices. First, a thin layer of metal powder is added on the substrate platform. Using a laser, the powder is melted in places and thus fused with the layers below. After cooling, the substrate platform is lowered, a new layer of metal powder is added and the next layer can be fused to the additively manufactured part. As the metal powder provides only limited support and heat extraction for upper layers, support structures must be added to the print in locations where critical overhangs are present [4]. When the print is finished, the support structures are removed and the print is post-processed (usually sandblasted), removing the remaining powder and improving the print's surface quality. Typical minimal layer thicknesses range from 20–60 μm, depending on the surface quality requirements and material [5]. The unfused metal powder can be retrieved and reused for future prints, making SLM a method of manufacture with little material waste. It is relatively affordable, fast and cost-efficient in small batch sizes, making it ideal for individualised applications [6].

Because SLM is fast and cheap, we can for the first time consider making "bespoke" individualised static mixers instead of a "one size fits all" solution. However, determining a suitable mixer geometry for each application in lab trials is prohibitively expensive. Simulations using Computational Fluid Dynamics (CFD) can be used to evaluate the mixing performance at far lower cost [7]. Different types of parameterised optimisation methods have been utilised for the optimisation of (dynamic) mixing elements. An investigation by Janßen et al. used an automated process to find the local optimum within a parameterised version of a dispersive mixer [8], while Hube et al. used a combination of free-form deformation and surface splines to parameterise and optimise rhomboid mixing elements, using a linear-elasticity-based mesh update method to deform the mixing element without remeshing [9]. Within the category of X-type static mixers, different configurations and designs are possible and subject to optimisation. Among the degrees of freedom of a static mixer are not only the number of mixing elements, but also the number of cross-bars over the width of the channel N_x, the number of parallel cross-bars per mixing element N_p and the angle between the bars θ. Previous research has shown that an optimal configuration for any X-type static mixer can be established via the design rule $N_p = (2/3)N_x - 1$ for $N_x = 3, 6, 9, \ldots$ [10,11]. It has also been shown that the X-type static mixer design is superior to most other static mixer designs, as it maximises mixing performance per pressure loss [12]. However, the design of static mixers is marked by the high amount of degrees of geometric freedom, as the parameterisation method described above only applies to symmetrical mixers with rectangular bars.

Apart from mixing, another goal can be the cooling of the melt. Conventional static mixers introduce considerable viscous dissipation as the melt is subjected to significant shear and elongation by the mixing bars. To counter-balance that, integrated oil channels enable simultaneous mixing and cooling of the melt, like in the P1 cooling mixer (Promix Solutions AG, Winterthur, Switzerland) [13].

This cooling transforms the mixing task from a somewhat isothermal flow problem to a heat transfer problem, often necessitating the use of simulation software. Past simulations of other types of static and dynamic mixers have been performed in the open source software OpenFOAM (OpenFOAM Foundation Ltd., London, UK) by Erb et al. [14], Alexias et al. [7,15] and Kettemann et al. [16]. In order to model material mixing, i.e., the transport of matter, 3D advection-diffusion equations can be used to model the transport of a passive scalar. While complex geometries such as the X-type static mixer are challenging even for state-of-the-art meshing softwares, the use of Immersed Boundary (IB) methods allows us to circumvent this problem. In IB methods, the whole computational domain is meshed as one domain, and subsequently, parts of the mesh are defined as belonging to the solid domain and made impassable to flow. This allows not only for a computationally cheap way of meshing the geometry, but also avoids re-meshing when the geometry changes. The OpenFOAM module *adjointShapeOptimizationFoam*, which was first introduced by Othmer et al. [17,18], takes advantage of this implementation. In this module, an adjoint optimisation algorithm using a gradient-based approach adjusts the boundaries between the fluid and solid domain by changing the extent of the solid domain according to local information on which parts of the domain are favourable or counterproductive for the overall target. While the original implementation was targeted towards minimal pressure loss of the fluid, other target functions are also possible. Previous research not published within the widely-shared OpenFOAM repositories includes the homogeneity of a passive scalar at the outlet [19,20] and also thermal optimisation models, e.g., with the goal of reaching a homogeneous temperature distribution or a specified target temperature [21–24]. While these optimisation algorithms yield good results, geometries that are compatible with additive manufacturing require additional checks and modifications [25].

In previous work at the Institute for Plastics Processing (IKV) at RWTH Aachen, an OpenFOAM simulation environment has been developed specifically for the simulation of plastic melt in extrusion processes. This includes the implementation of shear rate- and temperature-dependent material models, the integration of a 3D energy equation and a 3D advection-diffusion equation, as well as an immersed boundary method with material dependent specific heat capacities and diffusion coefficients [26–30]. This simulation environment has already been applied and validated in lab trials for the application and manual optimisation of static mixers [26–29]. In some of these applications, additive manufacturing has been successfully applied to introduce a higher degree of freedom [9,26,28,29]. In Figure 1 the resulting static mixer geometry of this research [26–28] is shown, where an X-type mixer has been modified to satisfy SLM manufacture requirements. In addition, three temperature control channels have been integrated. It has been shown that this static mixer geometry improves the ratio of thermal mixing and pressure loss by 50% and enables cooling of the melt by up to 10 °C, which is why we use it as the basis of the optimisation in this paper.

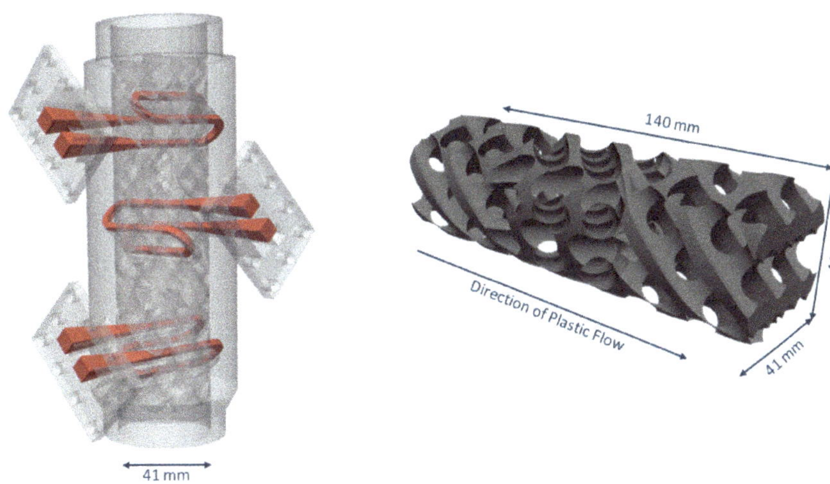

Figure 1. (**Left**) Static mixer with integrated temperature control channels. Adapted with permission from Ref. [28]. 2020, Institute for Plastics Processing (IKV) at RWTH Aachen. (**Right**) Static mixer suitable for SLM based on an X-type mixer geometry.

While the results shown above have been promising, the design and optimisation of this static mixer was largely crafted by hand and is therefore very inefficient, as it requires a large amount of skilled labour. The objective, then, is not only the automatic design of improved static mixers for plastics extrusion, but also the reduction of development time for new geometries. The goal of the research presented is the automatic optimisation of static mixers in OpenFOAM. Based on the module *adjointShapeOptimizationFoam*, new target functions are introduced specific to the optimisation of static mixers. In an additional step, the algorithm checks with each update of the mixer geometry whether the new geometry is additively manufacturable and automatically adjusts the geometry if necessary. This new algorithm is used to optimise a version of a static mixer with integrated oil channels for additional cooling using different target functions and oil temperatures, and the most promising results are manufactured using SLM and practically evaluated in lab trials.

2. Materials and Methods

First, a numerical model for the simulation and optimisation of the static mixers was developed. The presented model is specific to the needs of polymers, such as their high viscosity and strictly laminar flow regimes with according shear heating and simultaneous low heat capacity. Subsequently, optimisations were performed on pre-existing static mixer geometries which include integrated oil channels and influencing the optimisation routine. The best candidates were then additively manufactured using SLM and tested under lab conditions.

2.1. Numerical Model

While there are different types of IB methods available, one common method is to make the solid domain impassable to flow by introducing an additional term to the momentum equation, which includes an individual porosity α for each cell, which is set to zero for fluid domain cells and to a non-zero value for solid domain cells [17,31].

adjointShapeOptimizationFoam is based on a gradient-based optimisation algorithm using Lagrange multipliers [17]. In this algorithm, the state variables are solved in both a primal and an adjoint set of partial differential equations, the combination of which can subsequently be used to compute the sensitivities of the state variables $\mathbf{u}$, p and T. The topology of the static mixer modelled by the IB method is then optimised using the local sensitivities. As this method allows to compute the complete gradient using only two solver calls (primal and adjoint) and is independent of the search space dimension (i.e., the space spanned by the degrees of freedom), this algorithm is a computationally cheap way to obtain optimised static mixer topologies.

The flow of plastic melt in the static mixer is governed by the Navier-Stokes equations, which for this purpose have been extended by a 3D energy equation and a 3D advection-diffusion equation [30]. The set of primal equations of mass, momentum, energy and diffusion are therefore

$$\nabla \cdot \mathbf{u} = 0, \tag{1}$$

$$\nabla p_{norm} = \nabla \cdot (\mathbf{u} \times \mathbf{u}) - \nabla \cdot \tau_{norm} - \alpha \mathbf{u}, \tag{2}$$

$$0 = \nabla \cdot (\mathbf{u}T) + \nabla \cdot (D_T \nabla T) + \frac{\tau_{norm}}{c_p} : (\nabla \times \mathbf{u}), \tag{3}$$

$$0 = \nabla \cdot (\mathbf{u}c) + \nabla \cdot (D_c \nabla c), \tag{4}$$

where $\mathbf{u}$, p_{norm}, T and c denote the velocity, the pressure normalized to the density ρ, the temperature and the concentration, respectively. In addition, α is the individual porosity introduced by the immersed boundary method. D_T, D_c and c_p denote the local thermal diffusivity, the local diffusion coefficient and the local specific heat capacity, while τ_{norm} is the tensor of the shear stress normalized to the density ρ with ν denoting the kinematic viscosity:

$$\tau_{norm} = \nu((\nabla \times \mathbf{u}) + (\nabla \times \mathbf{u})^T). \tag{5}$$

The application at hand requires the distinction between three domains: plastic melt, solid steel and oil. Therefore, the material model was extended. Using the parameter α_{oil}, the viscosity was set individually for each cell. If a cell was flagged to contain plastic melt, the viscosity was modelled as a shear-thinning non-isothermal fluid using a Carreau-WLF model (see Equations (6) and (7)), while the viscosity for cells containing oil was set using a Newtonian model with a constant viscosity.

$$\nu = \frac{a_T \cdot A}{(1 + a_T \cdot B \cdot \dot{\gamma})^C} \tag{6}$$

$$\lg(a_T) = -\frac{8.86(T - T_s)}{101.6K + T - T_s} \tag{7}$$

The module *adjointShapeOptimizationFoam* was first described in [18]. It uses an immersed boundary method to optimise the shape of a flow canal for minimal total loss of pressure (PtLoss). This code was extended to also include an adjoint 3D energy equation. The resulting set of adjoint equations of mass, momentum and energy are as follows:

$$\nabla \cdot \mathbf{u}_a = 0, \tag{8}$$

$$\nabla p_{a,norm} = \nabla \cdot (\mathbf{u}_a \times \mathbf{u}_a) - \nabla \cdot \tau_{a,norm} - \alpha \mathbf{u}_a, \tag{9}$$

$$0 = \nabla \cdot (\mathbf{u} T_a) + \nabla \cdot (D_T \nabla T_a) + \frac{\tau_{a,norm}}{c_p} : (\nabla \times \mathbf{u}_a), \tag{10}$$

where the index a denotes the adjoint quantity and the adjoint shear stress tensor $\tau_{a,norm}$ is computed using the adjoint velocity and a viscosity calculated from an adjoint shear rate.

For the optimisation, the target functions for pressure loss and thermal homogenisation are incorporated in the outlet boundary conditions for pressure (Equation (11)), velocity (Equation (12)) and temperature (Equation (13)), where u_n and $\mathbf{u}_t$ refer to the normal and tangential velocity, respectively and T_{target} denotes the target temperature for the homogenisation. The weights w_p and w_T are used here to turn the cost functions on or off. In order to improve the numerical stability, the second order terms were neglected. For a full derivation of the equations we refer to [20,32].

$$p_{a,norm}\Big|_{Outlet} = \mathbf{u}\mathbf{u}_a + u_n u_{a,n} + T T_a + w_p \cdot ((-0.5\mathbf{u}^2) - u_n^2) \tag{11}$$

$$\mathbf{0} = u_{a,n}\mathbf{u}_t - u_{a,n}\mathbf{u}_{a,t} \tag{12}$$

$$0 = w_T \cdot |T - T_{target}| \tag{13}$$

After each iteration, the immersed boundary field is updated (Equation (14)) using a gradient-based approach, which uses the local sensitivities, where it is decided for each cell whether a fluid cell is favourable or counterproductive for the overall cost function. In this equation, α_0 and α_{max} are the individual porosities for fluid and solid, respectively, γ and λ are relaxation parameters, w_T is a weight turning thermal optimisation in the field update on or off and $\sigma_u = \mathbf{u} \cdot \mathbf{u}_a$ and $\sigma_T = T \cdot T_a$ denotes the local sensitivities with respect to velocity and temperature, respectively.

$$\alpha_{n+1} = \alpha_n(1-\gamma) + \gamma \cdot \min(\max(\alpha_n + \lambda_p \sigma_u + w_T \lambda_T \sigma_T, \alpha_0), \alpha_{max}) \tag{14}$$

After the immersed boundary field update an additional check is performed to ensure the geometry is still manufacturable using SLM. SLM has two main constraints [4,25,26]:

1. A cross section area of 1 mm^2 should be exceeded to prevent leakage and ensure mechanical stability.
2. A critical angle of 45° must be exceeded in every cell (see Figure 2 left). Otherwise support structures need to be added.

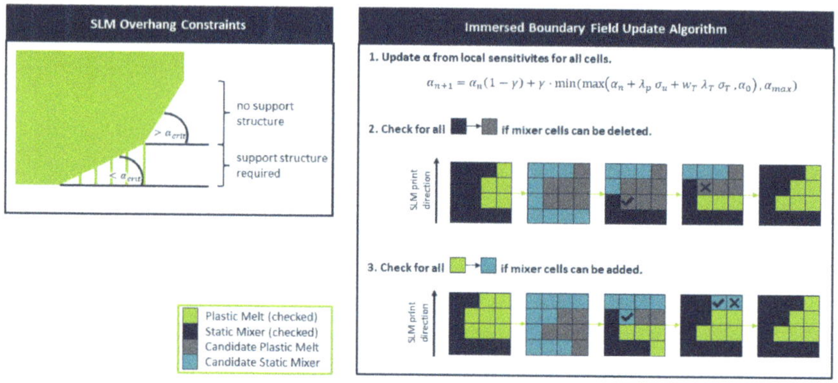

Figure 2. (**Left**) SLM overhang constraints regarding support structures. (**Right**) Schematic of the immersed boundary field update algorithm.

Constraint 1 is tackled with the introduction of an additional field $\alpha_{skeleton}$ which provides a minimal mixer structure that cannot be removed by the algorithm (see Figure 3). Cells that are flagged by this variable are automatically set to α_{max} and are exempt from all immersed boundary field updates. Similarly, all cells flagged as belonging to the oil channels by α_{oil} are exempt from the immersed boundary field update, ensuring that they are not removed in the process.

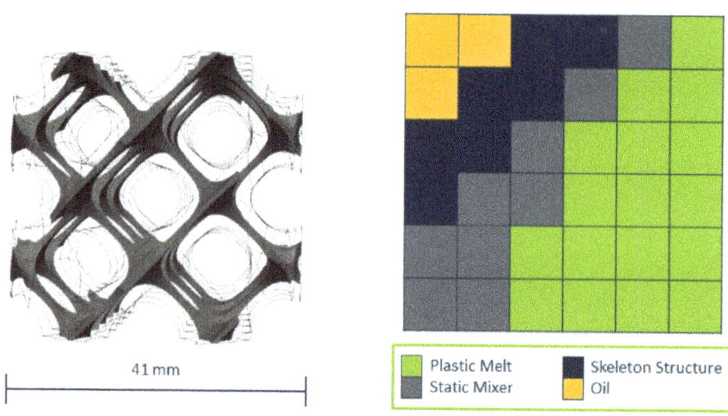

Figure 3. Immersed Boundary field with cells fixed as solid cells belonging to the static mixer or as fluid cells belonging to the oil domain.

For constraint 2, an additional algorithm is performed after each update (see Figure 2 right). All cells are checked individually in order of their print direction. If a cell changes from solid to fluid, it is checked whether there is a cell in the layer above that requires this cell for support, and if found, the cell has to stay solid. Similarly, if a cell changes its state from fluid to solid, it is checked to determine whether there is a sufficient support structure in the layer beneath, otherwise the change is rejected.

2.2. Simulative Optimisation

The reference geometry before optimisation is based on a SLM compatible version of an X-type mixer [26,29]. To maximize the contact area between the oil channels and the melt, a total of 8 oil channels were integrated, allowing for both longitudinal and radial heat exchange (see Figure 4 left).

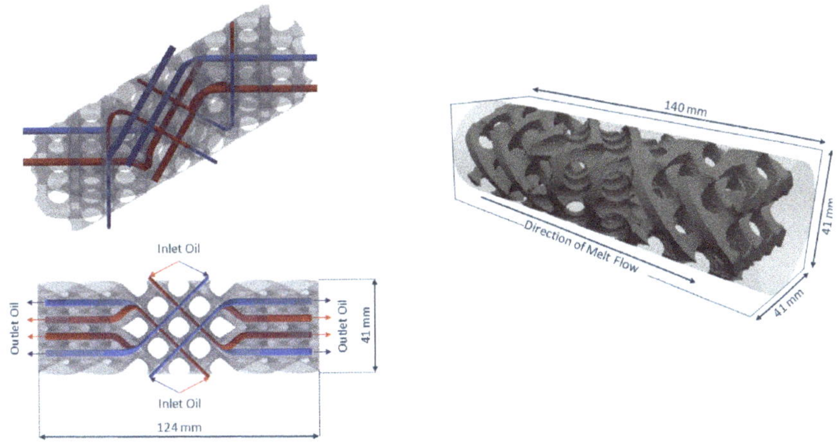

Figure 4. (**Left**) position and direction of flow of oil channels. (**Right**) Computational domain and reference geometry as immersed boundary field.

The whole computational domain (see Figure 4 right) consists of 3.26 million cells. The reference static mixer is comprised of 946,236 cells with 92,773 cells being flagged as $\alpha_{skeleton}$. The oil channels are modelled by 52,451 cells.

There are different types of thermal inhomogeneities that can typically occur in the extrusion process. One common temperature profile is caused by high dissipation in the extruder [1]; it is visualised in Figure 5 left and mathematically described in Equation (15), where r refers to the radius of the profile.

$$T(r) = (-1 \times 10^8 \times r^4 - 44430 \times r^2 + 31.615) \times 0.17990753 + 473.15 \quad (15)$$

As we intend to implement the experimental method described in [33], a non-uniform inlet concentration profile is used for validation puposes only. The concentration c in the advection-diffusion Equation (4) is used to model the transport of a passive scalar by setting the diffusion coefficient D_c to a very low value and prescribing a logistic function as a function of a coordinate perpendicular to the profile at the inlet (see Equation (16) and Figure 5 right):

$$c(z) = \frac{1}{1 + \exp(-2000 \cdot z)}. \quad (16)$$

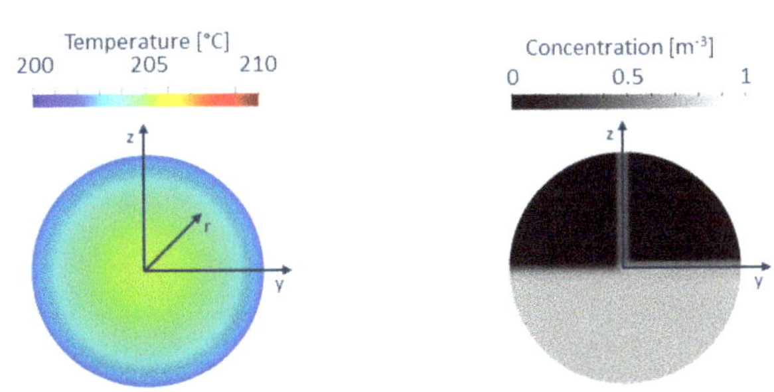

Figure 5. Temperature (**left**) and concentration (**right**) inlet profiles for simulations.

The remaining boundary conditions are displayed in Table 1. The u_{melt} was set to a value corresponding to a throughput of 100 kg/h and T_{melt} was set to 200 °C, while u_{oil} was set to a value corresponding to a throughput of 250 L/h for the whole oil circulation system. Previous research has shown that for a throughput of 100 kg/h the pressure loss is about 100 bar. As 20 bar pressure loss corresponds to approximately 1 °C increase in melt temperature and the best homogeneity is achieved if steel temperatures downstream of the screw tip follow this rule [34], the simulations were performed for two oil temperatures, one equal to and one 5 °C higher than the melt temperature.

Table 1. Boundary Conditions for CFD simulations.

	Inlet	Outlet	Wall	Inlet Oil	Outlet Oil
$\mathbf{u}$	u_{melt}	zeroGradient	noSlip	u_{oil}	zeroGradient
p_{norm}	zeroGradient	$1 \times 10^{-5}\,\frac{m^2}{s^2}$	zeroGradient	zeroGradient	$1 \times 10^{-5}\,\frac{m^2}{s^2}$
T	see Equation (15)	zeroGradient	T_{melt}	T_{oil}	zeroGradient
c	see Equation (16)	zeroGradient	zeroGradient	zeroGradient	zeroGradient
$\mathbf{u}_a$	$(-1\ 0\ 0)$	see Equation (12)	noSlip	zeroGradient	zeroGradient
$p_{a,norm}$	zeroGradient	see Equation (11)	zeroGradient	zeroGradient	$1 \times 10^{-5}\,\frac{m^2}{s^2}$
T_a	see Equation (15)	see Equation (13)	see Equation (13)	T_{oil}	zeroGradient

The three material domains are modelled as follows: The plastic melt is modelled as a non-isothermal shear-thinning fluid with a density $\rho = 736\,\text{kg/m}^3$, a specific heat capacity $c_p = 2900\,\text{m}^2/\text{s}^2\text{K}$, and a thermal diffusivity $D_T = 1.1997 \times 10^{-7}\,\text{m}^2/\text{s}$, with the Carreau-WLF parameters being A = 12.87 m²/s, B = 0.1871 s, C = 0.655 and $T_s = 237\,\text{K}$. The steel of the static mixer is modelled with a specific heat capacity of $c_p = 2900\,\text{m}^2/\text{s}^2\text{K}$ and a thermal diffusivity of $1 \times 10^{-5}\,\text{m}^2/\text{s}$. The oil is modelled with a specific heat capacity $c_p = 2220\,\text{m}^2/\text{s}^2\text{K}$, a thermal diffusivity $D_T = 5.415 \times 10^{-8}\,\text{m}^2/\text{s}$ and a constant viscosity $\nu = 0.0001\,\text{m}^2/\text{s}$.

First, a full simulation without optimisation was conducted as an initialisation. Subsequently, the optimisation was performed either for minimal pressure loss or for maximal thermal homogenisation. The resulting geometry was then simulated again using the original initial and boundary conditions to ensure that no artificial heating effects occurred in the optimisation process.

2.3. Validation Trials

The resulting geometries were post-processed in Meshmixer (Autodesk Inc., San Rafael, CA, USA), where minor mistakes in the geometry were fixed and the mixer geometry was merged with the outer mixer construction. Subsequently, the static mixers were manufactured using SLM. The resulting static mixers are displayed in Figure 6. The optimised static mixer on the right side differs from the reference static mixer on the left hand side by the decreased thickness of its 'bars', suggesting a reduction in pressure loss.

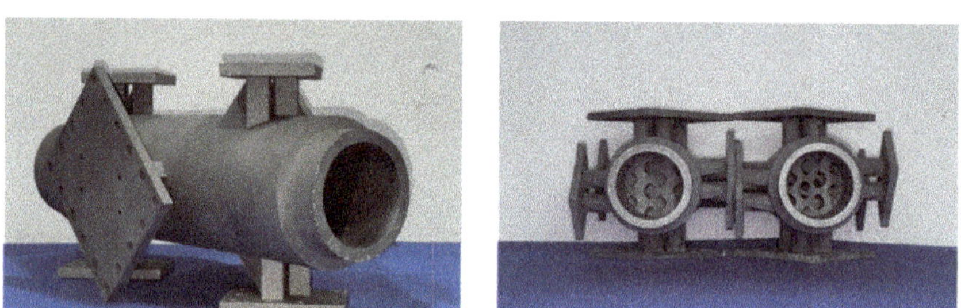

Figure 6. (**Left**) Manufactured static mixer with connectors for oil channels. (**Right**) Reference (**left**) and optimised (**right**) mixer geometries for validation trials.

To test the thermal as well as material mixing, a setup with two extruders was used. The validation trials were performed using a 60 mm main single-screw extruder and a

19 mm secondary single-screw extruder for the side feed. For an overall throughput of 20 kg/h, the rotational speed of the main extruder was set to 25 rpm, while the rotational speed of the side extruder was set to 20 rpm. Similarly, a throughput of 80 kg/h was achieved with rotational speeds of 99 rpm in the main extruder and 80 rpm in the side extruder. The material used was a blown film grade of HDPE (Hostalen GD 9550F, LyondellBasell GmbH, Wesseling, Germany), and the side extruder was fed with the same material together with 5% carbon black master batch material.

Both materials were processed at a nominal melt temperature of 220 °C, and the temperatures in the barrel zones were set up as a rising temperature profile (180 °C, 200 °C, 220 °C). All other melt-carrying parts of the setup had a nominal temperature of 220 °C controlled by heating bands. First, trials were run without additional cooling from the oil channels to test the performance of the static mixer alone. For the trials including the oil for additional cooling, the oil temperatures were varied between the nominal melt temperature (220 °C) and a temperature 5 °C higher than the nominal melt temperature (225 °C). Both upstream and downstream of the static mixer, a pressure sensor and an immersed temperature sensor were positioned to measure both the pressure loss of the static mixer and the radial temperature profiles before and after the mixer. The immersed temperature sensor measured the radial melt temperature at r = 0 mm, 7 mm and 14 mm, with r = 20 mm being the temperature at the barrel wall. For more details on the experimental setup, we refer the reader to [33].

For each operating point, 6 samples were taken with a 5 min difference between them to eliminate spontaneous fluctuations in the extrusion process.

2.4. Evaluation Criteria

In the simulations, the pressure loss (see Equation (17)) is evaluated using the average value at cut planes before and after the mixer as depicted in Figure 7 left.

$$\Delta p = p_{ave,before} - p_{ave,after} \tag{17}$$

The mixing characteristic numbers $e_{thermal}$ and e_{conc} are defined as in Equations (18) and (19), evaluating the relative change of the cumulative local inhomogeneities (see Figure 7 right). A value equal to zero means no change in the homogeneity and a value equal to one means an ideal maximal homogenisation.

$$e_{thermal} = \frac{\delta_{thermal,before} - \delta_{thermal,after}}{\delta_{thermal,before}} \quad \text{with } \delta_{thermal} = \sum_i |T_i - T_{ave}|, \tag{18}$$

$$e_{conc} = \frac{\delta_{conc,before} - \delta_{conc,after}}{\delta_{conc,before}} \quad \text{with } \delta_{conc} = \sum_i |c_i - c_{ave}|. \tag{19}$$

In the validation trials, the pressure loss is measured using a pressure sensor both before and after the static mixer. In addition, radial temperature profiles were measured both before and after the static mixer and the distribution of black colourant was evaluated as mean grey values μ_{grey}, where a high number is indicative of improved material mixing as a theoretical value of $\mu_{grey} = 255$ corresponds to a perfectly black extrudate.

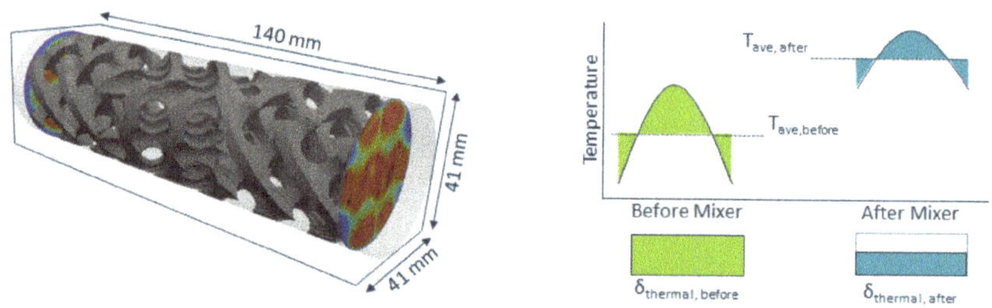

Figure 7. (**Left**) Positions of the cut planes relative to the static mixer for the evaluation of pressure loss. (**Right**) Graphical representation of $e_{thermal}$ and e_{conc} in simulations.

3. Results

First, the results of the numerical optimisation are presented and discussed. Subsequently, the results of the validation trials are shown.

3.1. Numerical Optimisation

Figure 8 depicts the pressure loss (left), thermal mixing (centre) and material mixing (right) for the two different oil temperatures. Both the optimisation strategy for minimal pressure loss as well as thermal mixing result in reduced pressure loss and improved thermal mixing relative to the reference geometry, while material mixing was barely affected. The best overall result was observed for the optimisation for minimal pressure loss with an oil temperature 5 °C above the melt temperature. In this scenario, material was removed from the mixing bars, therefore reducing the resistance to the flow as well as heating caused the additional shear. As a result, the pressure loss was reduced from 129 bar to 112 bar, while $e_{thermal}$ improved from 0.010 to 0.042 and e_{conc} barely changed (0.888 to 0.894).

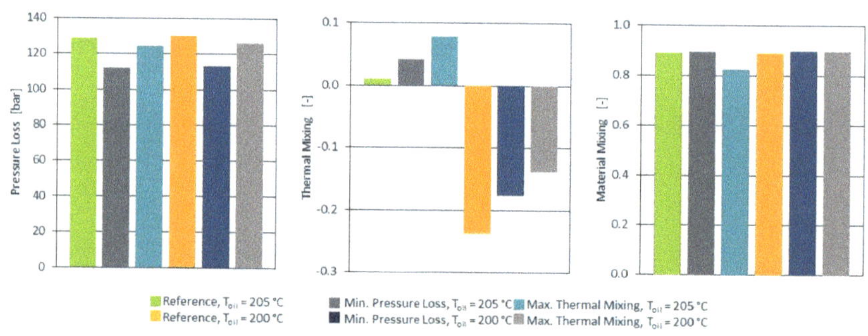

Figure 8. Pressure loss (**left**), thermal mixing (**centre**) and material mixing (**right**) for all simulations. Best results are seen for the higher oil temperature and optimisation for minimal pressure loss.

The temperature profiles before and after mixing for both the reference and the mixer geometry optimised for minimal pressure loss are displayed in Figure 9; the left side shows the the higher oil temperature and the right side shows the oil temperature at melt temperature. Looking at the temperature profiles in isolation, it appears that the characteristic number for thermal mixing $e_{thermal}$ decreased for the cooler oil temperature because the effect size of the cooling channels outweighs the effect size of the mixing bars. For the in-depth interpretation of this result, we have to employ the Graetz number. The Graetz number Gr defines the thermal regime in a domain, where small numbers (<1) indicate an equilibrium and the melt near the cylinder wall takes on the wall temperature and large numbers (>100) indicate that convection is the dominant thermal transport mechanism, the melt temperature increases linearly with the distance traveled and the wall temperature is irrelevant [35]. It is defined as

$$Gr = \frac{\mathbf{u} \cdot H^2}{D_T \cdot L}, \tag{20}$$

where H and L denote the height and length of the flow channel, respectively. For a throughput of 100 kg/h, a thermal diffusivity of 1.1997×10^{-7} m^2/s and a flow channel length of 120 mm and a distance between mixing bars of 10 mm, the Graetz number is greater than 200. A different way to visualise this is seen in Figure 10, where the temperature profile (scaled to [0, 1]) at the outlet has been multiplied by the local velocities (scaled to [0, 1]), thus indicating enthalpy convection downstream. It is apparent that, at these high throughputs, the residence time of the melt in the mixer is not long enough for significant heat exchange to occur.

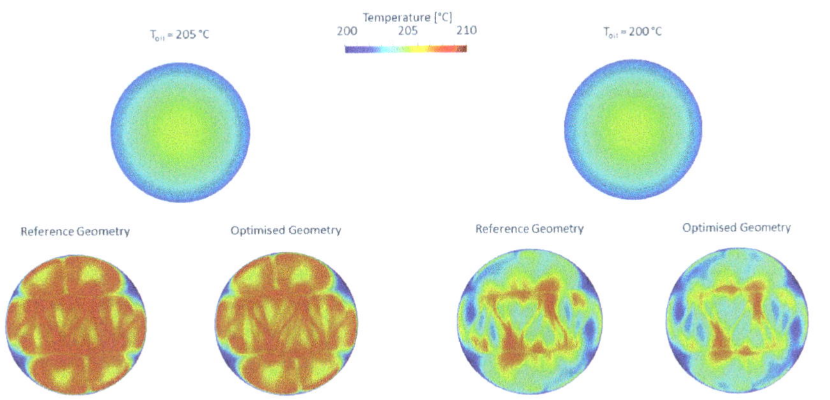

Figure 9. Thermal mixing for reference geometry and geometry optimised for pressure loss for an oil temperature 5 °C above melt temperature (**left**) and at melt temperature (**right**).

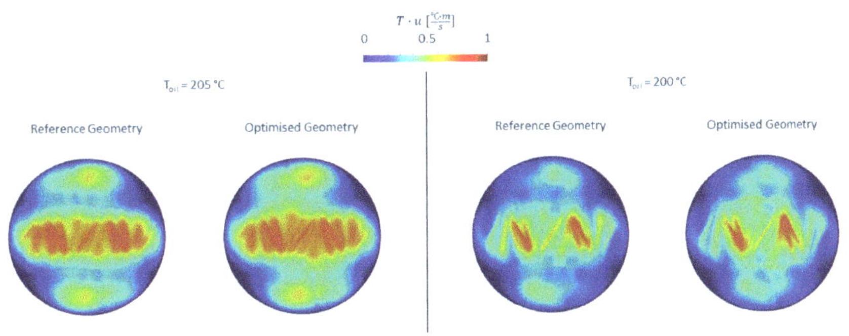

Figure 10. Temperature profile behind the mixer scaled for velocity for reference geometry and geometry optimised for pressure loss for an oil temperature 5 °C above melt temperature (**left**) and at melt temperature (**right**).

During the optimisation for either minimal pressure loss or maximal thermal mixing, the algorithm in both cases removed material from the mixer. To optimise for minimal pressure loss, the algorithm removed steel mainly towards the centre of the flow channel, where the impact of the thus reduced shear rate and flow resistance is the highest. This is consistent with previous findings, where optimisations of X-type mixers found a positive correlation between the number of mixing bars (i.e., reduction in total cross-section available for melt flow) and the pressure loss [10,36]. The results are also in agreement with [25] with respect to the changes to the geometry under the constraint of additive manufacturability. In addition, the thermal mixing improved, partly due to the reduction in the core temperature due to reduced viscous dissipation along the mixing bars [28], although these changes are small compared to the fluctuations and measurement uncertainties in the extrusion process. For thermal mixing on the other hand, the material removal was less concentrated in one location and a lot of material was removed closer to the cylinder wall as well. The removal of material here also explains the reduction in pressure loss as a secondary observation. Overall, the algorithm removed 6.7% of the static mixer geometry for minimal pressure loss and 12.3% for maximal thermal homogenisation. While the optimisation for thermal mixing was numerically successful, a closer investigation of the changes in the geometry found causes for concern regarding mechanical failure and leakage in locations not preemptively secured by the skeleton structure. For example, in some locations the mixer geometry surrounding a skeleton structure that was protecting an oil channel was completely removed and while this geometry is technically possible, the risk of failure was determined to be too high.

Based on these results, the reference geometry as well as the geometry obtained by optimisation for minimal pressure loss with an oil temperature 5 °C above the melt temperature (hereinafter referred to as *optimised*) were chosen for manufacture and experimental evaluation.

3.2. Validation Trials

The extruder setup used led to an increased temperature profile (see Figure 11 left), where the core of the melt reached temperatures as high as 250 °C. While both static mixers managed to reduce the core temperature and improve overall homogeneity, the maximum core temperature for the optimised mixer was approximately 6 °C lower compared to the reference mixer due to the decreased shear heating at the mixing bars. The pressure loss (see Figure 11 right) was reduced by 10% for the throughput of 80$\frac{kg}{h}$ (104.6 bar to 94.9 bar) and by 15% for 20$\frac{kg}{h}$ (36.45 bar to 31.3 bar).

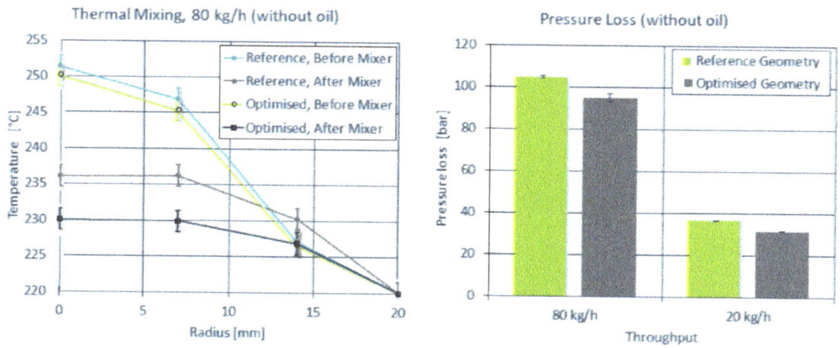

Figure 11. Thermal homogenisation and pressure loss of the reference and optimised mixer geometries. Note that the error bars in the temperature curves refer to the uncertainties in the temperature sensors as they outweigh the standard deviations in the recorded data.

For a direct comparison between simulation and experiment, the initial and boundary conditions of the extruder were recreated for OpenFOAM, corresponding to a melt temperature of 220 °C with a T_{max} = 250 °C. The temperature profiles measured in the lab trials are in acceptable agreement with the temperature profiles obtained by means of simulation (see Figure 12). While the simulation tends to overestimate the cooling in the reference mixer geometry, it correctly predicts the improved homogeneity for the optimised mixer compared to the reference mixer.

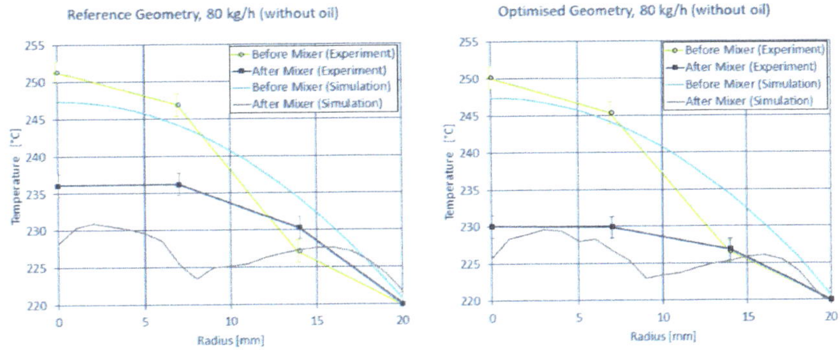

Figure 12. Simulated and measured temperature curves before and after the static mixer for the reference (**left**) and optimised (**right**) geometry.

During the trials, extrudate samples were taken, prepared and polished and subsequently photographed. An analysis of the colour distribution using ImageJ [37] and a *t*-test were performed to determine the statistical significance of the data. Two exemplary samples as well as the statistical analysis are shown in Figure 13. The *t*-test did not reject the null-hypothesis, suggesting no significant changes in the distributive mixing between the mixer geometries.

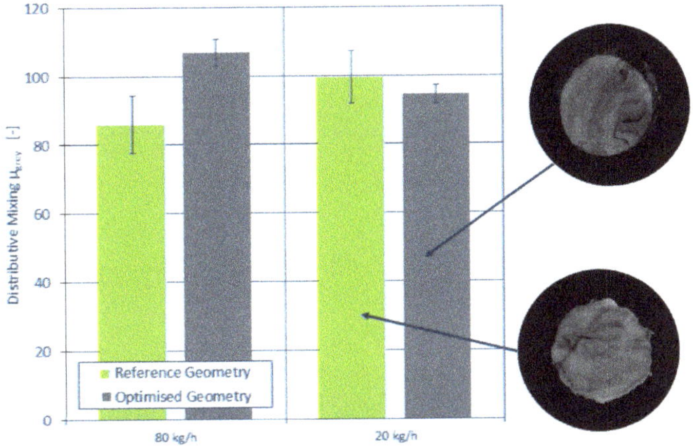

Figure 13. Distributive mixing capabilities of the reference and optimised mixer. High values of μ_{grey} are indicative of improved mixing.

In Table 2, the results of the simulation as it was performed during the optimisation, the simulation using the initial and boundary conditions as observed during the lab trials and the data measured during the lab trials are compared side by side. Both of the simulations come to similar conclusions regarding the 13% reduction in pressure loss and 3 °C reduction in maximum temperature at the outlet compared to their reference counterpart. While the lab trials measured a lower pressure loss reduction of only 9% and a larger temperature reduction of 6 °C, these results are still in acceptable agreement. The material mixing was not affected in either of the cases.

Table 2. Comparison between the results obtained by simulation and lab trials. Changes labeled as *Reduction* refer to the change with respect to their counterpart before the numerical optimisation.

	Simulative Optimisation	Validation (CFD)	Validation (Lab Trials)
T_{melt}	200 °C	220 °C	220 °C
$T_{max,inlet}$	205 °C	248 °C	250 °C
Reduction Δp	13% at 100 $\frac{kg}{h}$	13% at 80 $\frac{kg}{h}$	9% at 80 $\frac{kg}{h}$
Reduction $T_{max,outlet}$	3 °C	3 °C	6 °C
Material mixing	almost identical across reference and optimised geometries		

In addition, trials were run using the oil channels for additional cooling of the melt, once at the nominal melt temperature and once at a temperature 5 °C higher than the nominal melt temperature. The results are depicted in Figure 14. As it was the case with the numerical results, the influence of the oil channels on the temperature curves was negligible at this high throughput. Compared to the trials run without the active cooling by the oil, the difference between the reference and optimised geometry was reduced by 1–2 °C, putting the results closer to the results obtained by means of simulation.

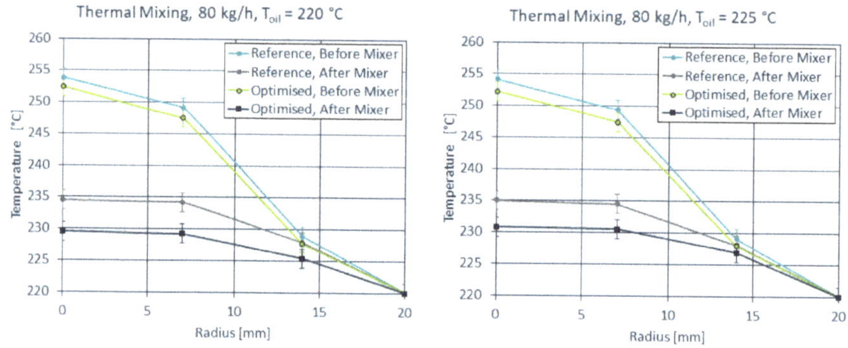

Figure 14. Temperature profiles of the reference and optimised geometries for both oil temperatures.

4. Discussion

The mixer geometries obtained by the automatic optimisation algorithm fulfilled the SLM constraints and required only minimal post-processing and general user input to obtain a printable and usable static mixer geometry. While the approach taken is quite similar to the approaches taken by Langelaar et al. [25,38,39], there are a few key differences. In their research, the focus is on structural optimisation and their code is also able to insert support structures and determine the best printing direction [39]. The algorithm presented here is not as sophisticated, but has its own advantages. First of all, because the additive manufacturability is checked in each iteration, the computational cost has to be very low. In addition, the features described by Langelaar et al. are not required for this specific use case, as the printing direction can be easily determined as the direction of extrusion and the static mixer geometry has to be manufacturable without additional support structures, as the main reason to use AM in this context is the integration of the oil channels in locations that are not accessible after printing (e.g., to remove those support structures). In this regard, the strategy for the optimisation of static mixers is closer to the approach taken by Leary et al. [40], where the added support structures are an inherent part of the final geometry. However, the algorithm presented still requires improvement in future research. The update of the immersed boundary field as described in Equation (14) is highly sensitive to the relaxation parameters, which also need to be modified if the throughput changes or different optimisation targets are used at the same time.

Both the optimisation strategy for minimal pressure loss and maximal thermal homogeneity were able to minimise their target functions. Furthermore, the other quality criteria (pressure loss, thermal mixing, material mixing) also improved even if they were not the focus of the optimisation strategy itself. While the optimisation for minimal pressure loss also led to a secondary improvement in the thermal mixing, the optimisation for maximal thermal mixing also resulted in a slight reduction in pressure loss. The latter is to be expected, as the algorithm has a tendency to remove more cells than it adds, and a reduction of the mixing bar size is associated with a decrease in flow resistance, resulting in reduced pressure loss. Another reason lies in the way the target function for thermal homogeneity (Equation (13)) is formulated. In the current form, the individual optimisation target of each cell at the outlet is to get as close as possible to the target temperature, which is set at the desired melt temperature. As a result, the algorithm is incentivised to minimise the shear heating in the static mixer, which is achieved by a reduction of the mixing bar size. Unfortunately, the current algorithm only checks for local manufacturability. It does not automatically account for mechanical stability as this is handled via the skeleton structure. The optimisation for thermal mixing has shown that the skeleton structure used in this case was not sufficient in order to ensure a manufacturable geometry. In addition, the optimised geometry also worked for off-design throughputs, as shown in the validation trials. A different application for this algorithm can be the optimisation of the overall length of the mixer, which has been shown to be another significant factor [10,28].

The experimental results in the validation trials confirm the results obtained by means of CFD. The simulations tended to underestimate the cooling effect of the static mixer (3 °C compared to the 4–6 °C measured in the validation trials, depending on the oil temperature). Even without the utilisation of the temperature control channels, the static mixers were able to both cool and homogenise the plastic melt. These results should be seen in the context of the previous research published in [26–28], where the performance of the SLM-compatible static mixer was compared against a commercially available static mixer. They found that their SLM-compatible static mixer geometry improves the ratio of thermal mixing and pressure loss by 50% and enables cooling of the meld by up to 10 °C. The initial geometry in the optimisation presented here was the result of that research, which implies that the method presented in this paper enables further optimisation compared to commercially available static mixers. Unfortunately, due to the low residence time at high throughputs, the thermal homogeneity could not be further improved by the oil channels. The temperature profiles after the mixer were the same regardless of the oil temperature, when accounting for measurement uncertainties. Future research should therefore focus on optimisation specifically for operating points at lower throughputs, where the chances for active cooling with the oil channels are improved. The material mixing did not fundamentally change between both mixer geometries, which is to be expected as the geometries were not different enough to change the distributive mixing behaviour.

One of the main advantages of the method presented is the reduced time (computational as well as skilled labour) for this method. While conventional optimisation methods need to run and analyse a minimum number of simulations and subsequently need to manually adjust the geometry and then re-simulate (or manufacture and validate in trials), and even automatic genetic optimisation algorithms need to evaluate a large amount of design points [3,8] the method presented only requires two to three simulations in total, i.e., initialisation, optimisation and an optional re-initialisation for evaluation purposes, resulting in less than 36 h of total run time. In addition, a lot less user input is required between the individual simulation runs, freeing the person performing the optimisation up to complete other tasks in the meantime. Therefore, automatic optimisation using Lagrange multipliers is a lot more efficient than manual and even simulation-assisted optimisation.

Future research should focus on the further development of the optimisation and the immersed boundary update algorithm. As the current IB update routine tends to favour the removal of solid cells instead of the addition of new solid cells, the extent to which new kinds of geometries can be created is very limited. In addition, a version where the skeleton structure can be moved to modify the position of the oil channels could enable the design of completely new structures. This future algorithm might take inspiration from [7], where a parameterised version of a static mixer modified the mixing geometry outside of a rigid geometry range. Also, an extension that checks the suitability for manufacture with conventional methods such as machining, assembly from sheets or casting instead of AM would be a useful addition and would broaden the field of applications for this method. A development of a user interface could improve industry adaptation of this method.

In the method presented, the initial geometry needs to already be suitable for AM, as the geometry would otherwise be dramatically altered to fit SLM constraints in the first iteration. To counteract that, coupling the algorithm to a routine that helps to improve the geometry's suitability for AM prior to optimisation, e.g., by adding transitional elements to horizontal bars to make them manufacturable will expand the fields of application for this optimisation method.

5. Conclusions

While static mixers are widely used for the thermal and material homogenisation of melt in extrusion, previous attempts at optimisation for reduced pressure loss and improved mixing had to be performed by hand and human experience. A new automatic optimisation method based on the open source software OpenFOAM was developed. With the utilisation of immersed boundary methods and the implementation of new target functions for improved homogeneity in the pre-existing routine, *adjointShapeOptimizationFoam*, along with an additional algorithm checking the suitability for additive manufacturing and adjusting the geometry during run-time, was presented.

The new algorithm was used to optimise an existing static mixer based on an X-type geometry with integrated oil channels maximising the heat exchange between oil and melt, enabling simultaneous cooling and mixing. Based on the results of these simulative optimisations, the best candidates were manufactured using SLM and experimental trials were run. Experimental validation has shown that with this optimisation algorithm, a pressure loss reduction of 10–15% was achieved and the core melt temperature was reduced by another 6 °C compared to the reference, improving the thermal homogenisation as well. These results are in good agreement with the results predicted by the simulations. While the main advantage of this method is the rapid optimisation taking the operating point into account, the trials also showed positive results in off-design operating points. This allows the low-cost design and manufacture of individualised static mixers.

Author Contributions: Conceptualization, M.S. and C.H.; methodology, M.S. and J.S.; software, J.S.; validation, J.S.; formal analysis, J.S.; investigation, J.S.; resources, C.H.; data curation, J.S.; writing—original draft preparation, J.S.; writing—review and editing, M.S. and C.H.; visualization, J.S.; supervision, M.S.; project administration, M.S.; funding acquisition, C.H. All authors have read and agreed to the published version of the manuscript.

Funding: The research project 21298 N of the Forschungsvereinigung Kunststoffverarbeitung has been sponsored as part of the "Industrielle Gemeinschaftsforschung und -entwicklung (IGF)" by the German Bundesministerium für Wirtschaft und Klimaschutz (BMWK) due to an enactment of the German Bundestag through the AiF. We would like to extend our thanks to all organizations mentioned.

Institutional Review Board Statement: Not applicable.

Informed Consent Statement: Not applicable.

Data Availability Statement: Data available on request.

Acknowledgments: The authors acknowledge the generous donation of raw material by Basell Sales & Marketing Company B.V. and would like to extend their thanks to these parties.

Conflicts of Interest: The authors declare no conflict of interest.

Abbreviations

The following abbreviations are used in this manuscript:

AM	Additive Manufacturing
CFD	Computational Fluid Dynamics
IB	Immersed Boundary
IKV	Institute for Plastics Processing
SLM	Selective Laser Melting
WLF	Williams, Landel and Ferry

References

1. Catherine, O. Effect of Typical Melt Temperature Non-Uniformity on Flow Distribution in Flat Dies. In Proceedings of the 70th Annual Technical Conference of the Society of Plastic Engineers (ANTEC), Orlando, FL, USA, 2–4 April 2012.
2. Gardiner, J.; Nguyen, X.; Genet, C.; Horne, M.D.; Hornung, C.H.; Tsanaktsidis, J. Catalytic Static Mixers for the Continuous Flow Hydrogenation of a Key Intermediate of Linezolid (Zyvox). *Org. Process. Res. Dev.* **2018**, *22*, 1448–1452. [CrossRef]
3. Neubrech, B. Thermo-Rheologische Optimierung Additiv Gefertigter Extrusionswerkzeuge. Ph.D. Thesis, Universität Duisburg-Essen, Duisburg, Germany, 2021. [CrossRef]
4. Lippert, R.B. Restriktionsgerechtes Gestalten Gewichtsoptimierter Strukturbauteile für das Selektive Laserstrahlschmelzen. Ph.D. Thesis, Gottfried Wilhelm Leibniz Universität Hannover, Hannover, Germany, 2018. [CrossRef]
5. Gebhardt, A. *Generative Fertigungsverfahren. Additive Manufacturing und 3D Drucken für Prototyping—Tooling—Produktion*; Carl Hanser Verlag: Munich, Germany, 2017.
6. Schleifenbaum, J. Verfahren und Maschine zur Individualisierten Produktion mit High Power Selective Laser Melting. Ph.D. Thesis, RWTH Aachen University, Aachen, Germany, 2012.
7. Alexias, P.; Giannakoglou, K.C. Shape Optimization of a Two-Fluid Mixing Device Using Continuous Adjoint. *Fluids* **2020**, *5*, 11. [CrossRef]
8. Janßen, M.; Schulz, L.; Vorjohann, F.; Schiffers, R. Automatisierte Optimierung von dynamischen Mischelementen. *Z. Kunststofftechnik/J. Plast. Technol.* **2022**, *18*, 24–61. [CrossRef]
9. Hube, S.; Behr, M.; Elgeti, S.; Schön, M.; Sasse, J.; Hopmann, C. Numerical design of distributive mixing elements. *Finite Elem. Anal. Des.* **2022**, *204*, 103733. [CrossRef]
10. Singh, M.K. Design, Analysis, and Optimization of Distributive Mixing: With Applications to Micro and Industrial Flow Devices. Ph.D. Thesis, Technische Universiteit Eindhoven: Eindhoven, The Netherlands, 2008. [CrossRef]
11. Singh, M.K.; Anderson, P.D.; Meijer, H. Understanding and Optimizing the SMX Static Mixer. *Macromol. Rapid Commun.* **2009**, *30*, 362–736. [CrossRef] [PubMed]
12. Meijer, H.E.; Singh, M.K.; Anderson, P.D. On the performance of static mixers: A quantitative comparison. *Prog. Polym. Sci.* **2012**, *37*, 1333–1349. [CrossRef]
13. Heusser, R. Cool gemischt: Intensives statisches Mischen mit effektivem Wärmeaustausch verbinden. *Kunststoffe* **2018**, *108*, 65–67.
14. Erb, T.; Geiger, K.; Bonten, C. Mixing processes—Influence of the viscosity model on flow calculations. *AIP Conf. Proc.* **2014**, *549*, 549–555. [CrossRef]
15. Alexias, P.; Giannakoglou, K.C. Optimization of a static mixing device using the continuous adjoint to a two-phase mixing model. *Optim. Eng.* **2020**, *21*, 631–650. [CrossRef]
16. Kettemann, J.; Bonten, C. Simulation of A Saxton-Mixer in High-Performance Extruders Using the Immersed Boundary Method. In Proceedings of the 78th Annual Technical Conference of the Society of Plastic Engineers (ANTEC)—Digital Edition, Online, 30 March–5 May 2020; pp. 851–857.
17. Othmer, C.; de Villiers, E.; Weller, H. Implementation of a Continuous Adjoint for Topology Optimization of Ducted Flows. In Proceedings of the 18th AIAA Computational Fluid Dynamics Conference, Miami, FL, USA, 25–28 June 2007; p. 77. [CrossRef]
18. Othmer, C. A continuous adjoint formulation for the computation of topological and surface sensitivities of ducted flows. *Int. J. Numer. Methods Fluids* **2008**, *58*, 861–877. [CrossRef]
19. Helgason, E.; Krajnovic, S. Implementation of an Adjoint-Based Optimization with Scalar Transport. In Proceedings of the ASME 2014 International Mechanical Engineering Congress and Exposition (IMECE2014), Montreal, QC, Canada, 14–20 November 2014. [CrossRef]
20. Helgason, E. Development of Adjoint-Based Optimization Methods for Ducted Flows in Vehicles. Ph.D. Thesis, Chalmers University of Technology, Göteborg, Sweden, 2015.
21. Mosca, R. *Implementation of New Boundary Conditions for External Flow Adjoint-Based Shape Optimization*; Student Work; Chalmers University of Technology: Göteborg, Sweden, 2020.
22. Mosca, R. Adjoint-Based Passive Optimization of a Micro T-Mixer. Ph.D. Thesis, Politecnico di Milano, Mailand, Italy, 2017.
23. Qian, X.; Dede, M.E. Topology optimization of a coupled thermal-fluid system under a tangential thermal gradient constraint. *Struct. Multidiscip. Optim.* **2016**, *54*, 531–551. [CrossRef]
24. Ruberto, E. An Adjoint Based Topology Optimization for Flows Including Heat Transfer. Ph.D. Thesis, Politecnico di Milano, Mailand, Italy, 2017.

25. Verboom, J. Design and Additive Manufacturing of Manifolds for Navier-Stokes Flow: A Topology Optimisation Approach. Master's Thesis, Delft University of Technology, Delft, The Netherlands, 2017.
26. Hopmann, C.; Schön, M.; Theunissen, M.; Meiners, W.; Shen, L. Simulative evaluation of the optimization potential of additively manufactured static mixing elements for extrusion. In Proceedings of the PROCEEDINGS OF PPS-33: The 33rd International Conference of the Polymer Processing Society, Cancun, Mexico, 10–14 December 2019; p. 020004. [CrossRef]
27. Hopmann, C.; Schön, M. Evaluation of Mesh Interface and Immersed Boundary Models for the Optimisation of Mixing Elements. In Proceedings of the 78th Annual Technical Conference of the Society of Plastic Engineers (ANTEC)—Digital Edition, Online, 30 March–5 May 2020.
28. Hopmann, C.; Schön, M.; Backmann, M.; Petzinka, F.; Leuchtenberger, L. Simulative optimisation of mixing and die technology in extrusion. In Proceedings of the 30th International Colloquium Plastics Technology 2020, Online, 11–12 March 2020; pp. 93–138.
29. Hopmann, C.; Sasse, J. Automatische Optimierung von additiv gefertigten statischen Mischern. *Blasformen Extrus.* **2021**, *17*, 5–7.
30. Hopmann, C.; Leuchtenberger, L.; Schön, M.; Wallhorn, L. Effect of Local Heat Pipe Cooling on Throughput Distribution and Thermal Homogeneity in a Binary Melt Pre-Distributor for Polyolefin Extrusion. *Polymers* **2022**, *14*, 2271. [CrossRef] [PubMed]
31. Khadra, K.; Angot, P.; Parneix, S.; Caltagirone, J.P. Fictitious domain approach for numerical modelling of Navier-Stokes equations. *Int. J. Numer. Methods Fluids* **2000**, *34*, 651–684. [CrossRef]
32. Hinterberger, C.; Olesen, M. Industrial application of continuous adjoint flow solvers for the optimization of automotive exhaust systems. In Proceedings of the CFD & Optimization 2011-069 an Eccomas Thematic Conference, Antalya, Turkey, 23–25 May 2011.
33. Hopmann, C.; Schön, M.; Reul, M.M.; Facklam, M. A Method for the Validation of Simulated Mixing Characteristics of Two Dynamic Mixers in Single-Screw Extrusion. *Polymers* **2020**, *12*, 2234. [CrossRef] [PubMed]
34. Hopmann, C.; Michaeli, W. *Extrusion Dies for Plastics and Rubber: Design and Engineering Computations*, 4th ed.; Hanser: München, Germany; Wien, Austria, 2016. [CrossRef]
35. Köpplmayr, T. Computational and Experimental Modelling of Microlayer Forming Processes. Ph.D. Thesis, Johannes Kepler Universtität Linz, Linz, Austria, 2014.
36. Liu, S.; Hrymak, A.N.; Wood, P.E. Laminar mixing of shear thinning fluids in a SMX static mixer. *Chem. Eng. Sci.* **2006**, *61*, 1753–1759. [CrossRef]
37. Schneider, C.; Rasband, W.; Eliceiri, K. NIH Image to ImageJ: 25 years of image analysis. *Nat. Methods* **2012**, *9*, 671–675. [CrossRef] [PubMed]
38. Langelaar, M. An additive manufacturing filter for topology optimization of print-ready designs. *Struct. Multidiscip. Optim.* **2017**, *55*, 871–883. [CrossRef]
39. Langelaar, M. Combined optimization of part topology, support structure layout and build orientation for additive manufacturing. *Struct. Multidiscip. Optim.* **2018**, *57*, 1985–2004. [CrossRef]
40. Leary, M.; Merli, L.; Torti, F.; Mazur, M.; Brandt, M. Optimal topology for additive manufacture: A method for enabling additive manufacture of support-free optimal structures. *Mater. Des.* **2014**, *63*, 678–690. [CrossRef]

polymers

Article

Experimental Investigation and Optimization of Turning Polymers Using RSM, GA, Hybrid FFD-GA, and MOGA Methods

Abdulrahman I. Alateyah [1,*], Yasmine El-Taybany [2], Samar El-Sanabary [2], Waleed H. El-Garaihy [1,3,*] and Hanan Kouta [2]

1 Department of Mechanical Engineering, College of Engineering, Qassim University, Unaizah 56452, Saudi Arabia
2 Department of Production Engineering and Mechanical Design, Port Said University, Port Fouad 42526, Egypt
3 Mechanical Engineering Department, Faculty of Engineering, Suez Canal University, Ismailia 41522, Egypt
* Correspondence: a.alateyah@qu.edu.sa (A.I.A.); w.nasr@qu.edu.sa (W.H.E.-G.); Tel.: +966-55-313-3322 (A.I.A.)

Abstract: The machining of polymers has become widely common in several components of industry 4.0 technology, i.e., mechanical and structural components and chemical and medical instruments, due to their unique characteristics such as: being strong and light-weight with high stiffness, chemical resistance, and heat and electricity insolation. Along with their properties, there is a need to attain a higher quality surface finish of machined parts. Therefore, this research concerns an experimental and analytical study dealing with the effect of process parameters on process performance during the turning two different types of polymers: high-density polyethylene (HDPE) and unreinforced polyamide (PA6). Firstly, the machining output responses (surface roughness (Ra), material removal rate (MRR), and chip formation (λc)) are experimentally investigated by varying cutting speed (v_c), feed rate (f), and depth of cut (d) using the full factorial design of experiments (FFD). The second step concerns the statistical analysis of the input parameters' effect on the output responses based on the analysis of variance and 3D response surface plots. The last step is the application of the RSM desirability function, genetic algorithm (GA), and hybrid FFD-GA techniques to determine the optimum cutting conditions of each output response. The lowest surface roughness for HDPE was obtained at v_c = 50 m/min, f = 0.01 mm/rev, and d = 1.47 mm and for PA6 it was obtained at v_c = 50 m/min, f = 0.01 mm/rev, and d = 1 mm. The highest material removal rate was obtained at v_c = 150 m/min, f = 0.01 mm/rev, and d = 1.5 mm for both materials. At f = 0.01 mm/rev, d = 1.5 mm, and v_c = 100 for HDPE, and v_c = 77 m/min for PA6, the largest chip thickness ratios were obtained. Finally, the multi-objective genetic algorithm (MOGA) methodology was used and compared.

Keywords: polymers; turning; surface roughness; MRR; chip formation; ANOVA; optimization; RSM; GA; MOGA

Citation: Alateyah, A.I.; El-Taybany, Y.; El-Sanabary, S.; El-Garaihy, W.H.; Kouta, H. Experimental Investigation and Optimization of Turning Polymers Using RSM, GA, Hybrid FFD-GA, and MOGA Methods. *Polymers* **2022**, *14*, 3585. https://doi.org/10.3390/polym14173585

Academic Editors: Ying Yan and Yiqi Wang

Received: 1 August 2022
Accepted: 26 August 2022
Published: 30 August 2022

1. Introduction

Nowadays, the demand for polymers in several industries and manufacturing areas, especially industry 4.0 technology, has increased owing to their unique physical, mechanical, and thermal properties. High specific strength and modulus, good damping properties, low density and weight, good corrosion resistance, mass production possibility, low friction coefficient, good thermal/electrical insulation, and the ability to be processed without external lubrication are the main advantages of polymers over metallic materials. As a result of these exceptional properties, polymers are used in numerous applications such as gears, bearings, rolling elements, structural components, and chemical and medical instruments [1–4]. However, polymers are considered as difficult-to-cut materials as they have few limitations over metals and alloys, such as the melting point of polymers being comparatively low, which leads to the easy softening of polymers when machining due to heat generated in the working area. Therefore, the applications of polymers at high working temperatures is not preferred [5]. In addition, the thermal expansion of polymers is ten times more than metals, which is one of the limitations that need to be considered in some applications [6].

Generally, primary injection molding processes i.e., injection, blow, compression, transfer molding, etc., are used for the manufacturing of large-scale polymeric products [7]. However, these methods can be somewhat restrictive and do not produce good dimensional accuracy or surface finish requirements. Thus, machining and post-processing are preferred in these cases [8–10]. High dimensional accuracy and superior surface quality are vital requirements of polymers products in precision machinery, electronics, medical, and optics applications. To attain these characteristics, polymers need to undergo machining operations such as turning, milling, and drilling [5–7].

Therefore, the machining of polymers has become the focus of interest for many researchers and many attempts have been made for the sake of understanding the polymers' performance and gaining knowledge of their machinability characteristics as well as obtaining the best conditions during machining operations using different statistical analysis tools.

Correa et al. [11] analyzed the effect of cutting speed, feed rate, and tool tip angle on the holes' dimensional deviations, circularity error, fiber pull out, and burr formation during the drilling of polyamide reinforced with 25% glass fiber. Kannan et al. [12] investigated the hole quality and thrust force during drilling of carbon fiber reinforced polymers (CFRPs). Gaitonde et al. [13] examined the surface roughness in the high-speed drilling of unreinforced polyamide (PA6) and reinforced polyamide with 30% glass fibers (PA66 GF30) by developing response surface methodology. Solymani et al. [14] conducted numerous drilling experiments on PA6 nanocomposites to obtain the thrust force values, used a particle-swarm-optimization-based neural network (PSONN) to create a predictive model, and compared the results with that of the conventional neural network.

To understand the micro-machinability of polyamide-6 (PA6) and glass-fiber-reinforced polyamide 6 (GFPA6), Kuram et al. [15] investigated the influences of various spindle speeds and feed rates on the cutting force, surface roughness, burr formation, and tool wear during the micro-milling of these materials. Other different kinds of polymers such as polymethylmethacrylate (PMMA), polyetheretherke tone (PEEK), and polyimide (PI) were experimentally tested in terms of surface roughness, burrs, and cutting chip characteristics in the high-speed micro-milling process [16]. From the micro-machinability point of view, better results were obtained from unreinforced PA6 compared to GFPA 6. Surface roughness was experimentally investigated and optimized and a predictive model based on an artificial neural network (ANN) was developed during the milling of (PA-6) nanocomposites [17] and slot milling of polypropylene materials [18]. In addition, the response surface methodology (RSM) technique was used to evaluate the effect of machining parameters on surface roughness during the milling of polyetheretherketones (PEEK) implant material [19]. Based on their optimization results, a better surface finish was obtained at the lowest level of feed rate. Verma et al. [20] applied a hybrid method during the milling of epoxy nanocomposites reinforced by graphene oxide/carbon fiber (G/CF) to optimize the material removal rate (MRR), cutting force (Fc), and surface roughness (Ra).

Other attempts were made to develop a model for surface roughness prediction and optimization when turning PA6 by applying RSM and analysis of variance (ANOVA) to the experimental data [21], as well as an ANN [22], and the fuzzy logic technique along with the Taguchi's orthogonal array [2]. Machinability aspects in terms of cutting force, power, and specific cutting pressure, when turning unreinforced (PA6) and reinforced (PA66 GF30) polyamides [23] and PEEK, unreinforced and reinforced (GF30), ref [24] using the RSM-based parametric analysis results, revealed that cutting force and power increased with cutting conditions, whereas specific cutting pressure decreased by increasing the feed rate. In the same context, Vaxev et al. [25] studied the influence of cutting speed and feed rate on cutting force components during the turning of PA66 GF30 by application of ANOVA and ANN. Aldwell et al. [26] analyzed the effects of turning parameters on the chip formation, cutting forces, and surface roughness in ultra-high-molecular-weight polyethylene, a common material used in biomedical applications. M. Kaddeche et al. [27] evaluated the surface roughness, cutting pressures, and temperatures during the turning of two grades of high density polyethylene (HDPE-100 and HDPE-80) extruded pipes and ANOVA was performed to establish predictive models. In addition, Hamlaoui et al. [28] experimentally investigated the influence of cutting speed, feed rate, and depth of cut on surface roughness and cutting temperature during the turning of tough polyethylene pipe material HDPE-100 using a Taguchi (L27) orthogonal array. A predictive model was established between input and output parameters via the RSM. The material removal rate was optimized along with different combinations of the process control parameters, such as the cutting speed, feed rate, and depth of cut using the Taguchi design and ANOVA methods during the turning of Teflon (PTFE) cylindrical rounds [29]. Chabbi et al. [30] applied the RSM and ANN methods when turning polyoxymethylene (POM C) polymer in order to investigate surface roughness, cutting force, and power. In addition, optimizations were accomplished using the desirability function (DF).

From the literature, limited work has been carried out to study the machinability aspects of polymers. Further, the relationship between the influencing factors and their effects on machinability are still not well-known. Among polymers, polyethylene (PE) and polyamides (PA) have been some of the most widely studied materials in polymer science due to their large number of applications. High density polyethylene (HDPE) pipes have been commonly used in recent urban networks for water conveyance due to their advantages in terms of setting up, joining, toxicological safety, and service lifetime [27]. Polyamides are used in applications where toughness, lubricity, and wear resistance are important, for example, in aircrafts, machines, marine applications, and automobile engine components such as sprockets, bearings, and gears [15]. Thus, this article aims at investigating, experimentally, the effects of cutting speed, feed rate, and depth of cut on surface roughness (Ra), material removal rate (MRR), and chip formation (λc) during the turning of two types of polymers: high-density polyethylene (HDPE) and unreinforced polyamide (PA6). These investigations are based on the concept of the full factorial design of experiment (FFD), which makes the experimental work not only effective but also efficient. Then, ANOVA and RSM are executed to analyze the process performance. After that, optimum cutting conditions are determined using genetic algorithm (GA) and hybrid FFD-GA approaches. A multi-objective genetic algorithm (MOGA) is used and compared as well. Figure 1 illustrates the flow of the experimentation strategy during this research.

2. Experimental Procedures

2.1. Materials and Measurement Methods

Two types of polymers, high-density polyethylene (HDPE) and Polyamides 6 (Nylon-pa), were used for the turning process. High-density polyethylene (HDPE) is a thermoplastic polymer produced from the monomer ethylene, as shown in Figure 2a. It is sometimes called "alkathene" or "polythene" when used for HDPE pipes with a high strength-to-density ratio. The Polyamides or Nylon class is one of the major engineering and high-performance thermoplastics classes because of its good balance of properties [28]. Polyamides contain repeating amide linkages (–CO-NH–), as shown in Figure 2b; they are formed by condensing identical units, copolymers, for different units. The two most common grades of Nylon are Nylon 6 (or PA6) and Nylon 66 (or PA66). The number relates to the number of methyl groups that are located on each side of the nitrogen atoms (amide groups). The main properties of these materials are given in Table 1. The workpiece specimens were cylindrical rods with diameters of 30 mm for HDPE and 35 mm for PA6, which were 300 mm in length and divided into six segments of 50 mm each, separated by 2 mm grooves.

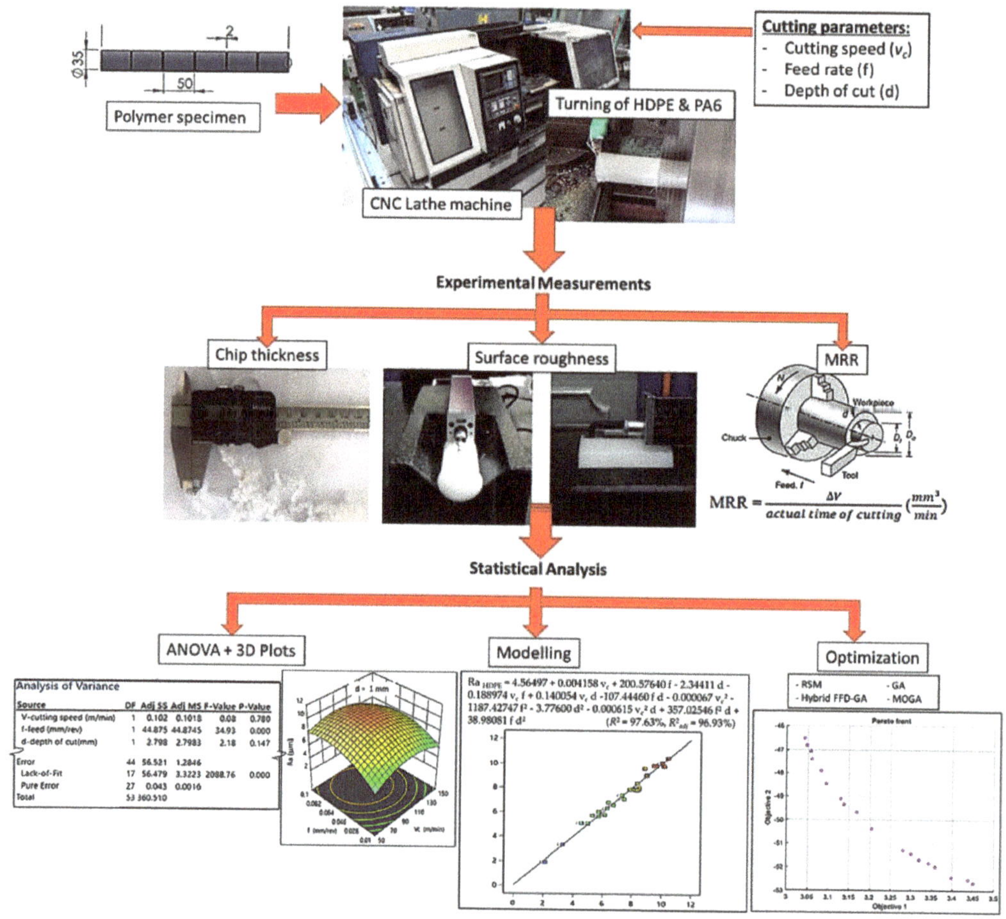

Figure 1. Flow chart of the experimentation strategy.

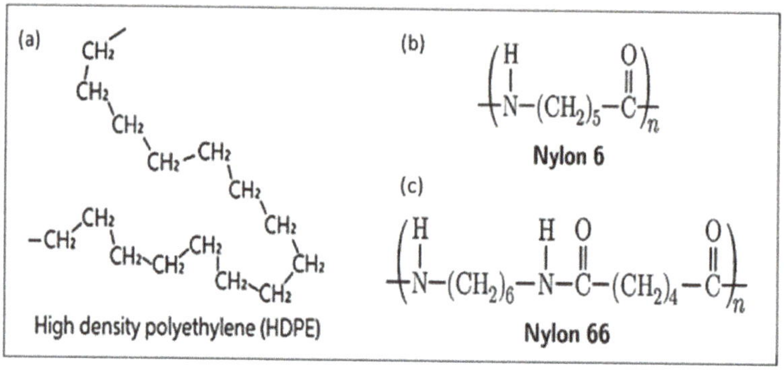

Figure 2. General structure of (**a**) Monomer ethylene, (**b**) PA6, and (**c**) PA66.

Table 1. Main properties of tested polymers.

Material	Physical Properties	Mechanical Properties		Thermal Properties	
	Density	HB	Tensile Strength	Thermal Conductivity	Melting Temperature
	(gm/cm^3)	(MPa)	(Mpa)	(W/km)	(°C)
HDPE	0.95~0.98	48.3	21	0.396	221
PA6	1.14	150	76	0.25	340

A Colchester 4000L CNC lathe machine was used with Fanuc OT commands. Dry tests were carried out using a cemented carbide insert (K10) tool (DCMT 11 T3 02-MF 1125) bolted on a tool holder SDJCR1616k11/L of size 16 × 16 mm. In the present study, the arithmetic mean roughness (Ra) were selected to express the surface roughness, since it is the most widely used surface roughness parameter in the industry to judge the surface quality. The surface roughness (Ra) was measured by means of a Mitutoyo Surftest SJ-201 roughness meter. The specimens were held on a V-block while a spirit level was used to ensure the proper alignment of the stylus motion. The measurements were conducted at four different points in the longitudinal feed direction with a 2.5 mm cut-off and 12.5 mm evaluation length; then, the average of arithmetic mean value (Ra) was calculated to represent the surface roughness. Actual MRR was calculated by subtracting the final volume of the specimen after turning (mm^3) from the initial volume of the specimen, then dividing it by the actual machining time (min). The chip thickness ratio (λc) is determined experimentally as the ratio between the deformed chip thickness (t_2), which was measured using an electronic LCD digital vernier caliper with an accuracy of 0.01 mm, and the undeformed chip thickness ($t_1 = f \cdot \sin\phi$), where f is the feed rate (mm/rev) and ϕ is the tool principal angle = 93°. The chips produced during turning were collected and the average deformed chip thickness was obtained by the average of ten readings at different locations of the same chips produced under constant conditions.

2.2. Experimentation and Data Collection

The turning experiments were planned using the full factorial design of experiments (FFD) consisting of three factors: cutting speed, feed rate, and depth of cut, with three levels, each with two replications, which led to a total number of 54 experiments. The values chosen were as follows: cutting speed v_c (50, 100, and 150 m/min), feed rate f (0.01, 0.05, and 0.1 mm/rev), and depth of cut d (0.5, 1, and 1.5 mm). Stat-Ease Design Expert (version 13.0.5) was used to generate the testing order as well as to assist in the statistical analysis of the experimental data. The experimental plan of the present investigation, as per FFD and their corresponding measured responses, is presented in Table 2.

Table 2. Full factorial design of experiment and the response results.

Test Order	Machining Parameters			Response Variables					
				HDPE			PA6		
	vc (m/min)	f (mm/rev)	d (mm)	Ra (μm)	MRR (mm^3/min)	λc	Ra (μm)	MRR (mm^3/min)	λc
1	50	0.01	1	4.528	1300	17.0238	2.3975	1700	39.0546
2	100	0.05	1	8.8405	5000	10.2143	8.0205	3845	7.81093
3	150	0.05	1.5	5.27075	10,450	9.6134	8.99725	10,962	7.81093
4	50	0.05	0.5	10.03475	1030	4.8067	9.88	834	11.0154
5	100	0.1	0.5	10.1365	3961	4.5056	11.14475	4046	3.3046
6	50	0.01	1.5	2.12075	2300	19.0266	3.4005	2543	52.0729
7	50	0.05	1.5	4.98275	3500	6.208	4.91625	3587	14.0196
8	150	0.1	1.5	6.1325	12,200	5.5085	8.8235	25,394	5.8081

Table 2. *Cont.*

Test Order	Machining Parameters			Response Variables					
				HDPE			PA6		
	vc (m/min)	f (mm/rev)	d (mm)	Ra (μm)	MRR (mm^3/min)	λc	Ra (μm)	MRR (mm^3/min)	λc
9	150	0.1	1	6.37425	9200	5.2072	9.91925	17,525	4.5063
10	150	0.1	0.5	7.845	9080	4.2058	10.2665	8750	4.1057
11	50	0.01	1	4.528	1300	16.0238	2.3975	1700	39.0546
12	50	0.05	1	8.49575	2480	5.2072	4.70925	2317	12.2171
13	150	0.05	1	8.38075	7720	8.61205	10.271	6975	7.8109
14	150	0.01	1.5	3.29975	7200	17.0252	4.144	7731	40.05607
15	100	0.01	1.5	4.981	5800	25.0364	3.1245	5900	55.0771
16	100	0.05	1	8.87125	5090	11.2143	7.923	3910	7.8109
17	100	0.1	1	8.958	7900	4.6064	10.256	6604	4.1057
18	100	0.1	0.5	10.179	3960	4.0056	11.1765	3125	3.3046
19	50	0.01	0.5	6.7405	673	11.0168	5.203	850	28.0392
20	50	0.1	1.5	5.7525	7932	5.8081	7.4535	4383	5.9082
21	50	0.1	0.5	9.476	2019	4.2058	9.595	1842	4.5063
22	50	0.1	1	8.48575	4904	4.9068	5.97225	2769	5.1071
23	100	0.01	1	7.47	3800	19.02803	5.8395	4680	41.0574
24	100	0.01	1.5	4.981	5800	26.0364	3.1245	5900	55.0771
25	50	0.1	0.5	9.6645	2020	5.2058	9.5435	1547	4.5063
26	50	0.1	1	8.08575	4952	4.6068	5.98725	2769	5.1071
27	100	0.01	1	7.47	3800	20.02803	5.8395	4680	41.0574
28	50	0.01	0.5	6.7405	673	12.0168	5.203	850	28.0392
29	50	0.01	1.5	2.12075	2300	18.2266	3.4005	2543	52.0729
30	150	0.05	1.5	5.32625	10,450	8.9134	9.02825	13,487	7.8109
31	150	0.01	1	5.7195	5000	16.0224	5.2	5150	29.0406
32	150	0.05	0.5	8.9865	5970	6.2086	9.66425	3506	6.8095
33	150	0.1	1	6.394	9230	5.7072	9.962	11,210	4.5063
34	150	0.1	1.5	6.15825	12,200	6.1085	8.87075	24,125	5.8081
35	100	0.1	1	9.0045	8020	4.9064	10.06275	4962	4.1057
36	100	0.05	1.5	8.20875	7930	11.6162	7.1545	6921	6.2086
37	50	0.1	1.5	5.70625	8088	6.3081	7.452	4641	5.9082
38	150	0.1	0.5	7.8625	9080	4.8058	10.26575	8578	4.1057
39	150	0.01	0.5	6.30775	3000	12.0182	5.0435	2500	25.03504
40	150	0.05	1	8.362	7730	8.11205	10.2755	6500	7.81093
41	50	0.05	0.5	10.03475	1030	5.4067	9.86975	396	11.0154
42	100	0.01	0.5	7.333	1500	16.0238	3.60275	1980	36.0504
43	100	0.1	1.5	8.3735	9900	5.2072	7.987	16,547	5.9082
44	50	0.05	1.5	4.98275	3517	6.0086	4.79075	3813	14.0196
45	150	0.05	0.5	8.9885	5970	6.8086	9.647	4375	6.8095
46	100	0.01	0.5	7.333	1500	17.0238	3.60275	1980	36.0504
47	100	0.05	1.5	8.2185	7800	10.6162	7.0805	6921	6.2086
48	150	0.01	1.5	3.29975	7200	18.0252	4.144	7731	40.05607
49	150	0.01	1	5.7195	5000	16.4224	5.2	5150	29.0406
50	50	0.05	1	8.49575	2490	5.00729	4.71075	2728	12.2171
51	150	0.01	0.5	6.30775	3000	13.0182	5.0435	2500	25.03504
52	100	0.1	1.5	8.3675	9900	5.70729	7.9625	15,470	5.9082
53	100	0.05	0.5	10.4005	3090	8.31093	7.20675	2348	5.20729
54	100	0.05	0.5	10.439	3060	7.8109	7.2955	2327	5.20729

Based on the experimental results, RSM was applied to analyze the results statistically to formulate the model, analyze, and develop an appropriate interaction between measured responses and input factors by regression modelling [31,32]. Regression models were obtained by the best interaction correlation between the input variables and the output responses into a non-linear equation. Finally, the optimal cutting conditions of turning HDPE and PA6 for the desired responses were attained by an application of the RSM desirability function, a genetic algorithm (GA), and hybrid FFD-GA techniques. The optimization target was set to the lowest possible Ra and the highest possible MRR and chip ratio.

3. Results and Discussion

3.1. Experimental Results

The surface characteristic of the machined part is affected by variations in the process input parameters (v_c, f, and d). Figure 3 shows the roughness surface profiles at different cutting conditions. It is observed that Ra deteriorated at v_c max and f max. Smooth surfaces were obtained at low v_c, f, and low-to-medium d. However, severe conditions that produced uneven surface finish were 150 mm/min, 0.1 mm/rev, and 1.5 mm, as shown in Figure 4a. Lower v_c of 50 mm/min, lower f of 0.01 mm/rev, and medium d of 1 mm was the combination that produced a smoother surface finish for PA6 of $\approx$2.4 µm, whereas d = 1.5 mm resulted in a smoother surface for HDPE of $\approx$2.12 µm, as shown in Figure 4b.

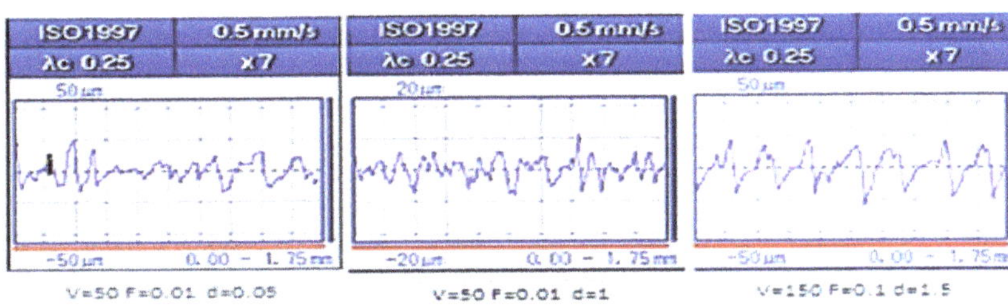

Figure 3. Surface profiles at different cutting conditions.

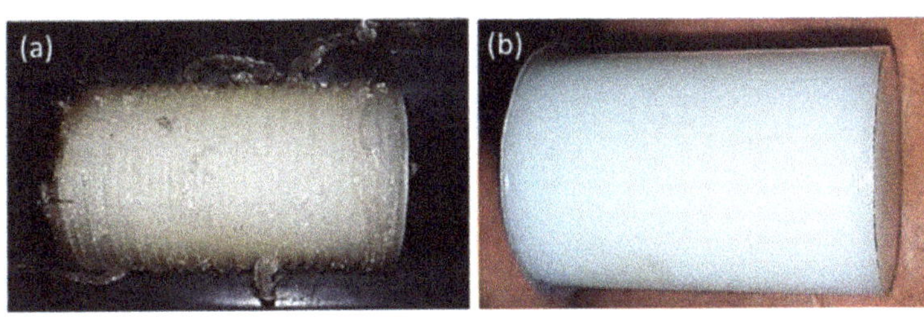

Figure 4. Photograph of the machined surface (**a**) worst surface, (**b**) best surface.

In addition to the surface roughness, the material removal rate also has an important role in the production of finished products. Higher MRR is preferred during machining in order to achieve high productivity, but with respect to the quality of the machined surface. As expected, it is obvious that, with an increase in v_c, f, and, d, MRR increases as the contact area between the tool and workpiece increases. However, the roughness increases as MRR increases and chip area increases where: MRR = $v_c \times f \times d$ mm^3/min and chip area = $f \times d$ mm^2.

The study of chip formation during machining operation has served a fundamental role to understand the mechanics of material removal. In addition, the determination of the chip thickness ratio is very important in the determination of material removal and surface roughness as it is one of the parameters that affect the shearing process and friction state at the entire chip/tool interface [23,33]. It is possible to recognize two main chip formation mechanisms: (I) continuous and (II) discontinuous chip. Based on these two types, the chips formed can be examined. Figure 5 shows the optical images of the collected chips formed at different v_c, f, and d for (a) HDPE and (b) PA6. It was observed that low v_c, f, and d, produce very long curly continuous chips, especially in case of HDPE machining. However, discontinuous chips are the common chip type formed during the machining of PA6. The transition between the two mechanisms is not sharp, as the chips appear, in some cases, highly stretched and with some melted filaments are observed. This behavior can be related to the fact that the machining of soft material produces helicoid continuous chips, as in the case of HDPE machining, which is more ductile and has relatively lower hardness compared to PA6. The high hardness of PA6 leads to discontinuous chip formation, which explains the higher values of surface roughness obtained in the case of PA6 compared to that of HDPE [34].

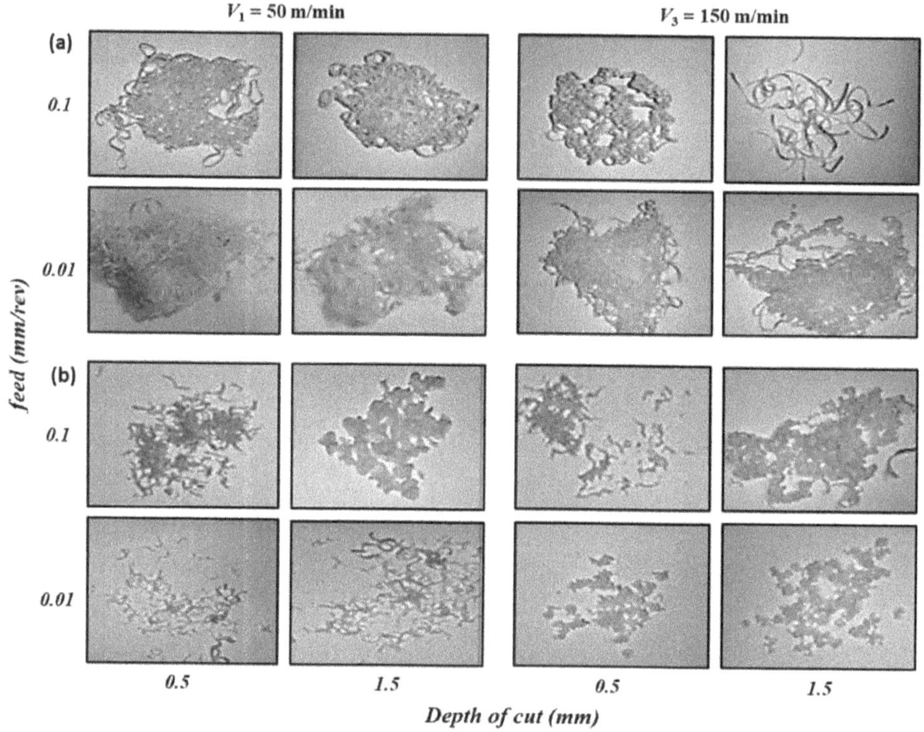

Figure 5. Optical images of chips obtained at different cutting conditions for (**a**) HDPE and (**b**) PA6.

3.2. Statistical Results

3.2.1. Regression Model

In this study, ANOVA was used to analyze the experimental data and determine the most significant impacts of input parameters (v_c, f, and d) on the output responses (R_a, MRR, and λc) by using the Design Expert software program (version 13.0.5, Stat-Ease, Inc., Minneapolis, MN, USA). Table 3 illustrates the summarized ANOVA results: F-value, p-value, lack of fit, adequate precision, R^2, adjusted R^2, and predicted R^2 at a confidence level of 95%. The p-values of all responses were less than 0.05 and all F-values were greater than four, which implies the adequacy of the predicted models and that the independent parameters, as well as the individual model coefficients and interaction terms, had a significant effect on the responses. It was found that the most significant factor on Ra_{HDPE} was the depth of cut with an F-value of 86.4, in comparison to v_c with an F-value of 289.33 on Ra_{PA6}. For both materials, HDPE and PA6, v_c had the greatest impact on MRR with F-values of 386.8 and 73.34, respectively. Finally, feed rate had a significant effect on λc for HDPE and PA6 with F-values of 2052 and 803.2, respectively.

Table 3. ANOVA summarized results of cutting responses.

	Response	F-Value (F > 4)	*p*-Value (*p* < 0.05)	Lack of Fit (*p* > 0.05)	Adeq Precision (ratio > 4)	R^2	R^2_{adj}	R^2_{pred}
HDPE	Ra	140.44	<0.0001	0.0001	47.3369	0.9763	0.9693	0.9623
	MRR	254.84	<0.0001	0.0001	55.6343	0.991	0.9871	0.9809
	λc	407.08	<0.0001	—	69.8882	0.9917	0.9892	0.9853
PA6	Ra	105.8	<0.0001	0.0001	32.9903	0.9846	0.9753	0.9601
	MRR	92.08	<0.0001	0.0001	41.9425	0.9642	0.9538	0.9397
	λc	238.39	<0.0001	—	46.0162	0.9823	0.9782	0.9725

The lack of fits for Ra and MRR for both materials, HDPE and PA6, were lower than 0.05, indicating that the model is insignificant. Chip thickness ratio (λc) had no lack of fit since the value of Sum of Squares (SS) of the regression model was larger than that of the residual. Moreover, the SS and mean squares (MS) of the pure error were equal to zero, so that there is no *p*-value and F-value for the lack of fit test. The signal to noise ratio (S/N) was computed with adequate precision to determine the model's validity. It is recommended that S/N ratio exceed four [23]. The obtained responses' adequate precision was greater than four and reveals that there was sufficient signal, and the model can be applicable to navigate the design space.

Many trails of regression transformation form and interaction between independent variables were examined in order to model the output responses. The relationship between the factors and the output parameters was modeled by quadratic regression. The best regression coefficient of determination (R^2) indicates that the models generated in the experimental research are statistically significant when R^2 is closer to 1 and can be used in predicting the response parameters with respect to the input control parameters [35].

The regression Equations (1)–(3) represent the predicted non-linear model of HDPE responses and (4)–(6) of PA6 responses, Ra, MRR, and λc, as a function of v_c, f, and d, with their associated determination and adjusted coefficients. In our study, R^2 and adjusted R^2 were very close to unity with values from 95.38–99.17%. Figure 6 illustrates the plot of the experimental values as a function of their corresponding predicted values of Ra, MRR, and λc for (a) HDPE and (b) PA6. By analyzing these figures, it became clear that there was a good agreement between experimental and predicted values, as most of the intersection points between them were very close to the median line, which confirms the effectiveness of the developed regression models.

$$Ra_{HDPE} = 4.56497 + 0.004158\, v_c + 200.57640\, f - 2.34411\, d - 0.188974\, v_c f + 0.140054\, v_c d - 107.44460\, f d - 0.000067\, v_c^2 - 1187.42747\, f^2 - 3.77600\, d^2 - 0.000615\, v_c^2 d + 357.02546\, f^2 d + 38.98081\, f d^2 \quad (R^2 = 97.63\%, R^2_{adj} = 96.93\%) \tag{1}$$

$$MRR_{HDPE} = 1564.63727 - 54.54011\, v_c - 84010.80298\, f - 583.17350\, d + 877.35515\, v_c f + 95.51891\, v_c d + 60491.89891\, f d + 0.326479\, v_c^2 + 5.81515E{+}05\, f^2 - 2269.31694\, d^2 - 619.97951\, v_c f d + 3.36686\, v_c^2 f - 0.544200\, v_c^2 d - 6908.98148\, v_c f^2 + 26.03333\, v_c d^2 + 94416.66667\, f^2 d + 4676.77596\, f d^2 \quad (R^2 = 99.10\%, R^2_{adj} = 98.71\%) \tag{2}$$

$$\lambda c_{HDPE} = -1.10886 + 0.355347\, v_c - 165.02460\, f + 4.09091\, d - 3.42531\, v_c f + 0.103812\, v_c d - 155.10604\, f d - 0.001872\, v_c^2 + 2627.13478\, f^2 + 0.025216\, v_c^2 f - 0.000514\, v_c^2 d - 14.16798\, v_c f^2 + 860.46391\, f^2 d \quad (R^2 = 99.17\%, R^2_{adj} = 98.92\%) \tag{3}$$

$$Ra_{PA6} = 41.30189 - 0.807919\, v_c + 357.63940\, f - 85.42634\, d - 2.44587\, v_c f + 1.78074\, v_c d - 212.57032\, f d + 0.003592\, v_c^2 - 3679.04630\, f^2 + 39.09850\, d^2 - 0.283895\, v_c f d + 0.007549\, v_c^2 f - 0.007744\, v_c^2 d + 52.69097\, v_c f^2 - 0.806673\, v_c d^2 + 1957.38194\, f^2 d + 39.52193\, f d^2 - 0.215384\, v_c^2 f^2 + 0.011114\, v_c^2 f d + 0.003432\, v_c^2 d^2 - 17.83417\, v_c f^2 d \quad (R^2 = 98.46\%, R^2_{adj} = 97.53\%) \tag{4}$$

$$MRR_{PA6} = -3991.74828 + 20.73222\ v_c + 2.09063E{+}05\ f + 7175.40968\ d - 1312.86239\ v_c\ f + 14.49852\ v_c\ d - 3.30948E{+}05\ f\ d - 1.15102E{+}06\ f^2 - 2551.52459\ d^2 + 1104.09016\ v_c\ f\ d + 11503.14815\ v_c\ f^2 + 6.66509E{+}05\ f^2\ d + 1.11560E{+}05\ f\ d^2 \quad (R^2 = 96.42\%,\ R^2_{adj} = 95.38\%) \tag{5}$$

$$\lambda c_{PA6} = 7.26518 + 0.462396\ v_c - 216.08394\ f + 29.83807\ d - 6.64209\ v_c\ f - 0.035717\ v_c\ d - 740.25858\ f\ d - 0.002598\ v_c^2 + 2381.11133\ f^2 + 0.037648\ v_c^2\ f + 4966.21195\ f^2\ d \quad (R^2 = 98.23\%,\ R^2_{adj} = 97.82\%) \tag{6}$$

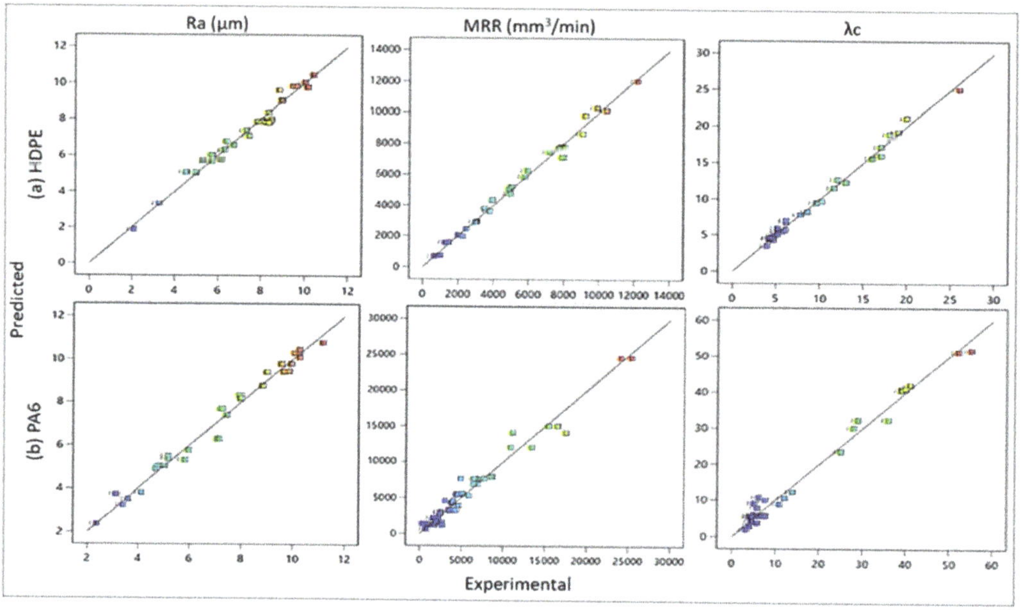

Figure 6. Comparison between experimental and predicted values of Ra, MRR, and λc for (**a**) HDPE and (**b**) PA6.

3.2.2. Effect of Cutting Parameters on Ra

Figures 7 and 8 show 3D response plots that were constructed based on the regression models in order to evaluate the change in Ra as a function of cutting parameters (v_c, *f*, and d) for HDPE and PA6, respectively.

For HDPE, as shown in Figure 7a, there is a proportional relationship between feed rate as well as cutting speed with Ra at constant depth of cut. It was observed that Ra increases rapidly with f and vc at certain depth of cut. At constant feed (Figure 7b), the increase in d produces better surface quality at low f; however, v_c does not show any significant changes in surface quality. Finally, it was concluded that Ra increases with increasing f and decreasing d at constant v_c, and that the lower roughness is obtained at low v_c (Figure 7c). As a result, there is a fair agreement between interaction plots and experimental results. The minimal Ra is obtained at the lowest value of v_c and f and highest d, which agrees with the best Ra $_{HDPE}$ ≈ 2.12 μm of experimental results obtained at 50 m/min, 0.01 mm/rev, and 1.5 mm.

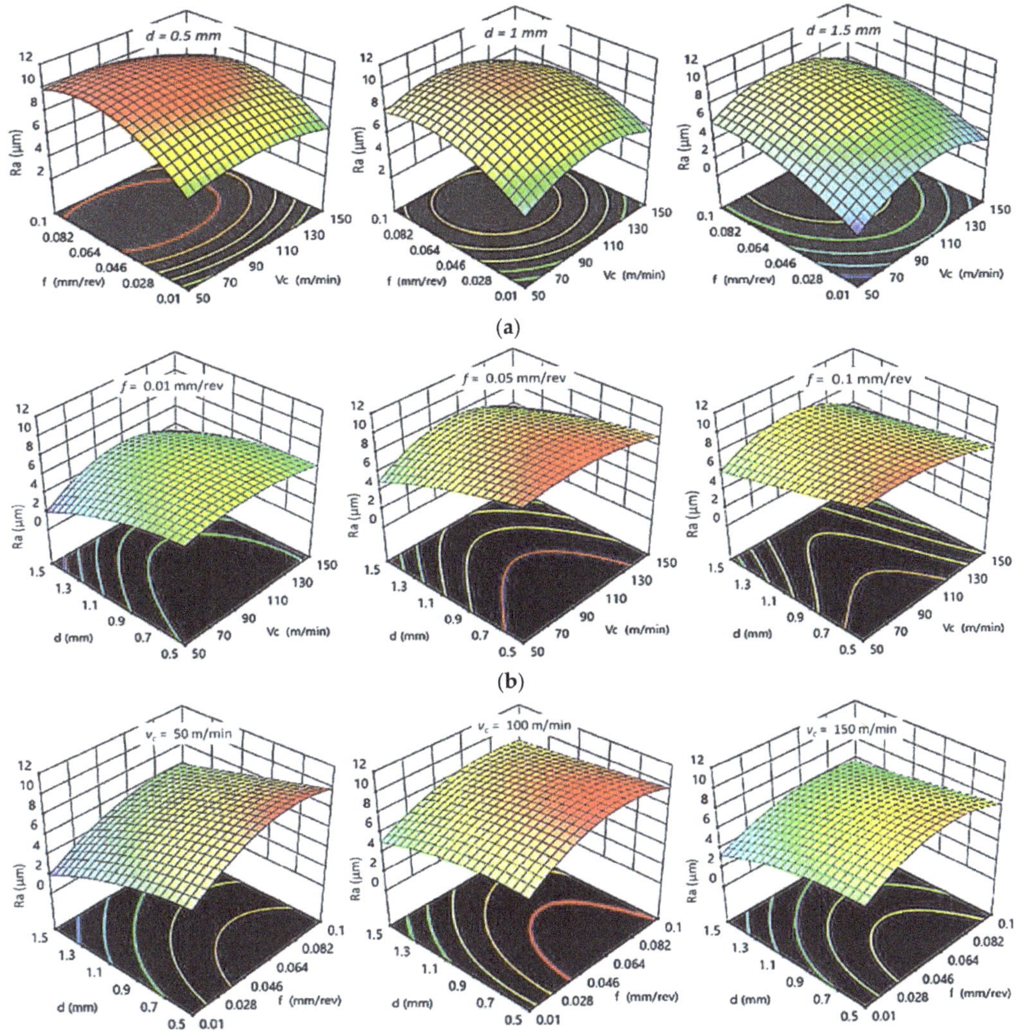

Figure 7. 3D response plots for surface roughness of HDPE. (**a**) Effect of cutting speed and feed rate on Ra $_{HDPE}$ at different depths of cut; (**b**) Effect of cutting speed and depth of cut on Ra $_{HDPE}$ at different feed rates; (**c**) Effect of feed rate and depth of cut on Ra $_{HDPE}$ at different cutting speeds.

For PA6, as shown in Figure 8a–c, it was concluded that the increase in both f and v_c result in increasing Ra while keeping d at constant level. However, by increasing d, Ra decreases until reaching its minimum value of 2.397 µm at 1 mm cutting depth. After that, Ra increases with further increasing of d. At constant v_c, d has the same influence on Ra as at constant f, which has a proportional relation with roughness, whereas better surface finish is found at low v_c. Accordingly, it is clear that the interaction plots results are compatible with the experimental results. The best surface finish is obtained at lowest v_c, f, and 1 mm depth of cut, agreeing with the experimental results, which gave the best Ra $_{PA6}$ $\approx$ 2.397 µm at 50 m/min, 0.01 mm/rev, and 1 mm.

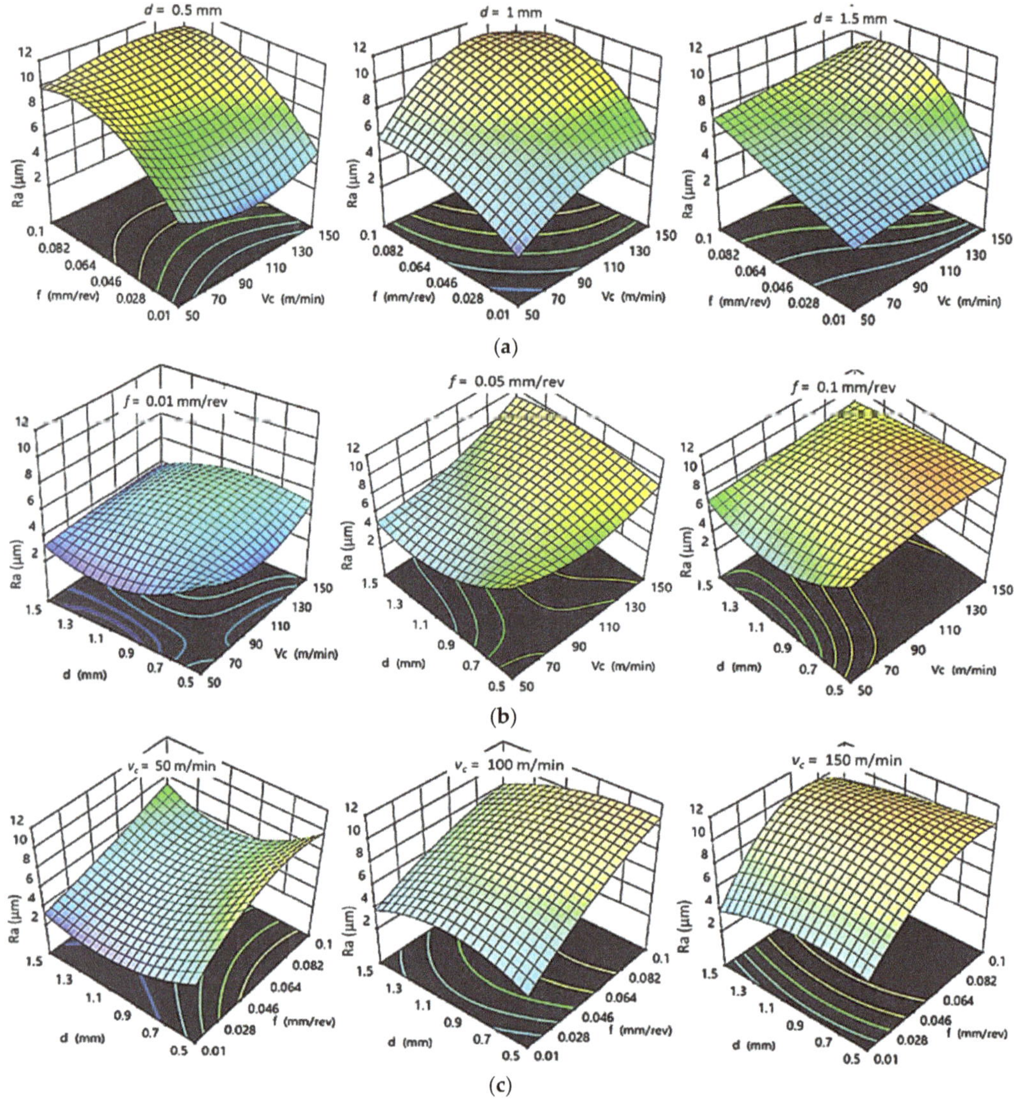

Figure 8. 3D response plots for surface roughness of PA6. (**a**) Effect of cutting speed and feed rate on Ra $_{PA6}$ at different depths of cut; (**b**) Effect of cutting speed and depth of cut on Ra $_{PA6}$ at different feed rates; (**c**) Effect of feed rate and depth of cut on Ra $_{PA6}$ at different cutting speeds.

3.2.3. Effect of Cutting Parameters on MRR

The effects of cutting parameters on MRR for HDPE and PA6 obtained by regression models are shown in Figures 9 and 10, respectively. It was noticed that there was a similar effect of cutting parameters on MRR for both HDPE and PA6. MRR increased with increases in v_c, f, and d. There is agreement between interaction plots and experimental results. The best MRRs of HDPE and PA6 obtained from interaction plots were found at high v_c, f, and d, which agrees with the experimental results, as best MRR $_{HDPE}$ = 12,200 mm^3/min and MRR $_{PA6}$ = 25,394 mm^3/min were obtained at 150 m/min, 0.1mm/rev, and 1.5 mm. In addition, it was noted that MRR $_{HDPE}$ is half MRR $_{PA6}$.

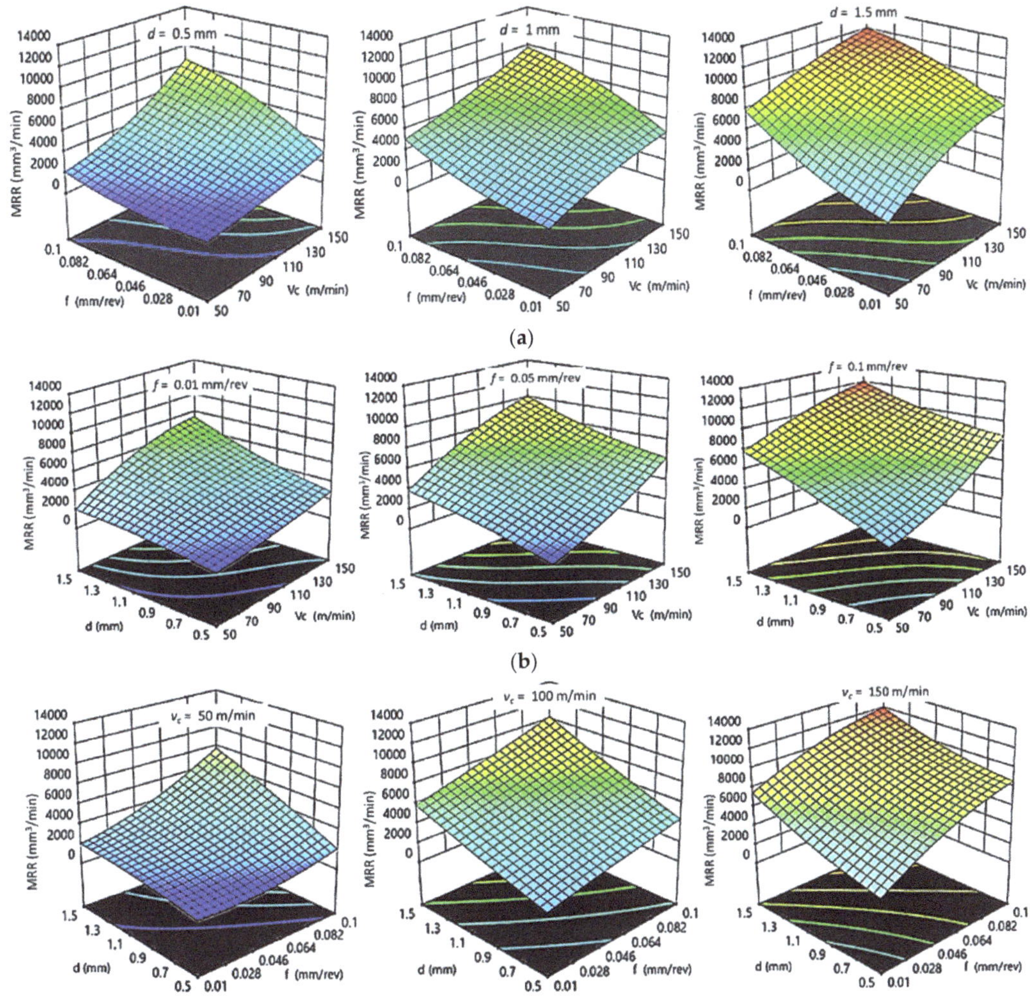

Figure 9. 3D response plots for material removal rate of HDPE. (**a**) Effect of cutting speed and feed rate on MRR $_{HDPE}$ at different depths of cut; (**b**) Effect of cutting speed and depth of cut on MRR $_{HDPE}$ at different feed rates; (**c**) Effect of feed rate and depth of cut on MRR $_{HDPE}$ at different cutting speeds.

3.2.4. Effect of Cutting Parameters on λc

Figures 11 and 12 illustrate the influence of changing cutting parameters obtained by regression models on the chip ratios (λc) for HDPE and PA6, respectively. The same behavior of chip ratio was remarked for both HDPE and PA6. At constant d, an increasing of f decreased the λc while an increasing of v_c rose the λc to its optimum value at 100 m/min; then, λc decreased again with a further increasing of v_c. The chip ratio also increased by an increasing of d, even at a fixed level of f or v_c. The higher the chip thickness ratio attained, the better the surface finish was. A higher λc for both HDPE and PA6, with values of 26.036 and 55.077, respectively, was achieved at v_c of 100 m/min, lower f of 0.01 mm/rev, and highest d of 1.5 mm, which confirms the experimental with the interaction plots results.

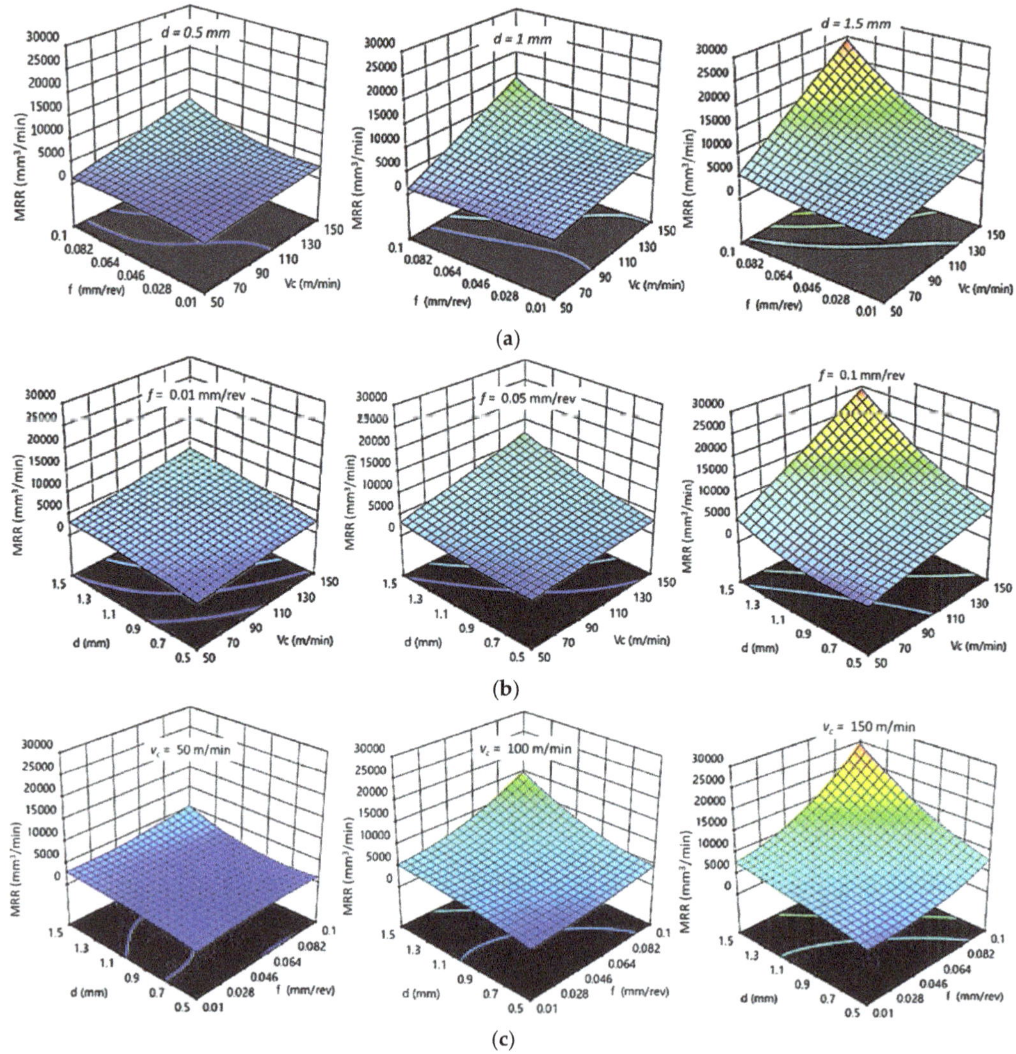

Figure 10. 3D response plots for material removal rate of PA6. (**a**) Effect of cutting speed and feed rate on MRR $_{PA6}$ at different depths of cut; (**b**) Effect of cutting speed and depth of cut on MRR $_{PA6}$ at different feed rates; (**c**) Effect of feed rate and depth of cut on MRR $_{PA6}$ at different cutting speeds.

3.3. Optimization Results

3.3.1. RSM Results

Based on the comprehensive analysis of each independent variable, the process optimization was developed to obtain the best parameter combination for HDPE and PA6 turning for the desire responses using the Design Expert software (version 13.0.5, Stat-Ease, Inc., Minneapolis, MN, USA). Figure 13a shows the optimized surface roughness (Ra) of turning HDPE and PA6 and corresponding conditions using the desirability function of the RSM. For all the following optimization findings the red dot and the blue dotes the indicated the cutting condition (v_c, f and d) and response (Ra, MRR and λc) respectively. As the machinability is better when Ra is low, so the solution destination was set to "Minimize", the optimization target was set to "In range", and the desirability function's predicted output was in the form of "smaller-is-better". For HDPE, the minimum surface roughness value of 2.1156 μm was predicted using the combination of v_c (A) = 50 m/min, f (B) = 0.01 mm/rev, and d (C) = 1.47 mm. For PA6, the minimum Ra $_{PA6}$ value of 2.376 μm was predicted using the combination of v_c (A) = 50 m/min, f (B) = 0.01 mm/rev, and d (C) = 1 mm.

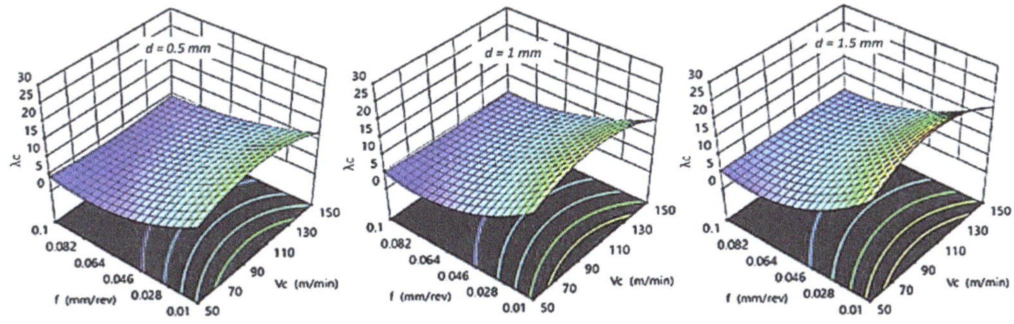

(a)

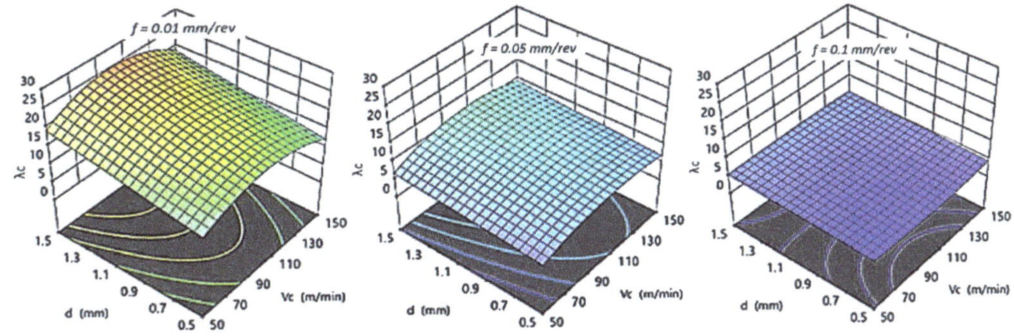

(b)

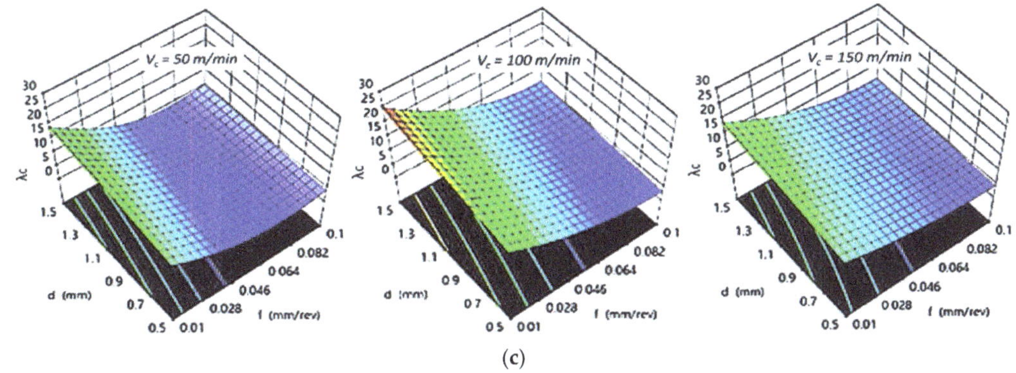

(c)

Figure 11. 3D response plots for chip ratio of HDPE. (**a**) Effect of cutting speed and feed rate on λc $_{HDPE}$ at different depths of cut; (**b**) Effect of cutting speed and depth of cut on λc $_{HDPE}$ at different feed rates; (**c**) Effect of feed rate and depth of cut on λc $_{HDPE}$ at different cutting speeds.

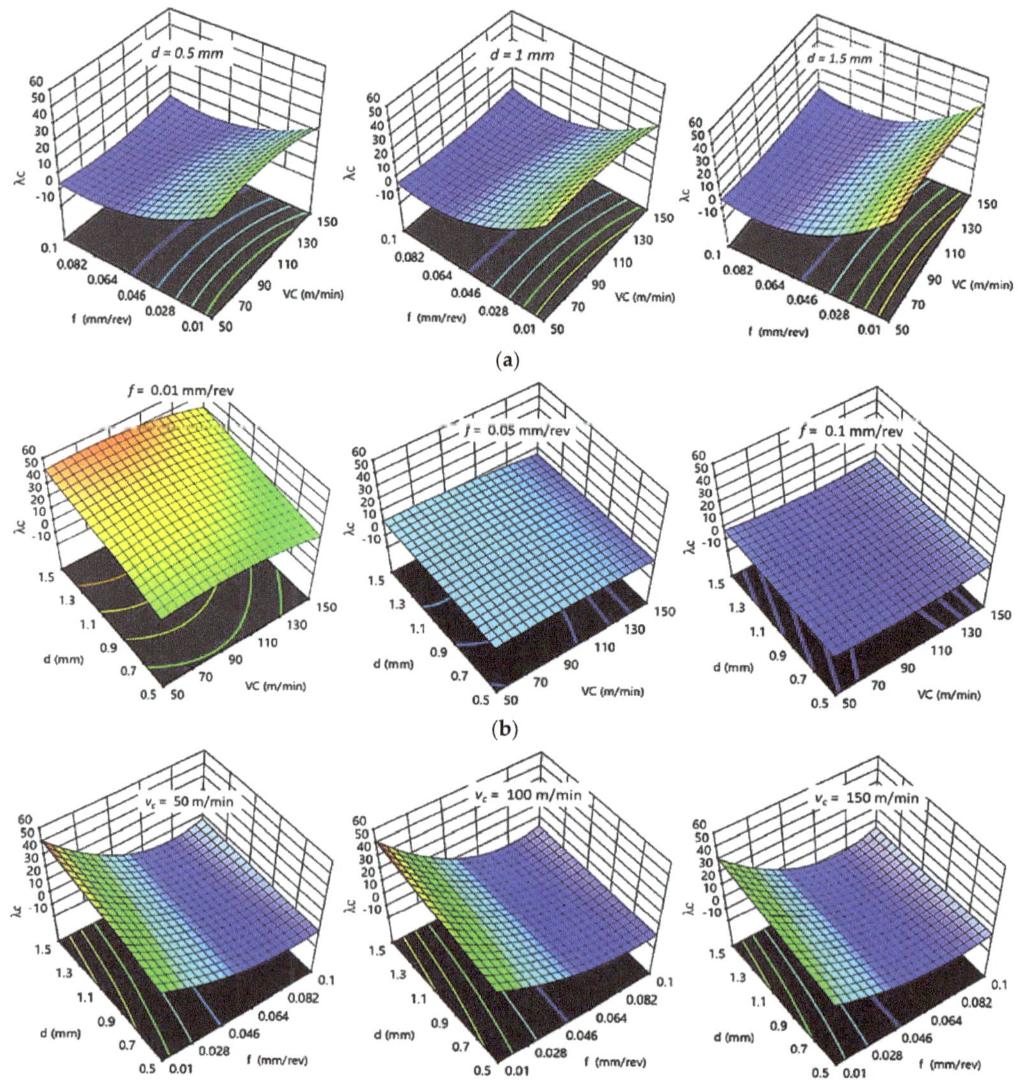

Figure 12. 3D response plots for chip ratio of PA6. (**a**) Effect of cutting speed and feed rate on $\lambda c_{\ PA6}$ at different depths of cut; (**b**) Effect of cutting speed and depth of cut on $\lambda c_{\ PA6}$ at different feed rates; (**c**) Effect of feed rate and depth of cut on $\lambda c_{\ PA6}$ at different cutting speeds.

Figure 13b presents the RSM optimization results of MRR and corresponding conditions. The optimization target was set to "In range", the solution destination was set to "Maximize", and the desirability function's predicted output was in the form of "larger-is-better". For HDPE, the optimal cutting conditions values were v_c (A) = 146.782 m/min, f (B) = 0.099 mm/rev, and d (C) = 1.496 mm for a maximum $MRR_{\ HDPE}$ value of 12206.8 mm^3/min. For PA6, the optimal cutting conditions values were v_c (A) = 150 m/min, f (B) = 0.1 mm/rev, and d (C) = 1.5 mm for a maximum $MRR_{\ PA6}$ value of 24,658.7 mm^3/min.

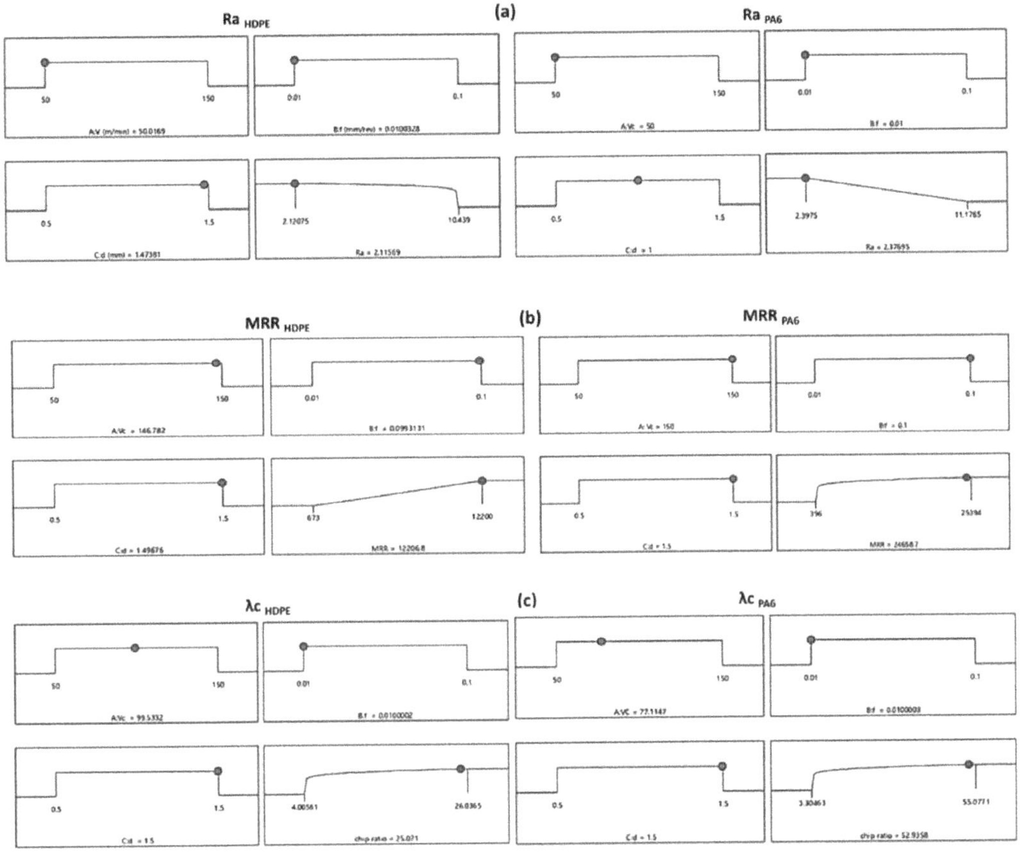

Figure 13. RSM optimization results of (**a**) Ra, (**b**) MRR, and (**c**) λc for HDPE and PA6.

The RSM optimization results of the chip ratio (λc) and their corresponding conditions are shown in Figure 13c. As the higher chip ratio obtained, the better surface finish is, so that the optimization target was set to "In range", the solution destination was set to "Maximize", and the desirability function's predicted output was in the form of "larger-is-better" characteristics. For HDPE, a maximum chip ratio value of 25.071 was attained at optimum cutting condition values of v_c (A) = 99.533 m/min, f (B) = 0.01 mm/rev, and d (C) = 1.5 mm. For PA6, the optimum cutting condition combination values were v_c (A) = 77.11 m/min, f (B) = 0.01 mm/rev, and d (C) = 1.5 mm for a maximum chip ratio value of 52.935.

For a confirmation test, the optimum cutting conditions and responses acquired by RSM are compared to the GA and hybrid FFD-GA results developed in the next section.

3.3.2. GA and Hybrid FFD-GA Results

A genetic algorithm (GA) was used to find out the optimum set of cutting independent variables that contribute to the lowest possible Ra and the highest possible MRR and λc. Equations (1)–(6) for each response were taken as the objective function and subjected to the cutting boundary conditions, v_c, f, and d, by using a genetic algorithm approach. Proposed objective functions can be expressed as follows:

Minimize (v_c, f, d)

Subjected to ranges of cutting conditions:

$50 \leq v_c \leq 150$ (m/min), $0.01 \leq f \leq 0.1$ (mm/rev), $0.5 \leq d \leq 1.5$ (mm)

For the GA optimization technique, Figures 14–16 present the performance of fitness value and run solver view generated from MATLAB and corresponding cutting conditions of the best Ra, MRR, and λc, respectively, for HDPE and PA6.

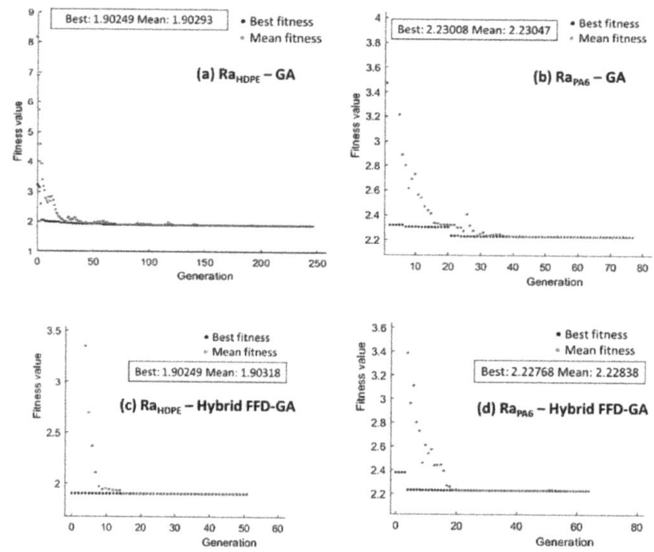

Figure 14. Optimum Ra of HDPE (**a**,**c**) and PA6 (**b**,**d**) by GA (**a**,**b**) and hybrid FFD-GA (**c**,**d**).

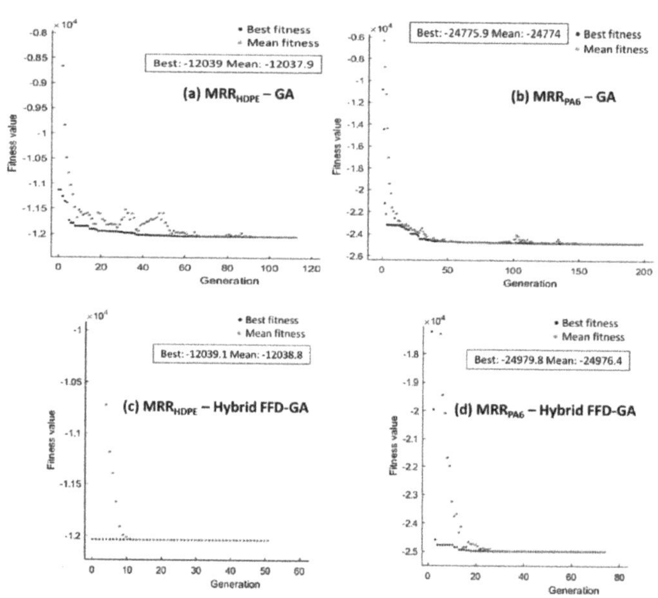

Figure 15. Optimum MRR of HDPE (**a**,**c**) and PA6 (**b**,**d**) by GA (**a**,**b**) and hybrid FFD-GA (**c**,**d**).

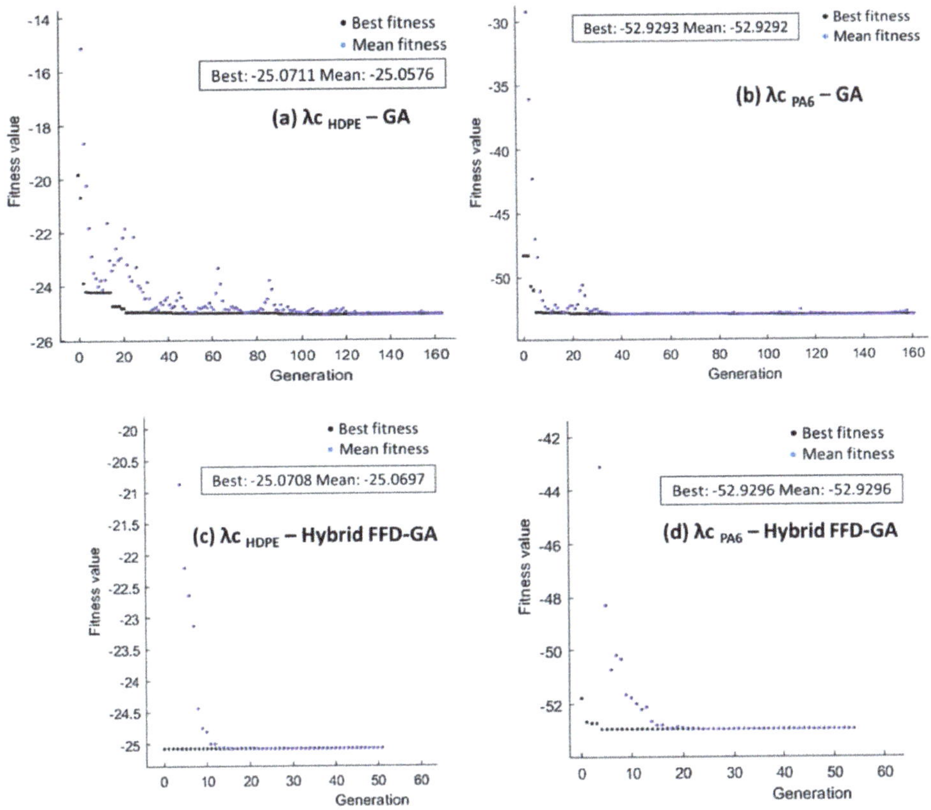

Figure 16. Optimum λc of HDPE (**a**,**c**) and PA6 (**b**,**d**) by GA (**a**,**b**) and hybrid FFD-GA (**c**,**d**).

The minimization of Ra proposed in Equations (1) and (4) was taken as the fitness function and subjected to the cutting boundary condition. The minimum value of Ra $_{HDPE}$ obtained by GA was 1.902 μm at v_c = 50 m/min, f = 0.01 mm/rev, and d = 1.5 mm (Figure 14a). In addition, the minimum value of Ra $_{PA6}$ was 2.23 μm, obtained by GA at v_c = 50 m/min, f = 0.01 mm/rev, and d = 1.139 mm (Figure 14b).

In order to improve the results generated from GA, a hybrid of full factorial design and GA (FFD-GA) was performed. Initial populations of hybrid FFD-GA based on FFD optimum cutting conditions were: v_c = 50 m/min, f = 0.01 mm/rev, and d = 1.5 mm, and v_c = 50 m/min, f = 0.01 mm/rev, and d = 1.5 mm for HDPE and PA6, respectively. The minimum Ra value for HDPE (Figure 14c) obtained by hybrid FFD-GA was 1.902 μm at 50 m/min, 0.01 mm/rev, and 1.5 mm. For PA6, the minimum Ra obtained by hybrid FFD-GA was 2.227 μm (Figure 14d) at 50 m/min, 0.01 mm/rev, and 1.139 mm.

The maximization of MRR proposed in Equations (2) and (5) was taken as the fitness function and subjected to the cutting boundary condition. The best values of MRR_{HDPE} and MRR_{PA6} by GA, shown in Figure 15a,b, were 12,039 mm^3/min and 24,775.9 mm^3/min, respectively, obtained at 150 m/min, 0.1 mm/rev, and 1.5 mm. Hybrid FFD-GA results of maximum MRR $_{HDPE}$ and MRR $_{PA6}$ were 12,039.1 mm^3/min and 24,979.8 mm^3/min, respectively, obtained at 150 m/min, 0.1 mm/rev, and 1.5 mm, as shown in Figure 15c,d.

The maximization of λc proposed in Equation (3) for HDPE and Equation (6) for PA6 were taken as the fitness function and subjected to the cutting boundary condition. As shown in Figure 16a,b, GA results indicated that the best value of λc $_{HDPE}$ was 25.0711 obtained at 99.4 m/min, 0.01 mm/rev, and 1.5 mm whereas λc $_{PA6}$ was 52.9293 obtained at 75.735 m/min, 0.01 mm/rev, and 1.5 mm. Hybrid FFD-GA results shown in Figure 16 c,d state that the maximum λc_{HDPE} = 25.07 was obtained at 99 m/min, 0.01 mm/rev, and 1.5 mm. In addition, the maximum value of λc_{PA6} = 52.92 was obtained at 75.665 m/min, 0.01 mm/rev, and 1.5 mm.

3.3.3. Multi-Objective Genetic Algorithm Optimization

The multi-objective genetic algorithm (MOGA) methodology was used to solve a mathematical model in which the input process parameters affects the output responses quality [36]. In the current study, multi-objective optimization using genetic algorithm (MOGA) has been used as the objective function in MATLAB 2020's GA Toolbox. The objective functions, fitness functions, are essential in GA to solve optimization problems and regression models were used to be the fitness function of the current optimization problem. The upper and lower boundaries were set based on turning input parameter values (v_c, f, d) and the number of variables was set to three. The MOGA parameters selected were as follows: initial population size was 50, optimization was achieved by setting intermediate crossover with a probability of 0.8 and constraint dependent mutation, the generation size was 300, the migration interval was 20, the migration fraction was 0.2, and the Pareto fraction was 0.35. The result of MOGA is the Pareto optimum, a non-dominated solution, which is a set of solutions that take into account all of the objectives while not losing any of them [37].

MOGA of Ra and MRR

The genetic algorithm, a non-traditional optimization technique, was used to utilize the minimum of Ra and maximum of MRR by considering both as multi-objective functions. For HDPE, the fitness functions of Ra and MRR were Equations (1) and (2), respectively. For PA6, the fitness functions of Ra and MRR were Equations (4) and (5), respectively. Table 4 lists the Pareto front points of Ra and MRR for HDPE and PA6 obtained by MOGA. Figure 17 presents the Pareto chart points of Ra (Objective 1) and MRR (Objective 2) for HDPE and PA6. It was noticed that high MRR produces rough surface. Consequently, the best surface finish can be achieved by sacrificing an increase in MRR. For HDPE, the best Ra (2.19 µm) had a minimal MRR of 2085 mm^3/min and the best MRR (12,024 mm^3/min) had the worst Ra with a value of 5.8 µm. For PA6, the best Ra (2.25 µm) had a minimal MRR of 2523 mm^3/min and the best MRR (24,967.4 mm^3/min) had the worst Ra with a value of 8.77 µm.

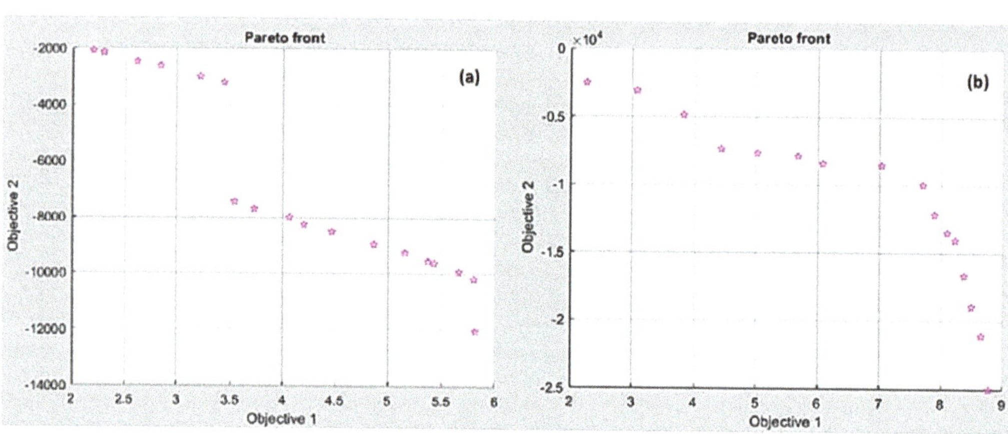

Figure 17. Pareto front chart of Ra and MRR for (**a**) HDPE and (**b**) PA6.

Table 4. Paterian points of Ra and MRR for HDPE and PA6.

No.	HDPE					PA6				
	Conditions			Responses		Conditions			Responses	
	v_c	f	d	Ra	MRR	v_c	f	d	Ra	MRR
1	50.10064	0.01263	1.49849	2.19748	−2085.9045	149.96604	0.09998	1.4999	8.77499	−24967.431
2	149.84927	0.09970	1.49925	5.80914	−12024.143	50.01438	0.01039	1.1419	2.25688	−2523.2370
3	149.57741	0.04108	1.49905	5.42572	−9610.6830	143.79544	0.016179	1.4831	5.01879	−7684.1109
4	149.57899	0.04663	1.49907	5.65631	−9929.6614	149.96604	0.09998	1.4999	8.77499	−24967.431
5	53.91927	0.01265	1.49807	2.61807	−2463.5907	106.11727	0.08151	1.4992	7.88962	−12127.141
6	61.51459	0.01341	1.49789	3.43465	−3201.2878	123.34297	0.09935	1.4989	8.49748	−18926.685
7	149.69267	0.05095	1.49899	5.798399	−10168.033	133.90288	0.09894	1.4990	8.6559	−21049.551
8	149.55036	0.02395	1.49895	4.462317	−8504.0965	91.28852	0.011075	1.4863	3.81448	−4866.6248
9	149.57155	0.030339	1.49898	4.865125	−8937.9101	97.71378	0.09749	1.4975	8.09410	−13485.296
10	148.61716	0.01027	1.49206	3.53617	−7438.4793	145.99145	0.011315	1.4382	4.42666	−7376.4625
11	50.67217	0.01287	1.4974	2.29414	−2150.3586	141.42424	0.02086	1.4867	5.68027	−7869.9244
12	59.24755	0.01333	1.49849	3.21338	−2992.0239	102.11577	0.09706	1.4873	8.22486	−14029.913
13	149.75849	0.02045	1.49892	4.19981	−8259.0439	113.96231	0.04970	1.4930	7.02449	−8528.3537
14	149.35865	0.03504	1.49762	5.15621	−9227.7809	103.23314	0.07284	1.4666	7.70323	−9950.7383
15	148.68008	0.01715	1.49639	4.0611	−7985.3721	114.12946	0.09763	1.4988	8.37259	−16651.477
16	55.16604	0.01355	1.49703	2.84483	−2612.9313	148.09737	0.02223	1.4918	6.07418	−8419.7041
17	148.95875	0.01319	1.49636	3.72502	−7693.2979	57.51669	0.01194	1.3257	3.07088	−3099.2571
18	149.58656	0.03989	1.49899	5.37042	−9539.5601	123.34297	0.09935	1.4989	8.49748	−18926.685

MOGA of Ra and λc

Minimal Ra and maximum λc were used as multi-objective functions using a genetic algorithm. Equations (1) and (3) were the fitness functions of Ra and λc for HDPE, respectively. Equations (4) and (6) were used to calculate the fitness functions of Ra and λc for PA6. Pareto front points of Ra and λc for HDPE and PA6 are listed in Table 5. Figure 18 shows Pareto chart points of Ra (Objective 1) and λc (Objective 2) for (a) HDPE and (b) PA6. It was revealed that the increase in chip ratio produces a rough surface. Therefore, the surface finish can be improved with the sacrifice of an increase in chip ratio. For HDPE, the best Ra (2.76 μm) had a minimal λc of 20.94 and the best chip ratio (25.01) had the worst Ra with a value of 5.01 μm. For PA6, the best Ra (3.04 μm) had a minimal λc of 46.5 and the best chip ratio (52.67) had the worst Ra with a value of 3.45 μm.

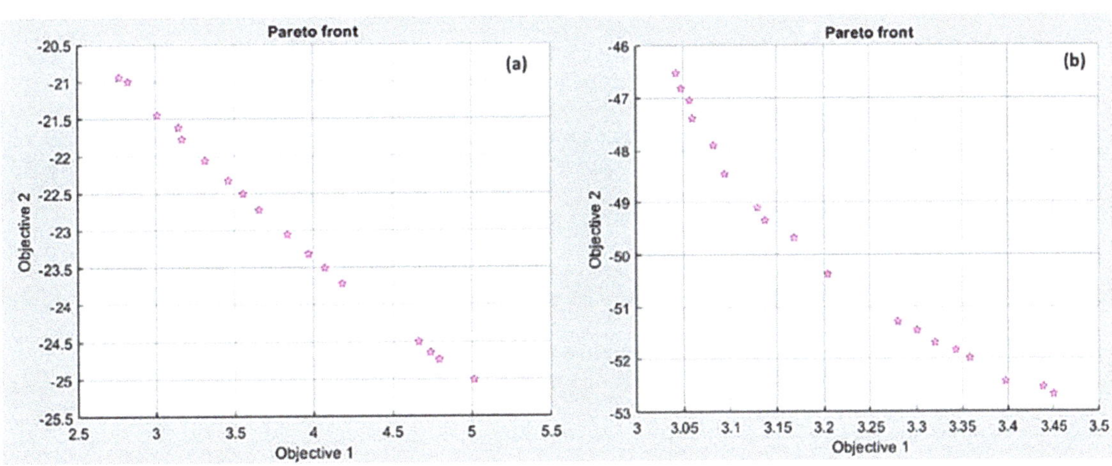

Figure 18. Pareto front chart of Ra and chip ratio for (**a**) HDPE and (**b**) PA6.

Table 5. Paterian points of R_a and λc for HDPE and PA6.

No.	HDPE					PA6				
	Conditions			Responses		Conditions			Responses	
	v_c	f	d	Ra	λc	v_c	f	d	Ra	λc
1	95.207315	0.010036	1.49981	5.011362	−25.011145	59.786200	0.010018	1.2251	3.04235	−46.53234
2	58.077290	0.010086	1.49961	2.765689	−20.945216	62.429872	0.010020	1.4997	3.39714	−52.42544
3	75.848055	0.010051	1.49959	4.180225	−23.717750	60.902873	0.010053	1.4783	3.34333	−51.83081
4	65.792705	0.010068	1.49921	3.459049	−22.332439	59.823397	0.010032	1.4107	3.20404	−50.37390
5	66.964083	0.010090	1.49970	3.552603	−22.511376	59.804024	0.010025	1.2395	3.04768	−46.82421
6	86.659306	0.010063	1.49946	4.739315	−24.649591	59.894496	0.010062	1.2933	3.08193	−47.90476
7	95.207315	0.010036	1.49981	5.011362	−25.011145	60.299039	0.010035	1.4611	3.30060	−51.45451
8	68.287165	0.010051	1.49958	3.653390	−22.729087	59.911380	0.010085	1.3803	3.16817	−49.68074
9	62.426334	0.010067	1.49972	3.167762	−21.768498	59.830667	0.010022	1.3606	3.13724	−49.34672
10	88.119805	0.010047	1.49973	4.793607	−24.743113	59.815306	0.010025	1.3183	3.09407	−48.46213
11	64.085893	0.010063	1.49973	3.311835	−22.057325	60.990353	0.010066	1.4862	3.35857	−51.98181
12	74.097835	0.010067	1.49970	4.069230	−23.506680	59.816572	0.010024	1.2670	3.05946	−47.39907
13	60.690432	0.010069	1.49968	3.011109	−21.453352	59.883487	0.010025	1.2501	3.05674	−47.05214
14	61.994241	0.010225	1.49958	3.146351	−21.619485	60.417895	0.010023	1.4711	3.32010	−51.68791
15	58.576330	0.010174	1.49958	2.823226	−21.003742	66.016942	0.010023	1.4953	3.43826	−52.53416
16	70.669659	0.010076	1.49965	3.833709	−23.058982	59.921783	0.010026	1.3488	3.12919	−49.10542
17	72.587857	0.010062	1.49967	3.967493	−23.320492	60.017329	0.010023	1.4536	3.27977	−51.29148
18	84.806305	0.010121	1.49975	4.664187	−24.504420	66.986410	0.010023	1.4999	3.45013	−52.67488

4. Conclusions

This study dealt with experimental investigations on some aspects of machinability such as surface roughness, material removal rate, and chip formation during the turning of two common types of polymers: HDPE and PA6, difficult-to-cut materials, using a cemented carbide insert (K10) tool.

Based on the experimental results, statistical analysis tools such as ANOVA and 3D response surface plots were evaluated. Moreover, RSM, GA, hybrid FFD-GA, and MOGA were developed to determine the optimal cutting parameters' setting, minimizing surface roughness and maximizing the material removal rate and chip thickness ratio. The following conclusions were drawn:

The most significant factor on Ra_{HDPE} is the depth of cut whereas cutting speed is most significant on Ra_{PA6}. For both materials, HDPE and PA6, cutting speed has the greatest impact on MRR and feed rate has a significant effect on λc for HDPE and PA6.

Ra increases with feed rate and cutting speed; however, the increase of depth of cut produces better surface quality. A high MRR of HDPE is found at high cutting speed, feed rate, and depth of cut, which affects the surface finish negatively. Better surface finish can be achieved at higher chip thickness ratio, which is attained by decreasing the feed rate while increasing both cutting speed and depth of cut.

Satisfactory correlations were acquired between experimental and predicted results for surface roughness criteria, material removal rate, and chip ratio.

The comparison of turning responses' values of experimental results obtained by FFD, optimization results by RSM, optimization results by GA, hybrid FFD-GA, and MOGA for HDPE and PA6 is summarized in Table 6.

Table 6. Summary results of HDPE and PA6 by turning process.

	Response		FFD	RSM	GA	FFD-GA	MOGA	
							Ra, MRR	Ra, λc
HDPE	Ra	Value	2.12075	2.11569	1.90249	1.90249	2.197	2.76
		v_c	50	50.0169	50	50	50.1	58.07
		f	0.01	0.0100328	0.01	0.01	0.012	0.01
		d	1.5	1.47381	1.5	1.5	1.498	1.49
	MRR	Value	12,200	12,206.8	12,039	12,039.1	12,024.143	
		v_c	150	146.782	150	150	149.84	
		f	0.1	0.0993131	0.1	0.1	0.099	
		d	1.5	1.49676	1.5	1.5	1.49	
	λc	Value	26.0365	25.071	25.0711	25.0708		25.01
		v_c	100	99.5332	99.425	99.032		95.2
		f	0.01	0.01	0.01	0.01		0.01
		d	1.5	1.5	1.5	1.5		1.499
PA6	Ra	Value	2.3975	2.3769	2.23008	2.22768	2.25	3.042
		v_c	50	50	50	50	50.01	59.78
		f	0.01	0.01	0.01	0.01	0.0103	0.01
		d	1	1	1.139	1.139	1.149	1.225
	MRR	Value	25,394	24,658.7	24,775.9	24,979.8	24,967.431	
		v_c	150	150	149.113	150	149.96	
		f	0.1	0.1	0.1	0.1	0.099	
		d	1.5	1.5	1.5	1.5	1.49	
	λc	Value	55.0771	52.9358	52.9293	52.9296		52.67
		v_c	100	77.1147	75.735	75.665		66.98
		f	0.01	0.01	0.01	0.01		0.01
		d	1.5	1.5	1.5	1.5		1.49

Author Contributions: Conceptualization, Y.E.-T., S.E.-S. and H.K.; methodology, Y.E.-T. and W.H.E.-G.; software, H.K.; validation, S.E.-S., H.K. and A.I.A.; formal analysis, Y.E.-T., S.E.-S. and H.K.; investigation, Y.E.-T.; resources, A.I.A.; data curation, W.H.E.-G.; writing—original draft preparation, Y.E.-T., S.E.-S. and H.K.; writing—review and editing, A.I.A. and W.H.E.-G. visualization, Y.E.-T.; supervision, Y.E.-T., S.E.-S. and H.K.; project administration A.I.A.; and W.H.E.-G. All authors have read and agreed to the published version of the manuscript.

Funding: This research received no external funding.

Institutional Review Board Statement: Not applicable.

Informed Consent Statement: Not applicable.

Data Availability Statement: All the raw data supporting the conclusion of this paper were provided by the authors.

Acknowledgments: The researchers would like to thank the Deanship of Scientific Research, Qassim University, for funding the publication of this project.

Conflicts of Interest: The authors declare no conflict of interest.

References

1. Bozdemir, M. The Effects of Humidity on Cast PA6G during Turning and Milling Machining. *Adv. Mater. Sci. Eng.* **2017**, *2017*, 5408691. [CrossRef]
2. Palanikumar, K.; Rajasekaran, T.; Latha, B. Fuzzy rule-based modeling of machining parameters for surface roughness in turning carbon particle-reinforced polyamide. *J. Thermoplast. Compos. Mater.* **2015**, *28*, 1387–1405.
3. Patel, P.; Chaudhary, V.; Patel, K.; Gohil, P. Milling of Polymer Matrix Composites: A Review. *Int. J. Appl. Eng. Res.* **2018**, *13*, 7455–7465.
4. Dehghan Manshadi, M.; Alafchi, N.; Tat, A.; Mousavi, M.; Mosavi, A. Comparative Analysis of Machine Learning and Numerical Modeling for Combined Heat Transfer in Polymethylmethacrylate. *Polymers* **2022**, *14*, 1996.

5. Gnatowski, A.; Gołębski, R.; Sikora, P. Analysis of the impact of changes in thermomechanical properties of polymer materials on the machining process of gears. *Polymers* **2021**, *13*, 28.
6. Mehdipour-Ataei, S.; Tabatabaei-Yazdi, Z. Heat Resistant Polymers. In *Encyclopedia of Polymer Science and Technology*; John Wiley & Sons: Hoboken, NJ, USA, 2015; pp. 1–31. ISBN 0471440264.
7. Wilczyński, K.; Wilczyński, K.J.; Buziak, K. Modeling and Experimental Studies on Polymer Melting and Flow in Injection Molding. *Polymers* **2022**, *14*, 2106.
8. Tushar, U.; Jagtap, H.A.M. Machining of Plastics: A Review. *Int. J. Eng. Res. Gen. Sci.* **2015**, *3*, 577–581.
9. Karataş, M.A.; Gökkayab, H. A review on machinability of carbon fiber reinforced polymer (CFRP) and glass fiber reinforced polymer (GFRP) composite materials. *Def. Technol.* **2018**, *14*, 318–326.
10. Sheikh-Ahmad, J.Y. *Machining of Polymer Composites*; Springer: Berlin/Heidelberg, Germany, 2009; ISBN 978-0-387-35539-9.
11. Corrêa, H.L.; Rodrigues, R.V.; da Costa, D.D. Machining process of glass-fiber-reinforced polyamide 6.6 Composite: Pathways to improve the drilling of recycled polymers. *Eng. Res. Express* **2020**, *2*, 015037.
12. Balan, A.S.S.; Kannan, C.; Jain, K.; Chakraborty, S.; Joshi, S.; Rawat, K.; Alsanie, W.F.; Thakur, V.K. Numerical modelling and analytical comparison of delamination during cryogenic drilling of cfrp. *Polymers* **2021**, *13*, 3995.
13. Gaitonde, V.N.; Karnik, S.R.; Rubio, J.C.; Abrao, A.M.; Correia, A.E.; Davim, J.P. Surface roughness analysis in high-speed drilling of unreinforced and reinforced polyamides. *J. Compos. Mater.* **2011**, *46*, 2659–2673.
14. Soleymani Yazdi, M.R.; Razfar, M.R.; Asadnia, M. Modelling of the thrust force of the drilling operation on PA6–nanoclay nanocomposites using particle swarm optimization. *J. Eng. Manuf.* **2011**, *225*, 1757–1771.
15. Kuram, E. Micro-machinability of injection molded polyamide 6 polymer and glass-fiber reinforced polyamide 6 composite. *Compos. Part B* **2016**, *88*, 85–100.
16. Yan, Y.; Mao, Y.; Li, B.; Zhou, P. Machinability of the thermoplastic polymers: Peek, pi, and pmma. *Polymers* **2021**, *13*, 69.
17. Moghri, M.; Madic, M.; Omidi, M.; Farahnakian, M. Surface Roughness Optimization of Polyamide-6/Nanoclay Nanocomposites Using Artificial Neural Network: Genetic Algorithm Approach. *Sci. World J.* **2014**, *2014*, 485205. [CrossRef]
18. Dhokia, V.G.; Kumar, S.; Vichare, P.; Newman, S.T.; Allen, R.D. Surface roughness prediction model for CNC machining of polypropylene. *J. Eng. Manuf.* **2008**, *222*, 137–153.
19. Raja Abdullah, R.I.; Yu Long, A.; Mohd Amran, M.A.; Kasim, M.S.; Mohd Hadzley, A.B.; Subramonian, S. Optimisation of Machining Parameters for Milling Polyetheretherketones (PEEK) Biomaterial. *Appl. Mech. Mater.* **2015**, *699*, 198–203.
20. Kumar, J.; Verma, R.K.; Mondal, A.K.; Singh, V.K. A hybrid optimization technique to control the machining performance of graphene/carbon/polymer (epoxy) nanocomposites. *Polym. Polym. Compos.* **2021**, *29*, S1168–S1180.
21. Mata, F.; Petropoulos, I.G.; Ntziantzias, J.P.D. A surface roughness analysis in turning of polyamide PA-6 using statistical techniques. *Int. J. Mater. Prod. Technol.* **2010**, *37*, 173–187.
22. Madić, M.; Marinković, V.; Radovanović, M. Mathematical modeling and optimization of surface roughness in turning of polyamide based on artificial neural network. *Mechanika* **2012**, *18*, 574–581.
23. Asghar, A.; Abdul Raman, A.A.; Daud, W.M.A.W. A comparison of central composite design and Taguchi method for optimizing Fenton process. *Sci. World J.* **2014**, *2014*, 869120.
24. Paulo Davim, J.; Reis, P.; Lapa, V.; Conceiçao António, C. Machinability study on polyetheretherketone (PEEK) unreinforced and reinforced (GF30) for applications in structural components. *Compos. Struct.* **2003**, *62*, 67–73. [CrossRef]
25. Fountas, N.A.; Ntziantzias, I.; Kechagias, J.; Koutsomichalis, A.; Davim, J.P.; Vaxevanidis, N.M. Prediction of Cutting Forces during Turning PA66 GF-30 Glass Fiber Reinforced Polyamide by Soft Computing Techniques. *Mater. Sci. Forum* **2013**, *766*, 37–58.
26. Aldwell, B.; Hanley, R.; O'Donnell, G.E. Characterising the machining of biomedical grade polymers. *J Eng. Manuf.* **2014**, *228*, 1237–1251. [CrossRef]
27. Kaddeche, M.; Chaoui, K.; Yallese, M.A. Cutting parameters effects on the machining of two high density polyethylene pipes resins: Cutting parameters effects on HDPE machining. *Mech. Ind.* **2012**, *13*, 307–316.
28. Hamlaoui, N.; Azzouz, S.; Chaoui, K.; Azari, Z.; Yallese, M.A. Machining of tough polyethylene pipe material: Surface roughness and cutting temperature optimization. *Int. J. Adv. Manuf. Technol.* **2017**, *92*, 2231–2245.
29. Raj, I.J.A.; Vijayakumar, P.; Kannan, T.; Kumar, P.; Ragavan, R.V. Design optimization of turning parameters of PTFE (Teflon) cylindrical rods using ANOVA Methodology. *Int. J. Appl. Eng. Res.* **2016**, *11*, 518–523.
30. Chabbi, A.; Yallese, M.A.; Nouioua, M.; Meddour, I.; Mabrouki, T.; Girardin, F. Modeling and optimization of turning process parameters during the cutting of polymer (POM C) based on RSM, ANN, and DF methods. *Int. J. Adv. Manuf. Technol.* **2017**, *91*, 2267–2290. [CrossRef]
31. Kilickap, E.; Huseyinoglu, M.; Yardimeden, A. Optimization of drilling parameters on surface roughness in drilling of AISI 1045 using response surface methodology and genetic algorithm. *Int. J. Adv. Manuf. Technol.* **2011**, *52*, 79–88.
32. Dadrasi, A.; Fooladpanjeh, S.; Gharahbagh, A.A. Interactions between HA/GO/epoxy resin nanocomposites: Optimization, modeling and mechanical performance using central composite design and genetic algorithm. *J. Braz. Soc. Mech. Sci. Eng.* **2019**, *41*, 63.
33. Davim, J.P.; Silva, L.R.; Festas, A.; Abrão, A.M. Machinability study on precision turning of PA66 polyamide with and without glass fiber reinforcing. *Mater. Des.* **2009**, *30*, 228–234. [CrossRef]
34. Shahabaz, S.M.; Sharma, S.; Shetty, N.; Shetty, S.D.; Gowrishankar, M.C. Influence of Temperature on Mechanical Properties and Machining of Fibre Reinforced Polymer Composites: A Review. *Eng. Sci.* **2021**, *16*, 26–46. [CrossRef]

35. Hazir, E.; Ozcan, T. Response surface methodology integrated with desirability function and genetic algorithm approach for the optimization of CNC machining parameters. *Arab. J. Sci. Eng.* **2019**, *44*, 2795–2809. [CrossRef]
36. Antil, P.; Singh, S.; Kumar, S.; Manna, A.; Katal, N. Taguchi and multi-objective genetic algorithm-based optimization during ecdm of sicp/glass fibers reinforced pmcs. *Indian J. Eng. Mater. Sci.* **2019**, *26*, 211–219.
37. Janahiraman, T.V.; Ahmad, N. Multi Objective Optimization for Turning Operation using Hybrid Extreme Learning Machine and Multi Objective Genetic Algorithm. *Int. J. Eng. Technol.* **2018**, *7*, 876. [CrossRef]

polymers
MDPI

Article

Effects of Melt Temperature and Non-Isothermal Flow in Design of Coat Hanger Dies Based on Flow Network of Non-Newtonian Fluids

Amin Razeghiyadaki [1], Dongming Wei [1,*], Asma Perveen [2], Dichuan Zhang [3] and Yanwei Wang [4,5]

[1] Department of Mathematics, School of Sciences and Humanities, Nazarbayev University, Nur-Sultan 010000, Kazakhstan; amin.razeghiyadaki@nu.edu.kz
[2] Department of Mechanical & Aerospace Engineering, School of Engineering and Digital Sciences, Nazarbayev University, Nur-Sultan 010000, Kazakhstan; asma.perveen@nu.edu.kz
[3] Department of Civil & Environmental Engineering, School of Engineering and Digital Sciences, Nazarbayev University, Nur-Sultan 010000, Kazakhstan; dichuan.zhang@nu.edu.kz
[4] Department of Chemical & Materials Engineering, School of Engineering and Digital Sciences, Nazarbayev University, Nur-Sultan 010000, Kazakhstan; yanwei.wang@nu.edu.kz
[5] Laboratory of Computational Materials Science, Center for Energy and Advanced Materials Science, National Laboratory Astana, Nur-Sultan 010000, Kazakhstan
* Correspondence: dongming.wei@nu.edu.kz

Abstract: In the design of coat hanger extrusion dies, the main objective is to provide a uniform flow rate at the die exit. Previously, a multi-rheology isothermal method model for coat hanger extrusion dies was developed to reach this objective. Polymer melts in extrusion dies commonly experience high shear rates. Viscous dissipation rooted by high shear rate may lead to significant temperature differences across the die. Due to temperature-dependency of viscosity, temperature differences may lead to nonuniform flow rates, which may significantly affect the flow rate at the die exit. As a result, a new design method is proposed to take into account the effects of temperature and viscous dissipation in the design of coat hanger dies. Although more non-Newtonian fluid rheology models can be adapted in the proposed study, as demonstration, temperature-dependent power-law and Carreau–Yasuda models are adapted in this study. Performances are compared with our isothermal method published earlier. In addition, the novel nonisothermal method is comprehensively examined where the effect of viscous dissipation is studied through Brinkman number of extrusion die. It is demonstrated that, for a low Brinkman number, both isothermal and nonisothermal design give similar flow uniformity level. However, for higher Brinkman numbers, the proposed nonisothermal method produces a design with more desirable velocity uniformity level along with a maximum improvement of 5.24% over the isothermal method. In addition, dependency of flow field on temperature, due to temperature-dependent viscosity, is studied, and it is demonstrated that fully-developed velocity profile changes as temperature increases along the flow channel. Moreover, the effect of the temperature sensitivity parameter in temperature-dependent non-Newtonian models is considered. It is demonstrated that the temperature boundary condition with the Biot number of 1.0 gives adequate results for lower values of the temperature sensitivity parameter.

Keywords: polymer processing; sheet die design; coat-hanger die; non-isothermal non-Newtonian fluids; constant shear-rate die; viscous dissipation; temperature effects

check for updates

Citation: Razeghiyadaki, A.; Wei, D.; Perveen, A.; Zhang, D.; Wang, Y. Effects of Melt Temperature and Non-Isothermal Flow in Design of Coat Hanger Dies Based on Flow Network of Non-Newtonian Fluids. *Polymers* **2022**, *14*, 3161. https://doi.org/10.3390/polym14153161

Academic Editors: Ying Yan, Yiqi Wang and Marcelo Antunes

Received: 4 July 2022
Accepted: 27 July 2022
Published: 3 August 2022

1. Introduction

The most challenging aspect of extrusion die design is providing a uniform exit velocity profile, hence no or reduced correction is required by adjusting bolts. Nonlinear behavior of viscosity of polymer melts and complicated multi-physics of extrusion dies make extrusion die design a challenging task. In addition, temperature rise due to viscous dissipation and temperature dependency of viscosity adds more complexity to the flow rate profile at die exit where a uniform profile is needed. Since the internal geometry of die and the process conditions determine optimal performance of extrusion dies, numerical methods are required to carry out inter-dependency of fluid flow and its dependency on temperature and nonlinear viscosity. Generally, a trial and error approach is adapted to test different designs to reach to an optimal design [1,2]. This approach requires a very high number of costly simulations to be carried out.

High computational cost of numerical methods limits numerical design optimization in engineering applications. Thus, introduction of simplified methods with low computational cost with efficiency is necessary. Different design methods based on analytical [3–5], semi-analytical [6,7] and numerical optimization were developed [8]. Winter and Fritz [3] proposed the first analytical design method for rectangular and circular-shaped sheeting extrusion dies. Degradation appears to be the main issue in extrusion of temperature-sensitive polymers. Awe et al. [4] proposed a shortened Winter–Fritz model to avoid this issue.

Semi-analytical methods based on a flow network method (FNM) are also reported in the literature. A flow network analysis, also known as a hydraulic-electric circuit analogy, uses the conventional concept of electric circuit theory for an analysis of fluid flow problems [9]. Some literature demonstrates [7,9] the application of this method for coat-hanger die design with different rheology models. Michaeli et al. [1] combined the finite element method with the network theory to find the optimum velocity distribution. Optimized designs for rectangular and circular shaped manifolds are achieved using power-law and Carreau–Yasuda rheology models. In addition, Yilmaz and Kadikkopru [7] developed a model based on FNM which successfully achieved optimal geometrical parameters in the manifold leading to uniform velocity profiles. Using a trial and error approach, Igali et al. [10] optimized flow distribution of sheeting extrusion die with finite element simulations. The above researchers did not consider temperature effects in their designs.

Due to temperature dependency of viscosity, temperature field can change flow field. As a result, the effect of temperature on flow uniformity was studied in literature. Lebaal et al. [11] optimized wall temperature and flow rate of a wire-coating die with Kriging interpolation and a sequential quadratic programming algorithm optimization method. Their study concluded with the possibility to design a coat-hanger wire-coating die for a different polymer range and flow rates. In another study, Lebaal et al. [12] studied the effectiveness of the response surface method and Sequential Quadratic Programming for optimization of wall temperature to gain the best velocity profile at die exit. Wu et al. [13] also developed a one-dimensional model that takes into account the effect of temperature variation and its effect on flow uniformity. In another study, Lebaal et al. [14] optimized the wall temperature profile of a coat-hanger extrusion die. By changing wall temperature, viscosity of the fluid in the vicinity of the wall can be changed and consequently flow rate can be manipulated. In a further study, Lebaal [12] presents a Kriging Swarm Optimization (KSO) algorithm for optimization of three geometrical design parameters to achieve a uniform velocity profile of a coat hanger die.

To the authors' best knowledge, no analytical or semi-analytical design method of a coat hanger die which takes into account the effects of temperature exists in literature. In our previous study, a new isothermal design method based on constant shear rate and uniform velocity assumptions was proposed [6]. It is known that the design of a coat hanger dies depends on wall shear rate, temperature and heat dissipation of polymer melt [15]. Therefore, a previous model is modified to take into account the effects of temperature and heat dissipation, and its subsequent effect on apparent viscosity of the polymer. This newly proposed model takes into account different non-Newtonian models. Based on our proposed model, new design curves are produced, and their performance is discussed through pressure, temperature and shear rate distribution.

2. Model and Methods

A sheeting extrusion die is made of two parts, the slit and the manifold, as shown in Figure 1. Due to symmetry, only half of the die is considered here. Since a uniform flow is desired at the exit of the die, the manifold has to be designed in such a manner that satisfies this goal. Manifold can be characterized by two geometrical parameters, i.e., distance from center of manifold to die exit $y(x)$ and radius $R(x)$ of the manifold. Flow rates in each manifold and slit segment are denoted by $Q_m(i)$ and $Q_s(i)$, respectively. Vertical and horizontal distance between two adjacent manifold segments are denoted by Δy and Δx, respectively. Length between to adjacent manifold nodes is denoted by $\Delta\zeta$ and is equal to $\sqrt{\Delta x^2 + \Delta y^2}$.

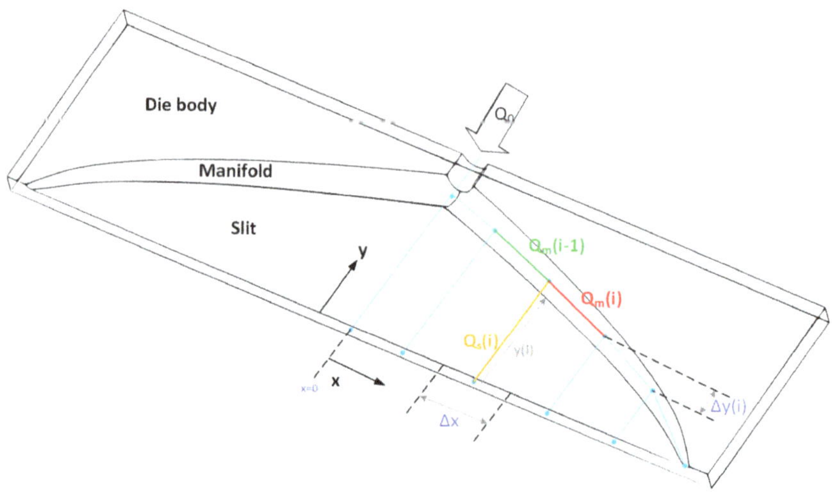

Figure 1. Schematic of an extrusion die.

2.1. Flow Network Method for Die Design

In our previous work [6], we proposed simplifying the die into a series of segments where fluid flows in each segment can be described analytically or numerically. As a result of segmentation, two continuous geometrical parameters reduce to two sets of discrete parameters. Two sets of equations are needed to be solved simultaneously to find these two geometrical parameters of extrusion. The following assumptions are made for the model proposed in this study:

- Steady-state, non-isothermal, incompressible flow;
- Streamlined flow;
- Uniform pressure and flow rates at die exit;
- Unidirectional and fully-developed flow in both manifold and slit;
- Constant wall shear rate in the manifold and the slit.

The constant wall shear rate assumption results in the following equation:

$$\dot{\gamma}_m(R) = \dot{\gamma}_s(y) \quad (1)$$

where $\dot{\gamma}$ is shear rate. Subscripts m and s refer to manifold and slit. Constant wall shear rate also results in the following equation known as the Winter–Fritz [3] equation:

$$\frac{dy}{dx} = -\left[\left(\frac{(dp/dy)_s}{(dp/d\zeta)_m}\right)^2 - 1\right]^{-1/2}; \tag{2}$$

in discretized form, Equation (2) can be written as follows:

$$(y_{i+1} - y_i) = -(x_{i+1} - x_i)\left[\left(\frac{\left(\frac{dp}{dy}\right)_s(i)}{\left(\frac{dp}{d\zeta}\right)_m(i)}\right)^2 - 1\right]^{-1/2} \tag{3}$$

$$(y_{i+1} - y_i) = -(x_{i+1} - x_i)\left[\left(\frac{\frac{\Delta p_{s,i}}{L_{s,i}}}{\frac{\Delta p_{m,i}}{L_{m,i}}}\right)^2 - 1\right]^{-1/2} \tag{4}$$

where $\Delta p_{s,i}$, $\Delta p_{m,i}$, $L_{s,i}$ and $L_{m,i}$ are pressure drops and corresponding length of the slit and the manifold, respectively. Since no analytical simple equation is available for pressure gradient of nonisothermal non-Newtonian fluid flows, a numerical method is adapted. Since uniform flow rates is assumed at die exit:

$$Q_s(i) = Q_s = \frac{Q_0}{N+1} \quad \text{where } 1 \leq i \leq N \tag{5}$$

where N and Q are the total number of segments and flow rate. Mass conservation for each node on manifold gives the following:

$$Q_m(i-1) = Q_m(i) + Q_s(i) \quad \text{where} \quad 1 \leq i \leq N \tag{6}$$

and

$$Q_m(N) = Q_s \quad Q_m(0) = Q_0 \tag{7}$$

from Equations (5) and (6); flow rates in both manifold and slit are calculated.

The algorithm of calculations is shown in Figure 2. The first step is initialization of parameters and calculation of flow rates in each segment. Next, shear rates, pressure drops and outlet temperature are calculated. With calculated shear rate, the radius of manifold is calculated by Equation (4). Lastly, with calculated pressure drops, new y-values are calculated by Equation (4). Calculations are repeated until convergence is reached. The convergence criterion is set to 10^{-5}. The convergence criterion is defined as follows:

$$max\left\{\left|\frac{y(i)^{\text{old}} - y(i)^{\text{new}}}{y(i)^{\text{old}}}\right|, \left|\frac{R(i)^{\text{old}} - R(i)^{\text{new}}}{R(i)^{\text{old}}}\right|\right\} < \varepsilon \quad 1 < i < N \tag{8}$$

Computational times were between 20 to 120 min on a core i9 personal computer.

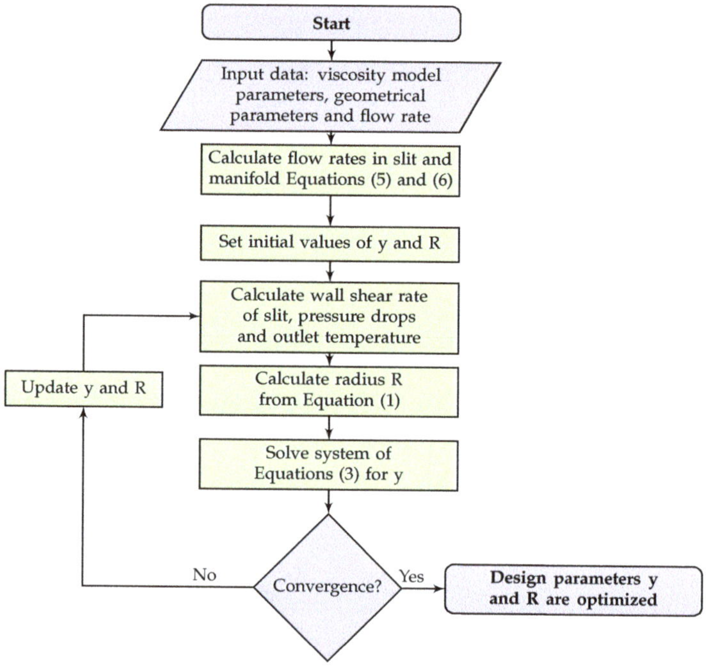

Figure 2. Algorithm used in the proposed flow network method for die design.

2.2. Non-Newtonian Models

For demonstration, temperature-dependent power-law and Carreau–Yasuda rheology models are considered. The shear-rate and temperature-dependent viscosity function of the Power-law model is as follows:

$$\eta(\dot{\gamma}, T) = m(T)\dot{\gamma}^{n-1} \tag{9}$$

where $m(T)$ is often referred to as the flow consistency index and n as the flow behavior index. The temperature dependence of the flow consistency index is given by

$$m(T) = m_0 \exp\left[-\alpha_{pl}(T - T_{\text{ref}})\right] \tag{10}$$

where α_{pl} is the temperature sensitivity parameter.

The Carreau–Yasuda model with temperature dependence is given as follows [16]:

$$\eta(\dot{\gamma}, T)/a_T = \eta_\infty + (\eta_0 - \eta_\infty)[1 + (a_T\dot{\gamma}\lambda)^a]^{(n-1)/a} \tag{11}$$

where $a_T = \exp[-\alpha_{cs}(T - T_{\text{ref}})]$ is often referred to as the shift factor; α_{cs} corresponds to the temperature sensitivity parameter.

2.3. Pressure Drop Calculation

In each segment of the manifold, the flow is assumed to be unidirectional. In addition, fluid flow is assumed to be laminar, hydrodynamically fully developed and thermally developing in channel segment direction. The non-Newtonian fluid enters at temperature T_0, and the effect of viscous dissipation is considered here. Flow in a slit segment can be assumed as a planar flow between two parallel plates and flow in a manifold segment can be assumed as pipe flow for a circular manifold, respectively. Polymer melts have high Prandtl number $\text{Pr} = c_p\mu/k_t$, where k_t, μ, and c_p are the thermal conductivity, apparent viscosity and heat capacity, respectively; and therefore velocity profile can be assumed to be fully-developed [17–22]. Due to coupling between velocity and temperature fields, energy and momentum equations are needed to be solved simultaneously. By solving the following equations, velocity and temperature profiles can be calculated for planar ($\kappa = 0$) and pipe flows ($\kappa = 1$):

$$\begin{array}{r l} \text{Momentum balance}: & -\frac{dp}{dx_2} + \frac{d\tau}{dx_1} = 0 \\ \text{B.C.1}: & \frac{\partial u}{\partial x_1} = 0 \quad \text{at}\, x_1 = 0 \\ \text{B.C.2}: & u = 0 \quad \text{at}\, x_1 = L_c \end{array} \tag{12}$$

$$\begin{array}{r l} \text{Energy balance}: & \rho c_p u \frac{\partial T}{\partial x_2} = \frac{k_t}{x_1^\kappa}\frac{\partial}{\partial x_1}\left(x_1^\kappa \frac{\partial T}{\partial x_1}\right) + \eta \left(\frac{du}{dx_1}\right)^2 \\ \text{B.C.1}: & T = T_0 \quad \text{at} \quad x_2 = 0 \\ \text{B.C.2}: & \frac{\partial T}{\partial x_1} = 0 \quad \text{at} \quad x_1 = 0 \\ \text{B.C.3}: & -k_t \frac{\partial T}{\partial x_1} = h(T - T_R) \quad \text{at} \quad x_1 = L_c \end{array} \tag{13}$$

where u, T, p, τ, x_1, x_2 and L_c are velocity, temperature, pressure, shear stress and transverse, axial coordinates and characteristic length (radius of manifold segment or height of slit segment), respectively. It is worth notiing that both Equations (12) and (13) are valid for both planar flow and pipe flows. In case of pipe flow, x_1 is the radius, while, for the case of planar flow, x_1 is the distance from the center plane between the two parallel plates. In other words, $x_1 = 0$ is axis of pipe or mid plane between the two parallel plates. Due to symmetry, boundary conditions at x_1 for both momentum and energy balance equations are set to zero gradient. Physical properties such as viscosity, density, constant pressure heat capacity and thermal diffusivity are denoted by μ, ρ, c_p and α_T, respectively. Apparent viscosity μ is a function of both shear rate and temperature. Shear stress and thermal diffusivity by definition are given as follows:

$$\tau = \eta(\dot{\gamma}, T)\dot{\gamma} \quad \alpha_T = \frac{k_t}{\rho c_p} \tag{14}$$

By solving these Equations (12) and (13), temperature, pressure drop and wall shear rate (velocity gradient at wall) can be calculated.

By non-dimensionalization of the momentum and the energy equations (Equations (12) and (13)) by the parameters in Table 1, we have for the momentum balance equation:

$$S = \frac{d\tilde{u}}{d\tilde{x}_1} = \frac{\tilde{x}_1}{\tilde{\eta}} \tag{15}$$

and for the energy balance equation:

$$\begin{array}{l} \tilde{u}\frac{\partial \tilde{T}}{\partial \tilde{x}_2} = \frac{1}{\tilde{x}_1^\kappa}\frac{\partial}{\partial \tilde{x}_1}\left(\tilde{x}_1^\kappa \frac{\partial \tilde{T}}{\partial \tilde{x}_1}\right) + \text{Br}\, S \tilde{x}_1 \\ \text{B.C.1}: \quad \tilde{T} = 1 \quad at \quad \tilde{x}_2 = 0 \\ \text{B.C.2}: \quad \frac{\partial \tilde{T}}{\partial \tilde{x}_1} = 0 \quad at \quad \tilde{x}_1 = 0 \\ \text{B.C.3}: \quad -\frac{\partial \tilde{T}}{\partial \tilde{x}_1} = \text{Bi}\, \tilde{T} \quad at \quad \tilde{x}_1 = L_c \end{array} \tag{16}$$

the Brinkman number, and the Biot number are defined as:

$$\text{Br} = \frac{\tau_w \dot{\gamma}_w L_c^2}{k_t(T_0 - T_R)} \tag{17}$$

$$\text{Bi} = \frac{h_{ext} L_c}{k} \tag{18}$$

the non-dimensional viscosity $\tilde{\eta}(\dot{\gamma}, T)$ for the power-law model (Equation (9)) and the Carreau–Yasuda model (Equation (11)) are defined in Tables 2 and 3, respectively, where shear rate at wall is calculated as follows:

$$\begin{aligned} \eta(\dot{\gamma}_w, T_w)\dot{\gamma}_w &= \frac{\Delta p}{2LR} \quad \text{for } \kappa = 1 \text{ (manifold segment flow)} \\ \eta(\dot{\gamma}_w, T_w)\dot{\gamma}_w &= \frac{\Delta p}{2LH} \quad \text{for } \kappa = 0 \text{ (slit segment flow)} \end{aligned} \tag{19}$$

the characteristic length L_c depends on the cross-section (radius r_0 for circular and half distance of plates $H/2$ for planar channel flow). Arpin et al. [23] suggested a Biot number of 1.0 as the temperature boundary condition of coat hanger dies. Our computational fluid dynamics (CFD) simulation confirmed this assumption. The algorithm for calculation of pressure drop is shown in Figure 3. Discretization results are as follows:

$$\frac{\tilde{u}(i+1) - \tilde{u}(i)}{\tilde{x}_1(i+1) - \tilde{x}_1(i)} = \frac{\tilde{x}_1(i)}{\tilde{\eta}(i)} \tag{20}$$

$$\begin{aligned} &\left[\frac{\tilde{u}(i)}{\Delta\tilde{x}_2} + \frac{2}{(\Delta\tilde{x}_1)^2}\right]\tilde{T}(i,j) + \left[\frac{2}{(\Delta\tilde{x}_1)^2} + \frac{\kappa}{\tilde{x}_1(i)\Delta\tilde{x}_1}\right]\tilde{T}(i-1,j) \\ &\quad + \left[\frac{2}{(\Delta\tilde{x}_1)^2} + \frac{-\kappa}{\tilde{x}_1(i)\Delta\tilde{x}_1}\right]\tilde{T}(i+1,j) = \frac{\tilde{u}(i)}{\Delta\tilde{x}_2}\tilde{T}(i,j-1) + \mathrm{Br}S_i^2 \end{aligned} \tag{21}$$

Table 1. Non-dimensionalization used in the energy equation.

Parameter	Dimension	Non-Dimensional
distance	x_1	$\tilde{x}_1 = x_1/L_c$
axial distance	x_2	$\tilde{x}_2 = x_2/(\mathrm{Pe}\,L_c)$
velocity	u	$\tilde{u} = u/(\dot{\gamma}_w L_c)$
temperature	T	$\tilde{T} = (T - T_R)/(T_0 - T_R)$

Table 2. Non-dimensionalized power-law viscosity function.

	$\tilde{\eta}(S, \tilde{T}) = a'_T S^{n-1}$
a_T	$m'_{0,pl} e^{-\alpha'_{pl}\tilde{T}}$
$m'_{0,pl}$	$e^{-\alpha_{pl}(T_R - T_{ref})}$
α'_{pl}	$\alpha_{pl}(T_0 - T_R)$

Table 3. Non-dimensionalized Carreau–Yasuda viscosity function.

	$\tilde{\eta}(S, \tilde{T}) = a'_T\,\tilde{\eta}_\infty + a'_T(\tilde{\eta}_0 - \tilde{\eta}_\infty)[1 + (a'_T\lambda' S)^a]^{\frac{n-1}{a}}$
$\tilde{\eta}_\infty$	η_∞/η_w
$\tilde{\eta}_0$	η_0/η_w
λ'	$\lambda\dot{\gamma}_w$
a'_T	$m'_{0,cs} e^{\alpha'\tilde{T}}$
$m'_{0,cs}$	$e^{\alpha_{cs}(T_R - T_{ref})}$
α'_{cs}	$\alpha_{cs}(T_0 - T_R)$

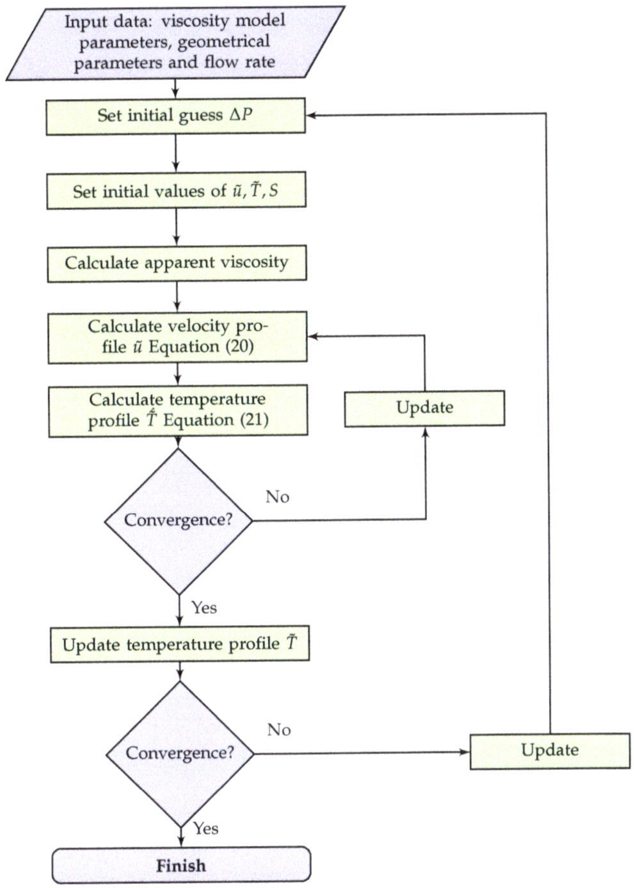

Figure 3. Algorithm of calculation of pressure drop.

2.4. *Verification and Validation by CFD*

CFD simulation is used as a tool to compare the results of the proposed model and those of the previous method [6]. The Ansys Fluent 2021.R2 finite volume CFD software package is adapted for this purpose. Computation domain is shown in Figure 4. Due to symmetry, only a quarter of coat-hanger dies are considered as the computation domain. The continuity, momentum and energy equations must be considered and solved simultaneously:

$$\nabla \cdot \mathbf{U} = 0 \tag{22}$$

$$\rho \mathbf{U} \nabla \cdot \mathbf{U} = -\nabla p + \nabla \cdot \tau \tag{23}$$

$$\rho c_p (\mathbf{U} \cdot \nabla) T = \nabla \cdot (k \nabla T) + \tau : \nabla \mathbf{U} \tag{24}$$

At the inlet of the computation domain, uniform velocity and temperature profile are applied. At the die exit, zero gauge pressure is assumed. Non-slip and Biot = 1 (Equation (13)) boundary conditions are applied to solid walls. A C++ user defined function is written to take into account viscous dissipation as a source term in the energy equation. For accurate representation of 3D geometry, a Python code in Ansys Spaceclaim 19.1 is written.

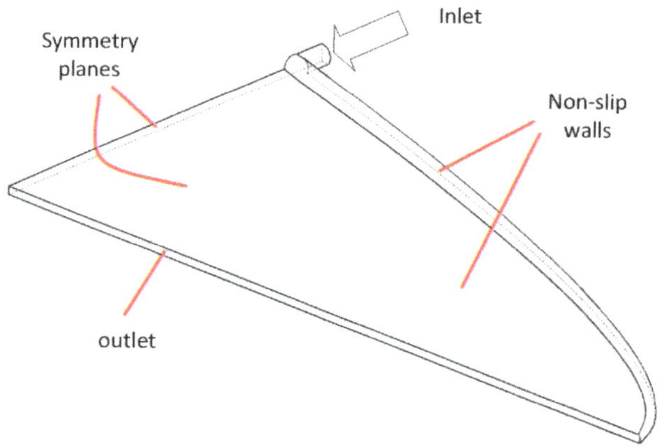

Figure 4. Computational domain and boundary conditions for CFD analysis.

3. Results and Discussion

3.1. Design Curves

In this work, values of the power-law model parameters are m_0 = 17,092 Pa·s^n, $n = 0.32$, $\alpha_{pl} = 0.011624\,\mathrm{K}^{-1}$, and $T_{\mathrm{ref}} = 473.15\,\mathrm{K}$; The Carreau–Yasuda model parameters are $\eta_0 = 8234\,\mathrm{Pa{\cdot}s}$, $\eta_\infty = 0\,\mathrm{Pa{\cdot}s}$, $\lambda = 0.129\,\mathrm{s}$, $n = 0.217$, $a = 0.468$, $\alpha_{cs} = 0.025\,\mathrm{K}^{-1}$ and $T_{\mathrm{ref}} = 473.15\,\mathrm{K}$. Figure 5 shows viscosity against shear rate of rheology models for both power-law and Carreau–Yasuda models.

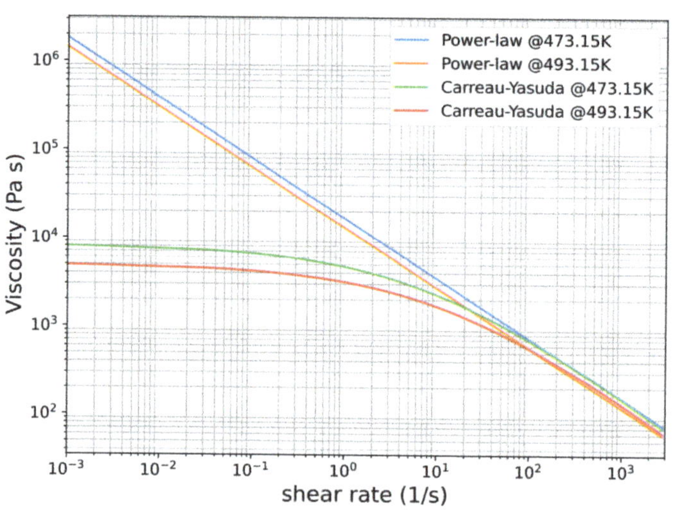

Figure 5. Viscosity versus shear rate of Power-law and Carreau–Yasuda at 473.15 K and 493.15 K, respectively.

Figure 6 depicts nonisothermal design curves for the power-law and the Carreau–Yasuda fluids. Input parameters for design and optimization are shown in Table 4. The isothermal power-law designs [6] for the otherwise same given parameters are also shown in Figure 6 as a comparison. Due to temperature changes across the die, calculated wall shear rate is non-uniform. This results in different radii as given by Equation (1). Design curves of y and radius for both power-law and Carreau–Yasuda fluids are almost the same. Calculated wall shear rates are in the range of 500 to 600 s^{-1}. In this range, both power-law and Carreau–Yasuda fluids behave in a similar manner as shown in Figure 5. Fluid flows through the manifold, it accumulates more energy from viscous dissipation and bulk temperature in manifold increases as shown in Figure 6c. The die has longer die land length at the center than its edges (Figure 1), thus fluid exits the slit at a higher temperature at the center than the edges as shown in Figure 6d–f, which depicts Brinkman numbers in the manifold and the slit, respectively. Flow rate decreases in the x-direction in the manifold, resulting in a decrease in the shear rate and subsequently Brinkman number in the manifold. As shown in Figure 6d, the slit temperature is deceasing from the center to the edge of the die, which results in high viscosity and higher Brinkman number, as shown in Figure 6f. For the given process conditions, the temperature across the die exit changes approximately 8 K as shown in in this figure.

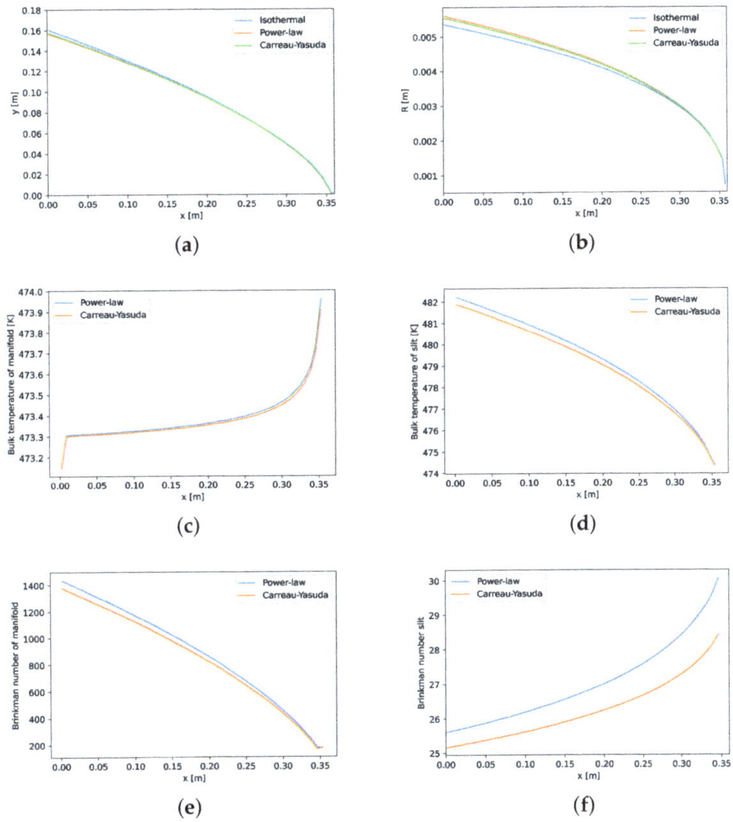

Figure 6. As a result of isothermal and isothermal designs: (**a**) die land length; (**b**) manifold radius; bulk temperatures of (**c**) manifold; and (**d**) slit and Brinkman number of (**e**) manifold and (**f**) slit for temperature-dependent power-law and Carreau–Yasuda fluid. See Table 4 for input parameters.

Table 4. Input parameters and physical properties for the design and optimization shown in Figure 6.

Parameter	Value
Flow rate at entry of die, Q_0	$5 \times 10^{-5}\,\mathrm{m^3 \cdot s^{-1}}$
Land height, h_s	1.5 mm
Half width of die, b	360 mm
Temperature at entry, T_0	473.15 K
Temperature of heater, T_R	463.15 K
Thermal conductivity, k_t	$0.16\,\mathrm{W \cdot m^{-1} \cdot K^{-1}}$
Density, ρ	$1150\,\mathrm{kg \cdot m^{-3}}$
Specific heat, c_p	$1800\,\mathrm{J \cdot kg^{-1} \cdot K^{-1}}$

3.2. *Effect of Heat Viscous Dissipation*

Effect of viscous dissipation can be quantitatively studied by the Brinkman number, which is defined as ratio of heat generation due to heat dissipation to heat transfer to wall. Since viscous dissipation across the die is changing, a new overall Brinkman number is defined based on Brinkman number of power-law fluid in the literature [24]. The overall Brinkman number of die for a power-law fluid is defined as follows:

$$\mathrm{Br_{die}} = \frac{m_0 b^2 {Q_0}^{n+1}}{k_t (T_0 - T_R)(b^2 h_s)^{n+1}}. \tag{25}$$

Four cases with different Br are defined as shown in Table 5. The only parameter that changes is the power-law consistency factor m_0. Values for consistency factors are 10^2, 5×10^3, 1×10^4 and 1.5×10^4, which corresponds to Brinkman numbers of 2.42, 122, 243 and 365.

Table 5. Parameters of cases for comparison in the study of the effect of viscous dissipation.

	Values
Power-law model parameter, n	0.296
Power-law model parameter, α_{pl}	$0.0045\ \mathrm{K^{-1}}$
Power-law model parameter, T_{ref}	503.15 K
Flow rate (Q_0)	$1.99\ \mathrm{m^3/h}$
Inlet temperature (T_0)	553.15 K
Heater temperature (T_R)	543.15 K
Half die width (b)	360 mm
Land height (h_s)	1.5 mm
Thermal diffusivity (α_T)	$1.57 \times 10^{-7}\ \mathrm{m^2/s}$

Figure 7 depicts the designs for all cases. It is worth mentioning that, for the isothermal method, all cases give the same design curves, since the isothermal method is independent of consistency factor. On the other hand, the nonisothermal method is affected by viscous dissipation, which is dependent on Brinkman number and subsequently on consistency factor, as shown in Equation (25). As shown in Figure 7, with an increase in Brinkman number, the radius of manifold increases while die land length y decreases. An increase in manifold radius leads to a lower shear rate and therefore lower heat generation due to viscous dissipation. Additionally, a lower die land length corresponds to a shorter travel length of polymer through a slit and hence a lower effect of heat viscous dissipation.

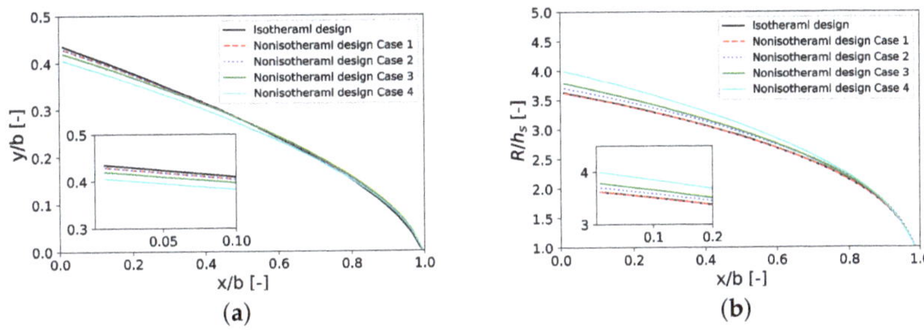

Figure 7. Nonisothermal designs for different cases (**a**) die land length y; (**b**) manifold radius R; see Tables 5 and 6 for the parameters used.

Table 6. Comparison of uniformity index ϕ obtained by CFD of different designs for different Brinkman numbers.

Design	Case 1	Case 2	Case 3	Case 4
m_0 [Pa·s^n]	100	5000	10,000	15,000
Br_{die}	2.42	122	243	365
ϕ of Isothermal design	0.269825	0.270569	0.285509	0.304196
ϕ of nonisothermal design (This work)	0.269825	0.278016	0.276205	0.288235

Table 6 also shows the velocity uniformity index of all cases based on CFD simulations. The uniformity level of velocity distribution is measured and defined by uniformity index as follows:

$$\phi = \frac{\sum_i^N |v_i - v_{ave}| dA_i}{A v_{ave}} \tag{26}$$

Figure 8 depicts improvement in velocity uniformity index of the nonisothermal design over the isothermal one. As shown in Figure 8a, with an increase in Brinkman number, more improvement in ϕ can be seen. This is due to the effect of viscous dissipation on temperature and subsequently on hydrodynamics of the extrusion die. For low Brinkman numbers, isothermal and nonisothermal designs result in similar die geometry and similar velocity distribution at the exit of the die. However, with an increase in viscous dissipation, improvement in the velocity uniformity index can be seen in almost linear fashion (Figure 8b). Thus, for fluids with temperature dependent viscosity, previous design methods are not sufficient and effects of viscous dissipation are necessary to be considered. Nonisothermal design for case 4 with Brinkman number 365 shows 5.24% improvement in velocity uniformity index compared to the isothermal design. Figure 9 shows isothermal and nonisothermal velocity profiles at the center line of the die exit for Brinkman numbers of 2.42 and 365. Both velocity profiles of nonisothermal and isothermal cases for both Brinkman numbers behave; similarly, there is a flat and uniform profile everywhere except a peak at the edge of the die where the manifold reaches the die exit. As shown in this figure, the value of peak increases as Brinkman number increases.

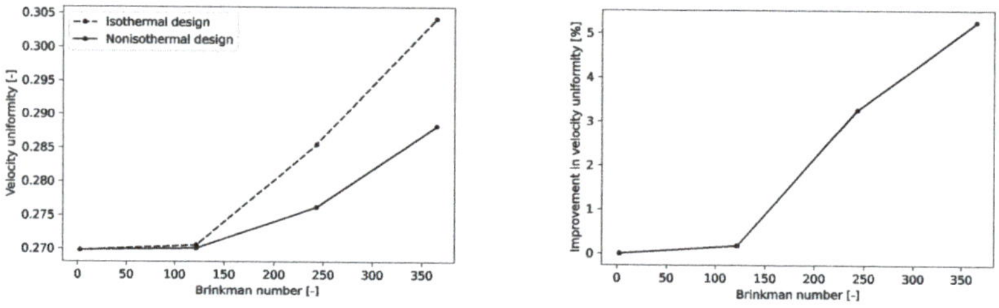

(a) (b)

Figure 8. Nonisothermal design performance and its comparison to previous design (**a**) isothermal and nonisothermal design velocity uniformity index and (**b**) percentage of improvement of uniformity of velocity profile; see Tables 5 and 6 for the parameters used.

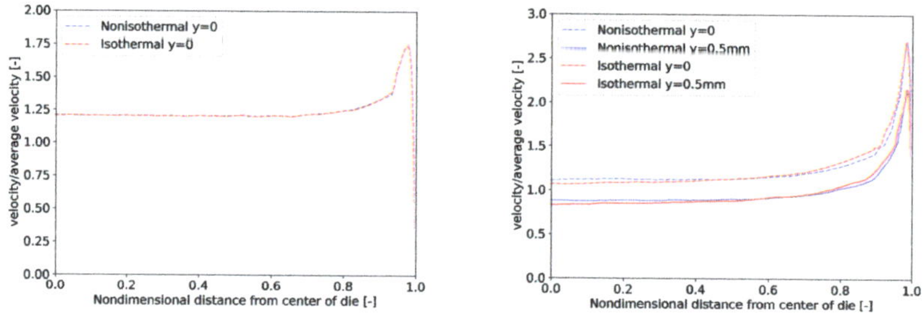

(a) (b)

Figure 9. Comparison of isothermal and nonisothermal designs velocity profiles (CFD) for (**a**) Case 1 with $Br_{die} = 2.42$ and (**b**) Case 4 with $Br_{die} = 365$. Dashed lines correspond to the center line at the exit, and solid lines correspond to the line half-way between the center and the slit wall.

3.3. *Effect of Temperature on Pressure Drop vs. Flow Rate Relation*

Any inaccuracy in calculation of pressure drops in Equation (4) results in inaccurate design curve (y). For simplicity, in this study, the velocity profile is assumed to be fully-developed while the temperature profile varies in the axial direction of a slit or manifold segment. Due to dependency of viscosity on temperature, the actual velocity profile is changing along the axial direction. Figure 10a shows a comparison of analytical (isothermal), one-dimensional, numerical (Equations (15) and (16)) and three-dimensional numerical (CFD) velocity profiles for a power-law fluid at one segment of the slit. Input parameters for all cases are given in Table 7. Due to dependency of viscosity on temperature, higher velocities are obtained by both numerical nonisothermal calculations. In opposition to the one-dimensional calculation, CFD simulation gives different velocity profiles at inlet and outlet of slit regions, as shown in Figure 10b. Despite the fact that the one-dimensional, numerical method is not an accurate representation of three-dimensional profiles, it provides better estimation than the isothermal method.

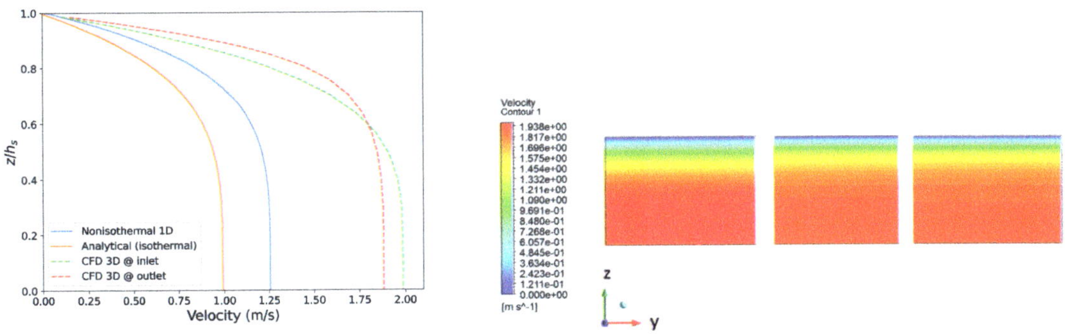

(a) (b)

Figure 10. Velocity distribution of a power law fluid in a slit segment: (**a**) velocity profiles obtained through isothermal analytical, one-dimensional, numerical and three-dimensional numerical (CFD) methods; (**b**) contour plot of velocity (CFD) at three different sections of channels. See Table 7 for values of the parameters used.

Table 7. Input parameters for comparison of the CFD, one-dimensional and analytical velocity profiles; results are shown in Figure 10.

	Values
Power-law model parameter, n	0.296
Power-law model parameter, m_0	1.5×10^4 Pa·s^n
Power-law model parameter, α_{pl}	0.0045 K^{-1}
Power-law model parameter, T_{ref}	503.15 K
Pressure drop (Δp)	40 MPa
Length of channel (L)	0.155 m
Inlet temperature (T_0)	553.15 K
Heater temperature (T_R)	543.15 K
Distance between two plates (L_c)	1.5 mm
Thermal diffusivity (α_T)	1.57×10^{-7} m^2/s
Thermal boundary condition	Bi = 1

3.4. Effect of Temperature Sensitivity Parameter

In the temperature-dependent viscosity model, the intensity of temperature dependency is determined by the temperature sensitivity parameter (α_{pl} or α_{cs}). The higher value of the temperature sensitivity parameter leads to higher dependency of viscosity on temperature. Therefore, it is paramount to study how the temperature sensitivity parameter affects design and subsequently flow distribution at die exit.

Design curves for temperature sensitivity parameters α_{pl} of 0.0045 K^{-1} and 0.029 K^{-1} are shown in Figure 11. Other physical properties and process conditions are given in Table 8. Figure 12 shows exit flow distribution of nonisothermal design at the die exit for both temperature sensitivity parameters. The CFD-obtained velocity uniformity indices for α_{pl} of 0.0045 K^{-1} and 0.029 K^{-1} are 0.280732 and 1.31557, respectively. This shows the importance of this parameter on the effectiveness of the proposed model.

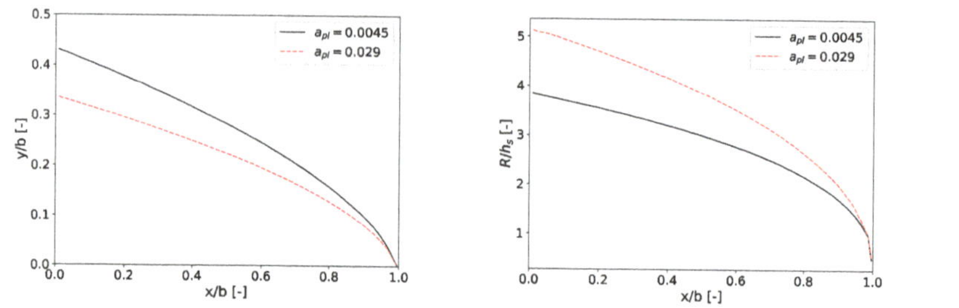

(a) (b)

Figure 11. Design curves for different temperature sensitivity parameters (**a**) die land length y; (**b**) manifold radius; see Table 8 for values of the parameters used.

Table 8. Process conditions and physical properties for comparison of the temperature sensitivity parameter as shown in Figures 11 and 12.

Quantity	Values
Power-law model parameter, n	0.296
Power-law model parameter, m_0	1.5×10^4 Pa·s^n
Power-law model parameter, T_{ref}	503.15 K
Flow rate (Q_0)	1.99 m^3/h
Inlet temperature (T_0)	553.15 K
Heater temperature (T_R)	543.15 K
Half die width (b)	360 mm
Land height (h_s)	1.5 mm
Thermal diffusivity (α_T)	1.57×10^{-7} m^2/s

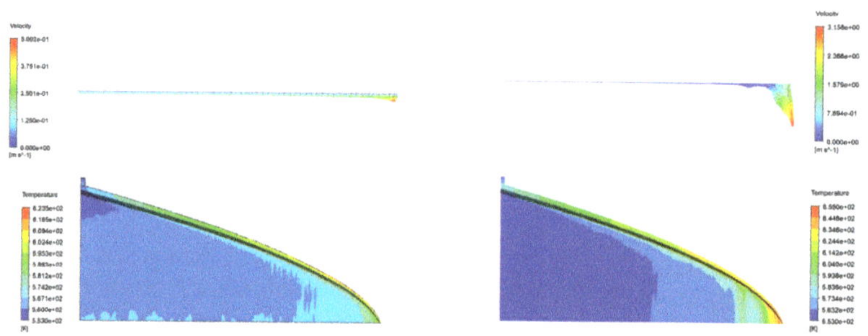

(a) (b)

Figure 12. Velocity distribution at die exit and temperature distribution at the surface of the die for temperature sensitivity parameter (**a**) $\alpha_{pl} = 0.0045$ K^{-1} and (**b**) $\alpha_{pl} = 0.029$ K^{-1}; see Table 8 for values of the parameters used.

In Figure 12, contours depict temperature distribution on the interior wall of the die. Temperature increases in the manifold results in lower viscosity and lower resistance to fluid flow, and therefore a higher percentage of polymer melts tends to flow through the manifold with higher velocity at the die exit (see Figure 12b). For high values of temperature sensitivity parameters, a better control of temperature rather than simplified assumption of Biot of 1.0 is required. It can be concluded that the temperature boundary condition with a Biot number of a numerical value of 1.0 gives only adequate results for lower values of temperature sensitivity parameters.

4. Conclusions

The previously proposed modified Winter–Fritz constant shear rate isothermal design is improved to take into account temperature dependency and effect of viscous dissipation. The new design is compared with the isothermal design through 3D CFD simulation. The temperature increase due to the viscous dissipation and its effect on hydrodynamics of extrusion die is illustrated with numerical simulation results. A Brinkman number is defined to quantify the effect of viscous dissipation in the extrusion die performance. It is shown that, for low to medium Brinkman numbers, the proposed design results in better velocity distribution at the exit of the die compared to previous isothermal design at a higher impact region of the Brinkman numbers. For example, when the Brinkman number is equal to 365, the proposed design results in 5.24% improvement in a velocity uniformity level at the die exit in the example. Assumption of a unidirectional fully-developed velocity profile along the channel is discussed, and it is shown that, due to temperature increase and dependency of viscosity, a fully-developed velocity profile tends to flatten. In addition, the effect of the temperature sensitivity parameter is also studied, and it is proposed to employ an adaptive temperature boundary condition that negates the effect of an excessive temperature increase in the manifold.

Author Contributions: Methodology, Investigation, Writing (original draft, review and editing): A.R.; Review and editing: A.P., D.Z. and D.W.; Conceptualization and Supervision: Y.W., A.P., D.Z. and D.W.; Funding acquisition: D.W. All authors are involved in discussions of the results. All authors have read and agreed to the published version of the manuscript.

Funding: This research was funded by Nazarbayev University under the Faculty Development Competitive Research Grant Program No. 021220FD4651, "Design & Development of Multiphysics Algorithm for Polymer Sheet Processing Die Design".

Institutional Review Board Statement: Not applicable.

Informed Consent Statement: Not applicable.

Data Availability Statement: Data will be available on request.

Conflicts of Interest: The authors declare no conflict of interest.

Nomenclature

Bi	Biot number
Br	Brinkman number
Br_{die}	overall Brinkman number
L	length of a segment
L_c	characteristic length (i.e., radius or height)
N	number of segments
Pr	Prandtl number
Q	flow rate
R	radius of manifold
S	non-dimensional shear rate
T	temperature
T_0	temperature at entry
T_{ref}	reference temperature in power-law and Carreau–Yasuda models
$\Delta\zeta$	length of a manifold segment
$\hat{T}$	non-dimensional temperature
$\hat{u}$	non-dimensional velocity
$\hat{x_1}$	non-dimensional transverse coordinate
$\hat{x_2}$	non-dimensional axial coordinate
$\mathbf{U}$	velocity vector
a_T	temperature dependence function of viscosity model
b	half width of die
c_p	heat capacity

h	heat transfer coefficient
T_R	temperature of heater
h_s	land height
k_l	thermal conductivity
m_0	power-law model consistency parameter
n	power-law index
p	pressure
u	velocity
x_1	transverse coordinate
x_2	axial coordinate
$y(x)$	distance from center of manifold to die exit
Greek letters	
α_T	thermal diffusivity
α_{cs}	Carreau–Yasuda model parameter
α_{pl}	power-law or Carreau–Yasuda model parameter
$\dot{\gamma}$	shear rate
ϵ	convergence criterion
η	viscosity
η_0	Carreau–Yasuda model parameter
η_∞	Carreau–Yasuda model parameter
$\hat{\eta}$	non-dimensional viscosity
λ	Carreau–Yasuda model parameter
μ	apparent viscosity
ϕ	flow uniformity index
ρ	density
τ	stress tensor
Subscripts	
0	at entry
ave	average
m	manifold
s	slit
w	evaluated at wall

References

1. Michaeli, W.; Kaul, S.; Wolff, T. Computer-aided optimization of extrusion dies. *J. Polym. Eng.* **2001**, *21*, 225–238. [CrossRef]
2. Michaeli, W.; Hopmann, C. *Extrusion Dies for Plastics and Rubber*; Hanser Publishers: Munich, Germany, 2016.
3. Winter, H.; Fritz, H. Design of dies for the extrusion of sheets and annular parisons: the distribution problem. *Polym. Eng. Sci.* **1986**, *26*, 543–553. [CrossRef]
4. Awe, T.J.; Eligindi, M.; Langer, R. Internal Design of Uniform Shear Rate Dies. *Morehead Electron. J. Appl. Math.* **2005**, 1–10. Available online: https://www.google.com.hk/url?sa=t&rct=j&q=&esrc=s&source=web&cd=&ved=2ahUKEwijhYzZxaf5AhX-mFYBHRfgBm8QFnoECA0QAQ&url=https%3A%2F%2Fscholarworks.moreheadstate.edu%2Fcgi%2Fviewcontent.cgi%3Farticle%3D1000%26context%3Dmejam_archives&usg=AOvVaw1Q6VmaMtFhflxsayPvSPR7 (accessed on 30 May 2022).
5. Igali, D.; Perveen, A.; Zhang, D.; Wei, D. Shear rate coat-hanger die using casson viscosity model. *Processes* **2020**, *8*, 1524. [CrossRef]
6. Razeghiyadaki, A.; Wei, D.; Perveen, A.; Zhang, D. A Multi-Rheology Design Method of Sheeting Polymer Extrusion Dies Based on Flow Network and the Winter–Fritz Design Equation. *Polymers* **2021**, *13*, 1924. [CrossRef] [PubMed]
7. Yilmaz, O.; Kirkkopru, K. Inverse Design and Flow Distribution Analysis of Carreau Type Fluid Flow through Coat-Hanger Die. *Fibers Polym.* **2020**, *21*, 204–215. [CrossRef]
8. Raju, G.; Sharma, M.L.; Meena, M.L. Recent methods for optimization of plastic extrusion process: a literature review. *Int. J. Adv. Mech. Eng.* **2014**, *4*, 583–588.
9. Oh, K.W.; Lee, K.; Ahn, B.; Furlani, E.P. Design of pressure-driven microfluidic networks using electric circuit analogy. *Lab Chip* **2012**, *12*, 515–545. [CrossRef]
10. Igali, D.; Perveen, A.; Wei, D.; Zhang, D.C.; Mentbayeva, A. 3D FEM Study of the Flow Uniformity of Flat Polypropylene Film/Sheet Extrusion Dies. *Key Eng. Mater. Trans. Tech. Publ.* **2020**, *841*, 375–380. [CrossRef]
11. Lebaal, N.; Puissant, S.; Schmidt, F.; Schläfli, D. An optimization method with experimental validation for the design of extrusion wire coating dies for a range of different materials and operating conditions. *Polym. Eng. Sci.* **2012**, *52*, 2675–2687. [CrossRef]
12. Lebaal, N.; Schmidt, F.; Puissant, S. Design and optimization of three-dimensional extrusion dies, using constraint optimization algorithm. *Finite Elem. Anal. Des.* **2009**, *45*, 333–340. [CrossRef]

13. Wu, P.Y.; Huang, L.M.; Liu, T.J. A simple model for heat transfer inside an extrusion die. *Polym. Eng. Sci.* **1995**, *35*, 1713–1724. [CrossRef]
14. Lebaal, N.; Puissant, S.; Schmidt, F. Application of a response surface method to the optimal design of the wall temperature profiles in extrusion die. *Int. J. Mater. Form.* **2010**, *3*, 47–58. [CrossRef]
15. Lebaal, N. Robust low cost meta-modeling optimization algorithm based on meta-heuristic and knowledge databases approach: Application to polymer extrusion die design. *Finite Elem. Anal. Des.* **2019**, *162*, 51–66. [CrossRef]
16. Amangeldi, M.; Wang, Y.; Perveen, A.; Zhang, D.; Wei, D. An Iterative Approach for the Parameter Estimation of Shear-Rate and Temperature-Dependent Rheological Models for Polymeric Liquids. *Polymers* **2021**, *13*, 4185. [CrossRef] [PubMed]
17. Bergman, T.; Lavine, A.; Incropera, F. *Fundamentals of Heat and Mass Transfer*, 7th ed.; John Wiley & Sons: Hoboken, NJ, USA, 2011.
18. Astarita, G. *Rheology: Volume 2: Fluids*; Springer: Berlin/Heidelberg, Germany, 2013.
19. Bird, R.B.; Stewart, W.E.; Lightfoot, E.N. *Transport Phenomena*, 2nd ed.; John Wiley & Sons: Hoboken, NJ, USA, 2006.
20. Santos, C.; Quaresma, J.; Lima, J. *Convective Heat Transfer in Ducts: The Integral Transform Approach*; Editora E-papers: Rio de Janeiro, Brazil, 2001.
21. Sato, S.; Oka, K.; Murakami, A. Heat transfer behavior of melting polymers in laminar flow field. *Polym. Eng. Sci.* **2004**, *44*, 423–432. [CrossRef]
22. Yang, H. Conjugate thermal simulation for sheet extrusion die. *Polym. Eng. Sci.* **2014**, *54*, 682–694. [CrossRef]
23. Arpin, B.; Lafleur, P.; Sanschagrin, B. A personal computer software program for coathanger die simulation. *Polym. Eng. Sci.* **1994**, *34*, 657–664. [CrossRef]
24. Etemad, S.G.; Majumdar, A.; Huang, B. Viscous dissipation effects in entrance region heat transfer for a power law fluid flowing between parallel plates. *Int. J. Heat Fluid Flow* **1994**, *15*, 122–131. [CrossRef]

MDPI
St. Alban-Anlage 66
4052 Basel
Switzerland
www.mdpi.com

Polymers Editorial Office
E-mail: polymers@mdpi.com
www.mdpi.com/journal/polymers

www.ingramcontent.com/pod-product-compliance
Lightning Source LLC
Chambersburg PA
CBHW061512210926
51850CB00008B/1896
9783725802500